Functional Mammalian Neuroanatomy

Functional Mammalian Neuroanatomy

with emphasis on dog and cat, including an atlas of dog central nervous system

Thomas W. Jenkins

B.S., M.S., Ph.D.

Associate Professor of Anatomy, College of Human Medicine, College of Veterinary Medicine, Michigan State University, East Lansing, Michigan

LEA & FEBIGER · PHILADELPHIA

1972

ISBN 0-8121-0354-8

Library of Congress Catalog Card Number 78-152028

Printed in the United States of America

TO MY FAMILY
From whom I was separated
while serving
the sentences of this book

Preface

There are many texts of human neuroanatomy from which to choose for study of the nervous system of man. Although there are atlases of dog and cat brains, to the best of my knowledge there is no text of functional neuroanatomy based on the dog or cat. Since the dog is selected as the standard small animal in most, if not all, schools of veterinary medicine, and it is very commonly used as an experimental animal in both veterinary and human medical teaching and research laboratories, it is intended that this book will serve a purpose in these areas. This book should be of value also for students of anatomy and paramedical disciplines who are interested in the principles of functional mammalian neuroanatomy and who may have access to dog (or cat) but not to human brains. Perhaps the industrial researcher who uses the dog and cat as experimental animals would find use for this book, which not only points out structures as found in an atlas but in addition gives some basic information regarding functions of structures as they apply to the dog and cat.

Although this is not a comparative text, the nervous system of the dog is compared with that of man and other mammals where appropriate. Due to the paucity of information on details of the central nervous system (CNS) of the dog in comparison with reports of cat experimentation, much of the text is based on the vast literature of the cat. An effort was made to make clear to the reader what species of animal was used for obtaining specific information reported.

Emphasis is placed on the similarities of

the gross and the microscopic structure of spinal cord and brain stem of dog and man; *i.e.,* the nuclei, pathways, and other structures in general share the same names and virtually the same anatomic relationships but with some differences in relative sizes. The main objective of this emphasis is to encourage the person trained only in human neuroanatomy to study brains from dogs. He will soon realize that the two mammals (man and dog) have brain stems that are quite similar in structure and function. The reverse also is true; if the veterinary neuroanatomist studies human spinal cord and brain stem, he will soon be amazed at the similarities in man and animals. If this objective is even partially achieved, perhaps this book may help to bridge the gap between veterinary and human neuroanatomy.

The book is divided into three parts: Part I, the text portion, begins with a brief phylogenetic survey and discussion of organization of the nervous system. Chapter 2 discusses development of the brain and gross anatomy of the brain of the adult dog. It presumably might be used in the laboratory in a basic study of gross anatomy, although it is not written as a dissecting guide. Chapters 6, 7, 9, and 17 discuss the functional anatomy of the peripheral nervous system. Chapters 10 through 16 cover the central nervous system (CNS), beginning with caudal levels of the spinal cord and progressing to the cerebral cortex. Chapters 18 and 19 concern the anatomy of the eye and ear, in addition to some functional and dysfunctional neuroanatomic considerations of the visual and vestibulocochlear systems. Part II is a summary of the functional anatomy of important nuclei, pathways, and other structures of the CNS. This is in brief form and arranged in alphabetical order for quick study with references to the Atlas for identification and to the text of Part I for more comprehensive information. Part III is an Atlas of the canine CNS, beginning at the caudal levels

of the spinal cord and progressing rostrally through the brain. The plates are photographs of actual brain sections stained by the Klüver and Barrera technique (*cf.* Chapter 3). The Atlas is arranged so that the student may turn the book and have a plate on the top page for self instruction of identification with the legend on the facing page for checking his answers.

It is with deep gratitude that I acknowledge the invaluable assistance that has been given by so many individuals in so many ways. I am indebted to Mrs. Edith Jones for her extraordinary devotion and expertise in preparation of the microscopic sections used for the Atlas. I gratefully acknowledge the contributions of Dr. Duane Haines, who authored Chapter 8, made helpful suggestions, and contributed greatly to the early stages of the Atlas. Special thanks are given to Mr. Robert Ewing for his special skill in preparing the illustrations and for being so cooperative. I gratefully acknowledge the photographic skill of Mr. Robert Paulson, who photographed the microscopic slides for the Atlas. I am indebted to many of my colleagues at Michigan State University for their valuable suggestions, criticisms, and help in many ways: Dr. Bruce Walker, Chairman of the Anatomy Department, and Dr. Arthur Foley of the same department; Dr. Richard Heisey of the Physiology Department for his assistance with the cerebrospinal fluid section of Chapter 5; and Dr. Waldo Keller, Chairman of Small Animal Surgery and Medicine, for his assistance with Chapter 18. In addition, I acknowledge the helpful suggestions of Dr. Thomas Fletcher, Department of Veterinary Anatomy, University of Minnesota.

I give thanks to my many students who have made suggestions on various parts of the book. I am grateful for the assistance of Mr. Dale Peerbolte, who worked very closely with me during the final periods of completion. Without his assistance in proofreading, indexing, preparation of the bibliography, and final work on the Atlas, completion of the book would have been delayed. I thank Mrs. Janeen Nixon and Mrs. Mary Grace for typing and retyping the manuscript.

I am indebted to the Pathology and Toxicology Research Unit of the Upjohn Company for financial assistance years ago for a project on dog brain which was an important factor in the evolution of this book. Special thanks are given to the Center for Laboratory Animal Resources at Michigan State University for assistance through N.I.H. Grant No. 5-P06-RR00366. Finally, I gratefully acknowledge the cooperation of the publisher, particularly Mr. Edward Wickland, Jr. (editor) and Mr. Thomas Colaiezzi (production manager), with whom I worked directly.

THOMAS W. JENKINS

East Lansing, Michigan

Contents

PART II

Summary of Internal Structures of the Brain and Spinal Cord, Including Locations, Connections, and Functions

PART III
Atlas of Central Nervous System of the Dog

Part I

Text

chapter 1

Evolution and Anatomic Divisions of the Nervous System

Evolution of the Nervous System

In consideration of the anatomic and physiologic organization of the highly developed mammalian nervous system, one should realize that as he studies the dog or human nervous system he is observing the final product of a long phylogenetic and ontogenetic history.

The survival of an individual organism and of its entire species depends on the ability to adapt successfully to environmental changes. Since active neuronal integration plays a primary role in such adaptation, and since not all species followed the same type of adaptation, specific functional anatomic differences are reflected in the variable nervous systems of different animals.

In progressing from lower to higher phyla, the comparative neuroanatomist emphasizes the increasing complexity of neuronal activity which is commensurate with the less stereotyped responses of higher animals to stimuli (environmental changes). Therefore the higher animals possess complex neuronal circuitry which permits greater functional variability.

It is generally agreed that man is the superior animal. This assumption is based primarily on man's possession of the greatest functional elaboration of the cerebral cortex, which permits him to reason and to communicate beyond the limits of any other animal. In the context of functional variability of higher animals, as discussed above, most people will concede that man is the most unpredictable in terms of how he will react to a given stimulus.

A brief survey of nervous mechanisms throughout the animal kingdom, from amoeba to man, reveals the pattern of anatomic changes that have occurred in order for animals to cope effectively with the increasingly more complex environmental changes and physiologic demands. Although functionally significant anatomic differences can be pointed to in the nervous systems of various animals, basic fundamental properties of protoplasm and neural mechanisms serve as common threads running throughout the entire animal kingdom.

Study of the anatomy and physiology of the nervous system in lower animals, the invertebrates, although it is simpler, therefore often adds to our understanding of the higher mammalian nervous system. In the brief discussion that follows, only a few phyla are cited to illustrate approximately where, in the phylogenetic scale, special characteristics of neural mechanisms have been introduced to permit a more sophisticated mode of living. Table 1-1 summarizes at a glance the pattern of the discussion.

Invertebrates

Animals without Neurons

Protozoa (unicellular animals) are able to survive without neurons, but not without manifesting nervous activity. That is, the protoplasm is capable of performing all of the essential life activities for an individual isolated cell. Throughout the animal kingdom, from the simplest to the most complex animal, the inherent general properties of protoplasm are maintained to some extent in all cells. The essential general properties of protoplasm are usually listed as: respiration, contractility, growth, reproduction, excretion, secretion, absorption, assimilation, irritability, and conductivity.

Although all of these general properties are essential for life of the protozoan, the last two listed, namely *irritability* and *conductivity*, are essential to nervous activity. As an example, the amoeba can react to noxious stimuli—can avoid heat, mechanical obstructions, and the like—primarily because of its irritability and conductivity. Because of its extremely limited anatomic and physiologic armamentaria, the reactions of the protozoan to stimuli are easily predicted.

Introduction of Neurons and Plexuses

It is in the phylum *Coelenterata* that we first identify true nerve cells (neurons), a nerve net (plexus), and the presence of synapses (junctions of two or more neurons). In these aquatic diploblastic (ectoderm and endoderm) animals we also first observe the tissue level of organization. In the best-known Coelenterates (*e.g.,* hydra, jellyfish, sea anemone) one can study the primitive

TABLE 1-1. Phylogenetic History of Development of the Nervous System

Invertebrates
    Unicellular animals without neurons
    Introduction of neurons and plexuses
    Centralization and cephalization
    Segmentation, and development of reflexes
    Specialization of sense organs
Vertebrates
    Notochord and neural tube
    Mammalian nervous system

neuronal system that is the precursor of plexuses and synapses found in higher animals. The synapse is primitive at this stage as based on the lack of dynamic polarity, that is, in coelenterates the impulses may traverse the synapse in either direction, whereas in higher animals the impulse is conveyed across the synapse in only one direction.

Centralization and Cephalization

Platyhelminthes (flatworms) is the next higher phylum of importance for introduction of neural advances. The increased complexities in organization are: (1) the introduction of a central nervous system composed of a brain and two ventral longitudinal nerve cords connected at intervals in a ladder-type formation; (2) the introduction of a head with sense organs; (3) formation of a final (third) germ layer (mesoderm) between the outside ectoderm and inside endoderm of the original diploblastic animal; (4) bilateral symmetry of the body; and (5) an organ-system level of organization. These anatomic-physiologic advancements make possible independent locomotion, more rapid and varied responses to stimuli, and a more varied behavior than that shown by coelenterates.

Segmentation, and Development of Reflexes

The phylum *Annelida* (e.g., the common earthworm) serves well to illustrate the advancements, over the lower phyla mentioned, of neural mechanisms, as well as advancements of other components of the body. The earthworm possesses a number of ganglia: two in the head region and one for each body segment. The ganglia in the body segments serve as integrative centers for receiving impulses via afferent (sensory) neurons and sending impulses peripherally via efferent (motor) neurons. The intricate segmental neural apparatus of the earth-

worm presumably possesses all the components of a *simple reflex arc* (Truex and Carpenter, 1969), so that the earthworm has the ability to respond segmentally and involuntarily to an appropriate stimulus.

There is evidence that *intrasegmental* and *intersegmental* reflexes exist in the 'lowly earthworm' and, although the mechanism is primitive, its plan is recognized as the basis for the more sophisticated neuronal interconnections in mammals as discussed in Chapter 7. The body segmentation (metamerism) of the earthworm suggests the pattern of metamerism of the mammalian body, in which it is best demonstrated in the spinal cord region, especially in the attachments of the thoracic nerves.

Specialization of Sense Organs and Development of Neuromuscular Apparatus

In the phylum *Arthropoda* (e.g., crayfish, insects, and spiders) the sense organs, especially the eye, are very well developed, and the plan is established that is followed in development of higher animals. The brain is more complex, as a compound ganglionic mass within the head, and functions as more than a relay center. The mechanism for coordination of movements of the limbs is established as a basic neuromuscular mechanism. The behavioral actions of insects are rather sophisticated, as exemplified by the well-known social orders of the honey bee.

This survey of invertebrates is very brief and far from complete. Further details of structure and function in the nervous systems of invertebrates may be obtained from the two-volume work by Bullock and Horridge (1965).

Vertebrates

Notochord and Dorsal Neural Tube

It is not until the highest animal phylum, *Chordata*, is reached that a dorsal neural tube (precursor of the spinal cord) is present and protected by a vertebral column. The notochord is not to be confused with any type of precursor or any entity directly related to the spinal cord. In general, the nervous system is derived from the ectoderm germ layer, and most authorities agree that the notochord is derived from mesoderm.

In the most primitive vertebrates the notochord is a well-developed fibrocellular cord lying ventral to the central nervous system and constitutes the chief axial supporting structure of the body (Patten, 1968). In the process of evolution the cartilaginous vertebrae are replaced by bony vertebrae and the notochord is compressed. In higher mammals a remnant of the notochord exists as an inconspicuous mark in the body of the vertebra and is more clearly identified in the central portion of the intervertebral disk as the nucleus pulposus.

The Mammalian Nervous System

The remainder of this book is devoted to discussion of some of the functional anatomic details of the mammalian nervous system, but the knowledge of a few basic concepts of body organization as related to the nervous system may help in understanding the neuroanatomic and neurophysiologic details presented in the succeeding chapters.

The same general properties of protoplasm that were mentioned in connection with protozoa are present in cells of multicellular animals, but not every cell has to maintain all of the properties at maximal levels, as does a unicellular animal. Various cells within the complicated mammalian body must become 'specialists' and perform certain duties better than their neighbors. The specialized duties are in addition to the other general protoplasmic functions. Animals are arranged on a phylogenetic scale not only according to their ancestry in terms of time but also according to the degree of cellular organization and division of cellular labor within the body tissues and organs.

Cells with a primary function referable to one property or to a limited number of the general properties of protoplasm are said to be *differentiated*, or *specialized*. As an example, muscle 'specializes' in contraction and conduction, gonads in reproduction, kidney tubules and gastrointestinal epithelium in absorption and assimilation, and glandular epithelium in secretion.

Neurons are highly differentiated because of their superior *irritability* and *conductivity*, the same general properties that permit the amoeba to respond to stimuli in a very simple stereotyped manner. To reiterate, despite the high degree of specialization with regard to these two properties, the other properties are present in the cells but may not exceed the level required for survival of the cells (neurons). Obviously, if all the other general properties of protoplasm were to disappear, the cells would die.

Interrelationship of the Nervous System and Other Body Parts

The function of the nervous system is to produce, by the generalized culmination of all the complex neuronal activity, the proper reaction by the organism to changes in its environment. This applies to changes in both the external and the internal environment. The functioning of the nervous system in this capacity is easily understood with reference to the external environment; its role is perhaps not so obvious in relation to the internal environment.

The anatomic organization of the mammalian body may be considered basically according to the following levels of complexity: (1) cell, (2) tissue, (3) organ, (4) organ system, and (5) organism. Note the similarities between the levels of organization of the individual mammalian body and the levels of organization of separate phyla as given in the general survey of the animal kingdom. Generally speaking, the external environment affects the entire organism (level 5). The *internal environment* is the intercellular fluid, which of course is present at the cellular level of organization (level 1).

The nervous system affects all five levels of organization. When there is a change in the external environment strong enough to affect the entire organism, the nervous system attempts to adjust the lower levels of organization so that the change is nullified or modulated at all levels, the ultimate effect being the attempt to maintain a constant internal environment. This maintenance of a constant internal environment is referred to as *homeostasis,* which is the primary function of the autonomic nervous system (see Chapter 9).

Each cell in the body is bathed in interstitial fluid (internal environment), which normally is chemically and physically compatible with the adjacent cells. When the interstitial fluid is changed physically or chemically beyond the endurance of the cells, the cells will change anatomically, and unless the process is reversed within a certain time the cells will die. As an example, if a particular solute is added to the interstitial fluid, it may become hypertonic with respect to the intracellular fluid. Then the cells become dehydrated, eventually become crenated, and die. Similarly, if a toxin enters the interstitial fluid the chemical constituents of the cells will change, and if the change progresses beyond a normal functional range the cells will be poisoned and killed.

The overall effect on the different levels of body organization depends on the number of cells that die as a result of deleterious changes in the internal environment. It is obvious that if certain vital areas of the body lose even a very small number of cells, death of the entire organism may result. However, other areas of the body may lose great numbers of cells with relatively little physiological embarrassment to a particular organ or the whole body. Since it is the autonomic nervous system that has a regulatory effect on all viscera and the internal environment, this system may determine the survival of the entire organism.

It is readily recognized that, in addition to the nervous system functioning as a regulatory mechanism, the endocrine system also functions in a similar manner. Most frequently these two systems act synergistically, the nervous system rather rapidly to produce the necessary compensatory effects, while the action of the endocrine system is characteristically slower. Again it is the autonomic nervous system that plays a major role in this regulation, owing to its action on the blood vascular system and on the endocrine glands.

In addition to the regulatory control exerted by the nervous system, as discussed previously, with major emphasis on the cellular level of body organization and internal environment, the nervous system acts as a functional coordinator for interaction at the system level (Fig. 1-1). As an example, one organ system of the body functionally affects any one or all of the other systems by means of the nervous system. One system may most directly affect itself rather than another by integration within the nervous system. As an example, muscle spindles in skeletal muscle are stimulated by muscular contractile change which initiates a proprioceptive impulse in the afferent neurons to be transmitted to the spinal cord, where the efferents are then stimulated to convey impulses peripherally back to the skeletal muscle (see Chapter 7).

Despite the wide scope of functional anatomic effects of neuronal activity throughout the body, there are only two types of cells, in addition to other neurons, which are stimulated by neurons: (1) muscle cells (skeletal, cardiac, and smooth), and (2) glandular epithelium. Therefore the systems indicated in Figure 1-1 respond to neuronal

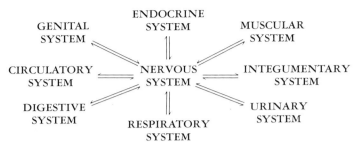

Fig. 1-1. Schema showing role of the nervous system as a mediator for intersystemic action throughout the body. See text for elaboration.

stimulation only by virtue of their muscular or glandular components, or both.

Anatomic Divisions of the Nervous System

The mammalian nervous system is divided anatomically into two main divisions: (1) The *central nervous system,* composed of the brain and spinal cord, is also referred to as the neuraxis. (2) The *peripheral nervous system* is composed of: (*a*) cranial and spinal nerves; (*b*) ganglia, both sensory and motor, (*c*) nerve endings, including receptor (sensory) and effector (motor) endings, and (*d*) the autonomic nervous system.

Despite the classical separation of the nervous system into central and peripheral divisions (systems), the two divisions are structurally and functionally interconnected. The peripheral nervous system transmits impulses toward the central nervous system via afferent (sensory) neurons and away from it via efferent (motor) neurons (see Chapter 7). Both of these functional types of neurons are present in a typical spinal nerve and in most of the cranial nerves. Based on the presence of both afferent and efferent fibers within the same nerve, most nerves are referred to as mixed. The basic organization of spinal nerves is discussed in Chapter 6.

In the cranial region the same basic plan is adhered to, but with some modifications.

The brain is a portion of the central nervous system, and the cranial nerves are components of the peripheral nervous system. However, the cranial nerves are modified in that each nerve does not have a dorsal and a ventral root. Some cranial nerves are purely sensory, and not all have sensory ganglia; that is, only nerves V, VII, VIII, IX, and X possess sensory ganglia. Details of cranial nerves are presented in Chapter 17.

The anatomic organization of the adult mammalian brain (*e.g.,* dog or man) is based on the five embryonic divisions which are easily demonstrated in embryos of submammalian species as well. The basic divisions are, from rostral to caudal: telencephalon, diencephalon, mesencephalon, metencephalon, and myelencephalon. The major components of each division are discussed in Chapter 2.

During study of successive chapters, one should keep in mind the basic plan of organization of the nervous system as it has evolved throughout the animal kingdom and as it develops within the individual mammal, as discussed in Chapter 2, and should remember that adult mammalian brains are basically similar in anatomic organization. The spinal cord and brain stem of the dog, as discussed in succeeding chapters, contain fundamentally the same tracts and nuclei as are found in the human. However, the telencephalon, particularly the cerebral cortex, of man is anatomically different; that is, the names and relationships of the gyri and sulci of dog and man are not interchangeable. This is not surprising, in that it is the exceptionally complex cerebral cortex of man that permits him to consider himself the 'superior animal.'

Bibliography

Bullock, T. H., and Horridge, G. A.: *Structure and Function in the Nervous Systems of Invertebrates.* San Francisco, W. H. Freeman and Co., 1965. Vols. I and II.

Patten, B. M.: *Human Embryology,* 3rd ed. New York, McGraw-Hill Book Co., Blakiston Division, 1968.

Truex, R. C., and Carpenter, M. B.: *Human Neuroanatomy,* 6th ed. Baltimore, Williams & Wilkins Co., 1969.

chapter 2

Development and Gross Anatomy of the Brain

Development of the Brain

Pertinent embryologic concepts are included throughout this text at points where such information will enhance the understanding of the anatomic or physiologic properties (or both) of the subject under discussion. Since the embryonic pattern is the 'blue print' for understanding the structure of the adult brain, a relatively extensive coverage of brain development is presented here. It will be seen that the divisions of the brain of the adult mammal (*i.e.,* dog, cat, man, and others) are the same as the divisions of the prenatal brain in terms of nomenclature, components, and relationships.

As one would expect, it is much easier to study the simple structure of the un-differentiated embryonic brain and the patterns of its development than to begin with study of the much more complex adult brain. Many of the relationships of the parts of the adult brain that are difficult to understand appear relatively simple when they are studied in the less complicated embryonic brain.

A search of the literature reveals that very little has been reported about the embryology of the canine nervous system. This paucity of information is not critical, because the embryonic pattern of development in all mammals is the same, to the depth of detail we wish to consider.

As a matter of fact, many of the descriptions of early development of the nervous system given below are characteristic of all vertebrates. From the standpoint of comparative anatomy, despite the common generalities of developmental patterns, a detailed study of neuroembryology of all animals reveals a rather complete story of phylogeny.

A comparative study of the prenatal and adult nervous systems of various animals also uncovers evidence of which environmental stimuli significantly influence the animals' mode of living. As an example, lower animals (rat, cat, and dog) depend much more on the sense of smell than does man. Therefore, the rhinencephalon (olfactory brain) is relatively much more elaborate in the animal brains than in human; consider, for instance, the relatively large size of the olfactory bulb and tracts in dog brain as compared with those in the human brain (see Fig. 2-14). In contrast, the neopallium of the brain in man is much more complex than that in the dog. The point of emphasis of these comparisons is that man depends on his great faculties of learning, memory, association, thinking, and communication for his survival; therefore the areas of the brain most necessary for such abstract phenomena are better developed in man than they are in animals.

Based on the above-stated principle that all mammals show the same developmental pattern, it is common practice to use pig embryos for study of the developmental processes as they occur in human embryos. Conception age differentials have been investigated with great precision; for example, as one studies the 5-mm pig embryo, he is studying an almost exact replica of a four-week-old human embryo; and a 10-mm pig embryo is an equivalent of a human embryo of the sixth week (Patten, 1968). The scarcity of detailed reports in the literature

pertaining to dog *per se* is therefore not a great hindrance, and much of the following discussion is based on the embryology of the pig (Patten, 1948) and human (Patten, 1968; Davies, 1963; and Thomas, 1968).

Development of the Neural Tube

It is not surprising that the nervous system, which is the most highly differentiated and complex system, is the first to begin, and the last to complete, development. Very soon after the three embryonic germ layers are established in sheetlike strata, the superficial outside ectoderm begins to thicken in the area of the future embryo as the first indication of a central nervous system.

The thickened ectoderm is the *neural plate* (Fig. 2-1 A), which soon develops a shallow, narrow indentation, the *neural groove* (Fig. 2-1 B), as the linear marking of the neuraxis in the future mid-dorsal line of the embryo. The neural groove deepens and its ectodermal walls thicken as the dorsal lips (*neural folds*) approach each other to close in the midline (Fig. 2-1 C). Figure 2-1 D illustrates that the *neural tube*, after the dorsal closure, is covered by the superficial ectoderm, from which the neural ectoderm of the tube becomes separated.

On each side, exterior to the neural tube, in the angle between the neural tube and the superficial ectoderm, there is a cluster of cells known as the *neural crest*. The neural crests give origin to the dorsal root (sensory) ganglia of the spinal nerves, the sensory ganglia of cranial nerves V, VII, VIII, IX, and X, and the autonomic ganglia. It is interesting to note that the same primordium gives rise to both sensory and motor (autonomic) ganglia. In addition it is thought that the *neurilemmal sheath* (of Schwann) *cells* of the peripheral neuronal processes arise from the neural crests.

The neural tube is the primordium for the whole central nervous system. The relevant details of spinal cord differentiation,

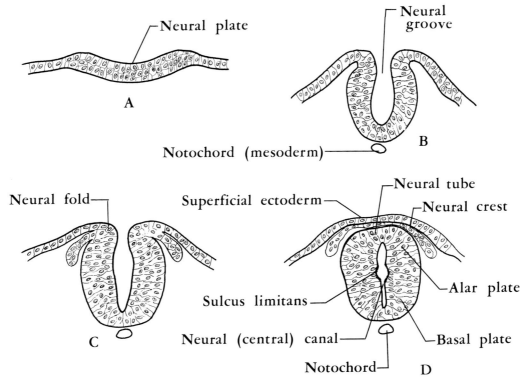

Fig. 2-1. The basic stages in the formation of the neural tube from ectoderm. A. The thickening of ectoderm as the *neural plate*. B. The indentation of the plate in the longitudinal axis of the future body, the *neural groove*. C. Deepening of the groove, with approximation of the neural folds (lips of the groove) in the dorsal midline. D. Closure of the groove dorsally to complete the *neural tube*. Between the dorsolateral surface of the tube and the superficial ectoderm the *neural crests* are formed.

particularly the histogenesis, are given in Chapter 10.

Differentiation of the Brain

It is obvious that the spinal cord area of the neural tube undergoes much less modification than does the region of the brain. Although the brain is highly differentiated and much more complex than the spinal cord, evidence of the basic neural tube is present in the adult brain as the brain stem. The following discussion will show that the cerebral hemispheres and the cerebellum are merely outgrowths of certain areas of the original neural tube.

Embryonic growth and development are inherently influenced by a phenomenon called *cephalization,* which basically means that the cephalic portion of the embryo possesses greater powers of proliferation and differentiation than does the caudal region. Examples of this phenomenon are much too numerous to list, but the concept should be remembered as we consider the development of the brain from the cephalic portion of the neural tube.

Brain Flexures, and Regional Differentiation

The rapid proliferation of the neural tube in the cephalic region occurs in an intra-cranial space that is inadequate to contain the linear tubular growth. Figure 2-2 illustrates that the neural tube area from which the brain is derived exhibits three primary brain regions: the *prosencephalon*, or *forebrain*, the *mesencephalon*, or *midbrain*, and the *rhombencephalon*, or *hindbrain*. The lumen of the neural tube continues into these regions and becomes modified to form the ventricles of the brain, as discussed later. The arrows in Figure 2-2 indicate the sites of the three points of bending in a dorso-ventral plane. The most cephalic bending is the *midbrain*, or *cranial, flexure*, the concavity of which faces ventrally. A second bending occurs at the caudal end of the rhombencephalon, approximately at the site of the future junction of the spinal cord and brain stem. Because of its location it is known as the *cervical flexure*, which, as illustrated in Figure 2-2 B, also has its concavity ventrally, like the midbrain flexure. Between the cranial and the cervical flexure is a bending of the neural tube in the rostral area of the rhombencephalon which opposes the other flexures in direction. Because this is the site of the future pons, this is named the *pontine flexure*. In the dog the cranial flexure begins at the 10-somite stage of development (17th day) and the cervical flexure at the fourth-somite level by the 20-somite stage (18th day); the pontine flexure is first visible at the 32-somite stage of development (21st day) (Houston, 1968).

Figure 2-2 C indicates that from the original three primary brain regions five divisions are derived in the following manner: The prosencephalon divides into a rostral *telencephalon* and a caudal *diencephalon*; the midbrain remains undivided as the *mesencephalon*; and the rhombencephalon divides into a rostral *metencephalon* and a caudal *myelencephalon*. These five embryonic divisions are retained as the basis for study of divisions and organization of the adult brain. Based on these five divisions, the following discussion is intended to describe the major features of differentiation of the neural tube wall and the lumen from caudal

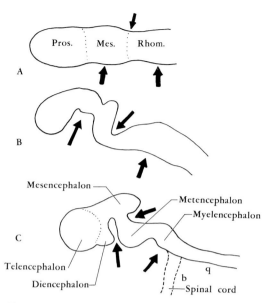

Fig. 2-2. Schematic illustration of the embryonic brain flexures. A. Lateral view of primitive three-vesicle stage (prosencephalon, mesencephalon, and rhombencephalon), with arrows indicating the three points of bending (flexure) which occurs in a dorso-ventral plane. From left to right, the three flexures are named: midbrain, pontine, and cervical flexure. B. Lateral view, showing directions of the flexures. C. The more advanced five-vesicle brain with the flexures shown. The cervical flexure is much less pronounced in quadrupeds, where the spinal cord (*q*) extends more horizontally, compared with that (*b*) in bipedal animals.

to rostral as an aid for understanding the organization of the adult brain. Discussion of specific details of anatomic components of each division are deferred to the second portion of this Chapter, on Anatomy. Table 2-1 gives the major details in outline form.

The Myelencephalon

The myelencephalon forms the medulla oblongata of the adult brain. The lumen of the neural tube in this region expands to form the caudal portion of the fourth ventricle.

TABLE 2-1. Derivatives of the Neural Tube

Primary Division	Subdivisions	Lumen	Major Derivatives
Prosencephalon	Telencephalon	Lateral ventricle + Rostral part of third ventricle	Cerebral cortex Basal nuclei Limbic system Rhinencephalon
	Diencephalon	Most of third ventricle	Epithalamus Dorsal thalamus Hypothalamus Subthalamus Metathalamus
Mesencephalon	Mesencephalon	Cerebral aqueduct (of Sylvius)	Tectum (Corpora quadrigemina) Tegmentum Cerebral peduncles
Rhombencephalon	Metencephalon	Fourth ventricle (rostral portion)	Pons Cerebellum
	Myelencephalon	Fourth ventricle (caudal portion)	Medulla oblongata
Remainder of Neural Tube	Spinal cord	Central canal	Spinal cord

Figure 2-3 shows that the myelencephalon is a direct cephalic continuation of the spinal cord. The very thin roof of the myelencephalic cavity is retained in the adult brain as the extremely delicate *posterior medullary velum.* Figure 2-3 also shows the auditory vesicle, the primordium of the internal ear which retains its relationship to the cranial nerves and ganglia.

Figure 2-4 illustrates how the mantle layer of the neural tube differentiates into four functional component representations of the spinal nerves, and how the lumen of the neural tube opens dorsally to form the fourth ventricle of the medulla and the mantle layer further differentiates to form the nuclei representation of the seven functional components of the cranial nerves.

Notice that the dorsal border of the neural tube lumen splits and opens with a lateral migration in a pivotal movement, with the ventral nuclei (somatic efferent) remaining stationary so that the sensory nuclei tend to be dorsolateral to the motor nuclei. The sulcus limitans is still visible in the adult and separates the sensory from the motor nuclei, as indicated in Figure 2-4. The ependymal cells remain in their original position and therefore line the fourth ventricle. Small blood vessels develop above the roof of the ventricle and push the roof partially into the ventricle to form the choroid plexus of the fourth ventricle.

With regard to the relation of gray and white matter in the spinal cord as compared with the medulla, the spinal cord retains the original neural tube relationship of mantle (gray substance) and marginal (white); in

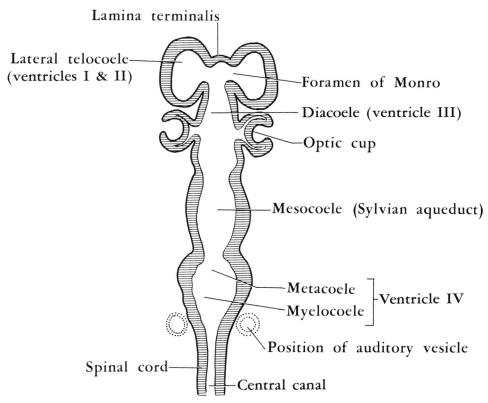

Lamina terminalis

Lateral telocoele (ventricles I & II)

Foramen of Monro

Diacoele (ventricle III)

Optic cup

Mesocoele (Sylvian aqueduct)

Metacoele ⎤
Myelocoele ⎦ Ventricle IV

Position of auditory vesicle

Spinal cord

Central canal

Fig. 2-3. Schematic illustration of the five-stage embryonic brain as if it were straightened out, showing the continuity of the ventricular system and the central canal of the spinal cord. (After Patten, 1948.)

the medulla, and the rest of the brain stem, this relationship is lost, so that there is more of an intermingling of gray (nuclei) and white (tracts). Grossly, the last seven cranial nerves are attached to the myelencephalic area, and therefore to the medulla of the adult brain.

The Metencephalon

The metencephalon forms the pons of the brain stem, the direct forward continuation of the medulla, and the cerebellum, which is the caudal superstructure attached to the brain stem. The lumen of this division (metacoele) joins the myelocoele without any line of demarcation so that both form the fourth ventricle, as illustrated in Figure 2-3.

The thin roof of the rostral portion of the fourth ventricle is known as the *anterior medullary velum*. The trigeminal (fifth cranial) nerve is attached to the pons. This is the largest cranial nerve, with two of its three divisions being purely sensory and the third division mixed. Considering the cranial nerves in their order from one to twelve, this is the first nerve to have a sensory ganglion, named the *semilunar*, or *Gasserian, ganglion*.

The Mesencephalon

The mesencephalon, or midbrain, shows the least deviation from the original neural tube of any division of the brain. Of the original three primary brain divisions, it is the only one that does not subdivide. It

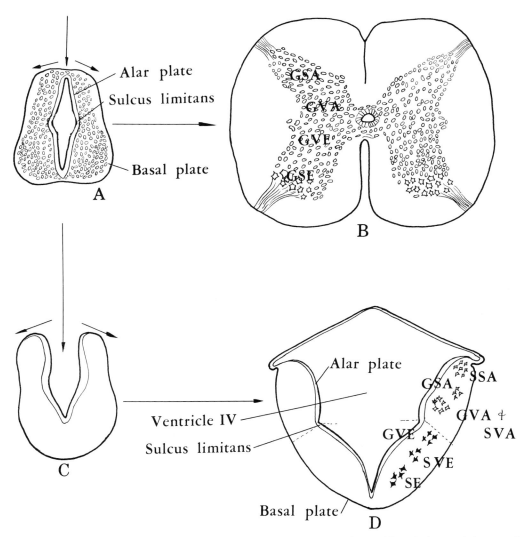

Fig. 2-4. Diagrammatic illustration of the differentiation of the alar and basal plates of the mantle layer of the neural tube (A). The four functional components of a typical spinal nerve are specifically represented in the spinal cord gray matter (B). Development from the neural tube to medulla (D) is more complicated and passes through an intermediate stage (C). The mid-dorsal area of the neural tube splits and the dorsal walls migrate laterally so that the somatic afferent components (which were located dorsally as in B) become located dorsolaterally in the medulla. The original neural canal (in A) becomes the fourth ventricle in the rhombencephalon and the sulcus limitans remains visible in the adult and continues to separate the sensory (alar plate) from the motor (basal plate) nuclei (D).

tends to retain a somewhat circular shape in cross section, with its floor being modified into two broad legs, the *cerebral peduncles,* which are separated by a triangular-shaped space, the *interpeduncular fossa.* The cavity of this region (mesocoele) remains small in diameter as the *aqueduct of Sylvius,* or *cerebral aqueduct* or *iter.* There is no apparent choroid plexus within this small aqueduct.

The *tectum* (roof) of the mesencephalon contains two pairs of rounded prominences, called the corpora quadrigemina. The two rostral elevations are the *rostral (superior) colliculi,* and the two caudal ones are the *caudal (inferior) colliculi.* In addition to the tectum as the dorsal portion in cross section, the *tegmentum* is the central portion, and the *peduncular area* is the ventral portion.

The motor nuclei of the third cranial nerve (oculomotor) are identifiable in a cross section through the rostral colliculi, and the motor nucleus of the fourth cranial nerve (trochlear) may be observed at the level of the caudal colliculi.

The Diencephalon

The diencephalon is the caudal division of the original prosencephalon which tends to remain near the midline as the telencephalon expands dorsally and laterally over it. The diacoele remains in the midline and flattened in a vertical plane to form almost all of the third ventricle of the adult brain. As shown in Figure 2-3, the diacoele connects caudally with the mesocoele and rostrally it joins the two telocoeles by means of the *interventricular foramina of Monro.* The roof of the diacoele becomes very thin, and small blood vessels from the outside push the roof into the cavity to form the choroid plexus of the third ventricle. From the caudal portion of the uppermost (dorsal) diencephalon region there appears a small evagination, the *epiphysis,* or *pineal body.*

Other outgrowths of the diencephalon which include extensions of the diacoele are illustrated in Figure 2-3: (1) the optic cups, which develop as bilateral outgrowths and are undifferentiated eyes; and (2) the *infundibulum,* which is the stalk of the hypophysis, or *pituitary gland,* which is attached to the ventral surface of the hypothalamus, the most ventral division of the diencepha-

lon. In addition to the infundibulum, the hypothalamus contains the *optic chiasm,* which is the crossing of some of the fibers of the optic nerves on the ventral brain surface; the *tuber cinereum,* which is a rather inconspicuous area surrounding the infundibulum; and the paired *mammillary bodies,* which are the most caudal hypothalamic structures visible on the ventral brain surface. The appearance of the above-named hypothalamic structures in the adult brain is shown in Figure 2-5.

Divisions of the Diencephalon

The diencephalon is divisible into five portions. Details of the functional neuroanatomic aspects are considered in Chapter 15, but the divisions are named here for proper orientation, as they appear in both the fetal and the adult brain. (1) The *hypothalamus* has been named above as the portion visible from the ventral aspect of the brain. (2) The *subthalamus* lies caudal and lateral to the hypothalamus. (3) The *thalamus proper,* or *dorsal thalamus,* which contains many specific nuclei and therefore develops from a great proliferation of neuroblasts in the mantle layer of the embryonic diencephalon, is separated from the hypothalamus ventrally by the *hypothalamic sulcus.* The two thalami are connected in the midline by a conspicuous structure known as the *massa intermedia,* or *adhesio interthalamica.* (4) The most dorsal, relatively thin portion of the diencephalon is the *epithalamus.* In addition to the epiphysis, or pineal body, this division contains the *habenula* and the *posterior commissure.* (5) The *metathalamus* is composed of the medial and lateral geniculate bodies.

Diencephalon–Brain Stem Relationship

The diencephalon is sometimes included as a portion of the brain stem (Truex and Carpenter, 1969; Peele, 1961; Meyer, 1964); other authorities, however, do not consider the diencephalon to be a component of the brain stem (Patten, 1968). Both the telencephalon and the diencephalon are derivations of the original prosencephalon. In the adult brain the intimacy of the two divisions is apparent, both structurally and functionally. It may seem logical to consider the diencephalon as separated from the brain stem, based on its close relationship with the telencephalon in the embryonic and in the adult brain.

In this text the diencephalon is considered to be within the brain stem as the most rostral portion, followed caudally in order by the mesencephalon, the pons, and the medulla oblongata. The following four reasons seem to be the cardinal ones for considering the diencephalon within the brain stem: (1) The third ventricle (lumen of the diencephalon) is a single cavity located in the midline in direct line with the cerebral aqueduct and fourth ventricle (Fig. 2-3). (2) In the adult brain the lamina terminalis is the rostral boundary of the third ventricle (see Fig. 2-7). (3) The sulcus limitans is generally considered to be represented in the adult diencephalon as the hypothalamic sulcus, whereas there is no representation of the sulcus limitans in the telencephalon. (4) The thalamus is the relay center for the long ascending (sensory) pathways, which traverse the brain stem and synapse in the thalamus with a terminal neuron that passes to the cerebral cortex (telencephalon).

The Telencephalon

The telencephalon evolves from the most rostral portion of the neural tube with lateral evaginations, called lateral telencephalic vesicles. As illustrated in Figure 2-3, the original telocoele also consists of two portions: the medial part which forms the rostral end of the third ventricle, and the lateral vesicles which ultimately form the lateral ventricles. As alluded to previously, the lamina terminalis marks the rostral termination of the neural tube (Fig. 2-3) and forms the rostral border of the third ventri-

cle in the adult (Fig. 2-7). The two portions of the telocoele are connected by the interventricular foramen of Monro in the embryo (Fig. 2-3); the interventricular foramina of Monro remain in the adult as the connections between the third and the lateral ventricles. Through these foramina the choroid plexus of the third ventricle continues into the lateral ventricles to form the large choroid plexuses of those ventricles. As they are in the fourth ventricle, the choroidal blood vessels in the lateral ventricles are separated from the cerebrospinal fluid by an intervening layer of ependymal cells.

The cerebral hemispheres enlarge greatly and grow rostrally, dorsally, and caudally to cover the diencephalon and the mesencephalon. The temporal lobes continue to extend on each side of the brain stem in a ventro-rostral arch to attain the adult relationship. The telocoele follows the arching growth, thus accounting for the complicated curved shape of the lateral ventricles. Since the diencephalon remains stationary as a pivot-point and the telencephalon arches around it as described, the structures of the hemispheric wall (e.g., the fornix and hippocampus) follow the C-shaped growth pattern and therefore conform to the curvature of the ventricles (see Fig. 2-15).

Development and Organization of Cerebral Cortex

During the early developmental stages the surface of the cerebral hemispheres is smooth, without much specific differentiation. In passing, it may be noted that the telencephalon of less complicated submammalian and lower mammalian brains (e.g., rat) remains smooth, and such brains are classified as lissencephalic. The cortex of the brain of higher mammals is convoluted and contains specific gyri (elongated elevations) and sulci (grooves between the gyri). The apparent reason for the more complex brain of higher animals possessing a convoluted cortex is to allow space for more cell bodies of sensory, motor, association, commissural, and projection neurons, by the tremendous increase in surface area and hence in the superficial gray cortex containing the cell bodies. As alluded to previously, the specific gyri and sulci patterns vary more among mammals than does any other portion of the brain, so that one can not easily interchange nomenclature for features of the cortex among different mammals (*e.g.,* man and dog), as can be done in comparative studies of the brain stem, which follows a more uniform pattern of architecture.

Phylogenetically there are three cortices which show relative degrees of dominance. In fishes and amphibians the pallium (cortex) of the telencephalon serves primarily to receive olfactory impulses, although the pallium structurally is unlike that found in mammals (Brodal, 1969). Typical of a conservative organ that saves structural evidence of its phylogenetic development, the brain of higher mammals reveals the suppression of the once-dominant olfactory brain, the archicortex and paleocortex, by the recent neocortex. The *archicortex* is represented by the hippocampus, and the *paleocortex* by the piriform area. These cortices have been pushed ventrally with the archipallium folded inside ventrally and medially by the force of the *neocortex* (*neopallium*), the dorsal cortex, which is best developed in higher mammals and includes most of the surface of the hemisphere.

The ontogenetic aspects of the differentiation of the cortex illustrates the significance of olfactory localization in that the first fissures to develop are the rhinal and hippocampal fissures (Davies, 1963). These separate the olfactory from the nonolfactory cortex.

In summary of the organization of the brain as discussed above for the developmental states, and as an introduction to the following section on the gross anatomy of the adult brain, Table 2-1 outlines the divisions of the brain and their major contents and derivatives.

Gross Anatomy of the Adult Brain

The five embryonic brain divisions discussed under Development in the first part of this Chapter will constitute the major headings for this discussion of the gross anatomy of the adult brain. Before considering each division and the organization of the intrinsic structures, one should remember that in very basic terminology the brain may be thought of as consisting of the brain stem plus two superstructures that are attached. The *brain stem* includes the myelencephalon, the pons (of the metencephalon), the mesencephalon, and the diencephalon. Obviously the two superstructures are the telencephalon and the cerebellum (of the metencephalon).

The following discussion will pertain only to the gross anatomic aspects of each division. Microscopic observations of specific small nuclei and pathways, with their functions, are considered in separate chapters.

The Myelencephalon

The Medulla Oblongata

The spinal cord continues rostrally through the foramen magnum, where it expands without a sharp line of demarcation to become the medulla oblongata, the sole derivative of the caudal portion of the primitive rhombencephalon. Figure 2-5 shows the caudal limit of the medulla to be approximately at the level of the first cervical nerve; the rostral limit is the upper border of the trapezoid body at the caudal edge of the pons, a site which is logically named the pons-medulla junction. The medulla lies within the posterior fossa of the skull, where it rests on the basilar por-

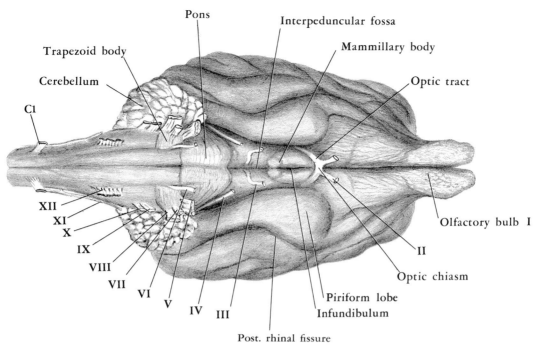

Pons

Interpeduncular fossa

Trapezoid body

Mammillary body

Cerebellum

Optic tract

C1

XII
XI
X
IX
VIII
VII
VI
V
IV
III

Olfactory bulb I

II

Optic chiasm

Piriform lobe
Infundibulum

Post. rhinal fissure

Fig. 2-5. Ventral view of the dog brain and cranial nerves.

tion of the occipital bone. Almost all of the medulla is covered dorsally by the cerebellum (see Fig. 2-7).

The visible sulci on the surface of the spinal cord are continued rostrally into the medulla and may aid as landmarks in describing the gross anatomy of the medulla.

The Ventral Area of the Medulla (*Fig. 2-5*)

Near the caudal end of the medulla, at the level of the rootlets of the hypoglossal nerve, the ventral median fissure is interrupted by rather inconspicuous interdigitating fibers that obliquely cross the midline between the pyramids as the *decussation of the pyramids*. These are the decussating corticospinal fibers that cross in the lower medulla to enter the lateral funiculi of the spinal cord as the lateral corticospinal (pyramidal) tracts.

CRANIAL NERVES

From the ventral aspect of the brain (Fig. 2-5) one can observe the attachments to the medulla of the last seven pairs of cranial nerves. As stated previously, microscopic observations and functions are discussed in separate chapters. The intramedullary nuclei and the course of the fibers of these cranial nerves are considered in Chapters 11 and 12. Their peripheral distribution and functional importance are discussed in Chapter 17.

Near the caudal limit of the medulla at the level of the pyramidal decussation, immediately lateral to the pyramids, several rootlets emerge to form the *hypoglossal nerve* (cranial nerve XII). Notice that the longitudinal craniocaudal arrangement of hypoglossal fibers is in direct line with the ventral roots of the spinal nerves.

The eleventh cranial nerve, the *accessory*

nerve, has two divisions, based on origin of their fibers from the central nervous system. The roots of the *spinal* division originate from the ventral gray column of the upper (usually upper four) cervical levels. As shown in Figure 2-5, the cervical contributions form a nerve which is parallel to the lateral surface of the spinal cord and passes cephalically through the foramen magnum to join the second portion, the *bulbar* division. This portion of the eleventh cranial nerve is responsible for the name 'accessory,' because it emerges from the ventrolateral surface of the medulla immediately caudal to the vagal fibers, which it joins as an accessory to the vagus nerve.

The *vagus nerve* (X) is attached to the side of the medulla by a series of root filaments in close apposition rostrally to those of the ninth cranial nerve (glossopharyngeal). The rootlets consolidate to form the vagus nerve, which proceeds distally as it is joined by the accessory nerve on its caudal side. In routine methods of removing the brain from the skull, the vagus nerve is usually cut proximal to the jugular ganglion, which is just about at the level of the jugular foramen through which the nerve passes to enter the petrobasilar fissure. Extracranially the vagus has a second sensory ganglion, the nodose, which is much larger. The details of functional significance are discussed in Chapter 17, but for the gross anatomic importance the vagus (X) and the glossopharyngeal (IX) nerve discussed below each has two sensory ganglia; the superior (proximal) in each case is immediately inside the skull, and the inferior (distal) is outside the skull. These two nerves are attached to the lateral side of the medulla in direct line with the dorsal roots of the spinal nerves. Although the accessory nerve (XI) has no ganglion in the adult, it is closely associated with the ninth and tenth cranial nerves, and all three pass through the same foramen.

The *glossopharyngeal nerve* (IX) is attached to the medulla by rootlets in the same man-

ner as the vagus. Because of the linear arrangement of the rootlets of the two nerves, it is often not possible to positively distinguish the rootlets of one from those of the other, except for the fact that those of the glossopharyngeal nerve are rostral to those of the vagus. As mentioned previously for the vagus nerve, in routine brain removal the glossopharyngeal nerve is cut proximal to the merging of the rootlets into a solid nerve, above the two sensory ganglia.

For the gross anatomic attachment of the rostral three cranial nerves attaching to the medulla (cranial nerves VI, VII, and VIII), the trapezoid body may be used as a point of reference. As illustrated in Figure 2-5, the facial (VII) and vestibulocochlear (VIII) nerves are very close together near the lateral border of the trapezoid body, directly ventral to the flocculus of the cerebellum.

The eighth cranial, or *vestibulocochlear,* nerve (formerly called auditory, acoustic, or statoacoustic) is divisible, as the proper name suggests, into a vestibular and a cochlear nerve. Grossly, however, the two nerves are tightly bound together to compose the eighth cranial nerve. This is the only cranial nerve that does not leave the skull; it extends only the short distance between the medulla oblongata, and the inner ear within the petrosal portion of the temporal bone. Each division of this nerve has a ganglion. With the aid of a dissecting microscope one may see the components of the *spiral ganglion* of the cochlear nerve within the modiolus of the cochlear portion of the inner ear. The *vestibular,* or *Scarpa's, ganglia* may be located (with difficulty) within the pyramid of the temporal bone near the ventral portion of the internal auditory meatus as the nerve passes to the vestibular portion of the inner ear. It should be remembered that in routine methods of brain removal the ganglia are not present on the portion of the nerve that remains attached to the medulla. With the aid of a dissecting microscope, however, one may identify the *ventral and dorsal cochlear nuclei* as slight swellings on the caudal edge of the nerve that is actually on the lateral surface of the restiform body immediately caudal to the transected cerebellar peduncles following removal of the cerebellum. The dorsal cochlear nucleus must be observed from the dorsal view.

The *facial nerve* (VII) is attached to the medulla in very close association with the vestibulocochlear nerve.

Specifically, the facial nerve is at the rostral border of the trapezoid body, ventral (medial) to the vestibular nerve and directly caudal to the large trigeminal nerve, which is attached to the pons. The physical closeness of the seventh and eighth cranial nerves is emphasized by the fact that the two nerves are enclosed in a common sheath as they pass together through the internal auditory meatus.

The intermedius, or glossopalatine nerve, which is a part of the seventh nerve and grossly separable in man and other species, is not grossly identifiable in the dog (McClure, 1964). The *geniculate ganglion,* the sensory ganglion for the facial nerve, is located within the facial canal distal to the internal auditory meatus and obviously is not visible on the short attachment of the nerve to the medulla as shown in Figure 2-5.

The sixth cranial nerve, the *abducens,* attaches to the trapezoid body medial to the seventh and eighth nerves, in a groove between the lateral edge of the pyramidal tracts and the trapezium. This nerve is closer to the ventral median fissure in a direct rostro-caudal line with the hypoglossal nerve already discussed and the oculomotor nerve which is attached to the ventral side of the mesencephalon. The functional significance of this is that these three nerves have the same functional components, as described in detail in Chapter 17. Unlike the seventh and eighth cranial nerves, the abducens has no sensory ganglion anywhere along its course.

The Dorsal Area of the Medulla

On the dorsal side the visible sulci and tracts of the spinal cord are continued rostrally into the medulla (Fig. 2-6). The medial fasciculus gracilis and the larger lateral fasciculus cuneatus shortly diverge to form a V-shaped caudal end of the fourth ventricle, the *obex* (Fig. 2-6). In general description this marks the point of junction between the central canal of the spinal cord and the fourth ventricle of the medulla. At about the level of the obex, the nuclei gracilis and cuneatus are present within the terminal ends of the respective fasciculi. Grossly, one sees a swelling on only the fasciculus cuneatus, the *cuneate tubercle,* which is much larger than the gracilis nucleus because it contains the cuneate nucleus plus the lateral (accessory) cuneate nucleus. The cuneate tubercle appears to continue rostrally into the caudal cerebellar peduncle or restiform body. On the dorsal side the external arcuate fibers that join the restiform body appear to obliterate the rostral portion of the cuneate tubercle, as illustrated in Figure 2-6. On the lateral side the external arcuate fibers cover the spinal tract of the trigeminal nerve, which may be identified grossly.

THE FOURTH VENTRICLE

The fourth ventricle may be exposed for study from the dorsal view by removal of the cerebellum. This is best accomplished by horizontally transecting the three cerebellar peduncles on both sides. Recall that this diamond-shaped cavity is the cavity of the embryonic rhombencephalon, which includes the pons as well as the medulla, with no anatomic line of demarcation between the two areas. Therefore, the fourth ventricle lies between the pons and medulla ventrally and the cerebellum dorsally. It is continuous rostrally with the cerebral aqueduct (of Sylvius) and caudally with the central canal of the spinal cord. From the func-

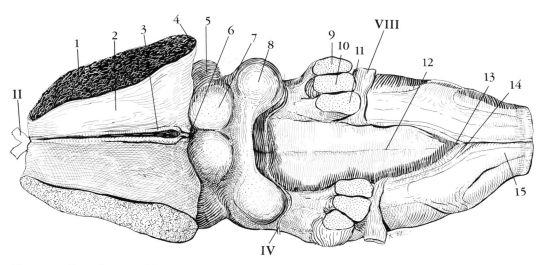

Fig. 2-6. Dorsal view of brain stem (dog). 1, cut surface between cerebrum and brain stem; 2, thalamus; 3, stria medullaris of thalamus; 4, lateral geniculate body; 5, medial geniculate body; 6, pineal body; 7, rostral colliculus; 8, caudal colliculus; 9, middle cerebellar peduncle; 10, caudal cerebellar peduncle; 11, rostral cerebellar peduncle; 12, dorsal median sulcus in rhomboid fossa; 13, obex; 14, fasciculus gracilis; 15, fasciculus cuneatus; II, optic nerve; IV, trochlear nerve; VIII, vestibulocochlear nerve.

tional standpoint it is not very important to consider the fourth ventricle as leading into the central canal of the spinal cord, because it probably makes little difference whether or not cerebrospinal fluid flows this route. From the standpoint of flow of the fluid, the importance of the connection of the fourth ventricle with the subarachnoid space and particularly with the cisterna cerebellomedullaris (cisterna magna) should be emphasized. Considering the cerebral aqueduct and the central canal of the spinal cord as the rostral and caudal points, respectively, of the diamond-shaped cavity, there are two bilateral evaginations, called the *lateral recesses,* which define the other two points of the diamond. The points of the lateral recesses are not closed, but remain open as the *foramina of Luschka,* through which cerebrospinal fluid may pass from the fourth ventricle to the subarachnoid space outside of the brain. Grossly, this can be identified by observing from the lateral view

that the choroid plexus of the fourth ventricle passes over each restiform body to emerge through the foramen of Luschka and become visible on the lateral surface directly below the flocculus of the cerebellum at the level of the vestibulocochlear nerve. These gross anatomic relationships will again be apparent in cross sections of the medulla shown in figures in Chapter 11 and the plates in the Atlas. As discussed in Chapter 5, a third exit of cerebrospinal fluid from the fourth ventricle, the *foramen of Magendie* in the posterior medullary velum, is usually mentioned. In agreement with McGrath (1960), this foramen is considered to be absent in the dog.

The floor of the fourth ventricle, or rhomboid fossa, is formed by the dorsal surfaces of the pons and the medulla oblongata without any exact line of demarcation between the two. A hand lens or dissecting microscope is useful for study of small anatomic markings on the floor, which may be correlated with the serial cross sections of the brain stem as illustrated in Chapters 11 and 12 and in the Atlas portion of the book.

The *median sulcus* (Fig. 2-6) separates the fossa into two symmetrical halves. The floor is not flat, and it blends gradually with the sloping walls, particularly in the vestibular area. The *sulcus limitans* is visible, but somewhat inconspicuous as it follows the base of the curved vestibular area, which marks the location of the vestibular nuclei. The vestibulocochlear nerve can be followed grossly into the area, and the swelling of the dorsal cochlear nucleus is visible on the dorsal surface of the restiform body at the level of the lateral recess of the ventricle.

At the transverse level immediately rostral to the lateral recess, the median sulcus is bordered on each side by a slight swelling, the *facial colliculus* (Fig. 2-6). This is at the level of the internal genu of the facial nerve, as illustrated in Plate 14 (see Atlas). At the level of the caudal portion of the vestibular swelling on the lateral wall, on each side of the midline (median sulcus) is a very inconspicuous rounded swelling, the site of the *hypoglossal nucleus.* Because of the triangular-shaped area resulting from the closeness of the obex and the walls of the ventricle diverging grossly from it, the names of the various structures include the word 'trigone.' Therefore the site of the hypoglossal nucleus described above is referred to as the *hypoglossal trigone.* Immediately below and slightly rostral to the obex, there is a small delicate-looking nodule, the *nucleus of the vagus nerve.* This too extends to the lateral walls and forms a triangle, and therefore is called the *vagal trigone.*

The rostral portion of *the roof of the fourth ventricle* is covered by the *rostral (anterior) medullary velum* (Figs. 2-6, 2-7), which is a very thin sheet of white matter extending from the cerebellum to the caudal level of the midbrain as it spans the two rostral (superior) cerebellar peduncles (brachia conjunctiva). At the rostral attachment of

the velum on the caudal side of the inferior colliculi of the midbrain, the very thin *trochlear nerves* may be located, emerging from the lateral sides of the velum to pass laterally and ventrally around the cerebral peduncles. Recall that the trochlear nerve (IV) is the only cranial nerve that emerges from the dorsal side of the brain stem, after the fibers have decussated within the rostral (anterior) medullary velum (Plate 17).

The caudal portion of the roof of the fourth ventricle, the *caudal (posterior) medullary velum,* is very thin and consists of only a layer of ependymal cells supported on the outside by the thin vascular pia mater (Ranson and Clark, 1959). This constitutes the *tela choroidea,* from which vascular tufts, covered by epithelium, invaginate into the cavity and form the *choroid plexus* of the fourth ventricle. The choroid plexus not only invaginates rostrally into the ventricle, but, as discussed above, the plexus may be observed from the lateral surface of the medulla as it emerges through the foramen of Luschka at the lateral recess on the surface of the restiform body directly below the flocculus of the cerebellum.

The Metencephalon

The metencephalon is derived from the rostral portion of the embryonic rhombencephalon. This division of the brain further differentiates into one ventral portion, the pons, a portion of the brain stem, and the cerebellum, the caudal superstructure referred to in the introduction to this section of the Chapter.

The Pons

From the *ventral view* of the brain the pons appears as a band of transverse fibers interposed between the trapezoid body of the medulla oblongata caudally and the cerebral peduncles (crura cerebri) of the mesencephalon rostrally (Fig. 2-5). As illustrated in the figure, the greatest rostro-caudal

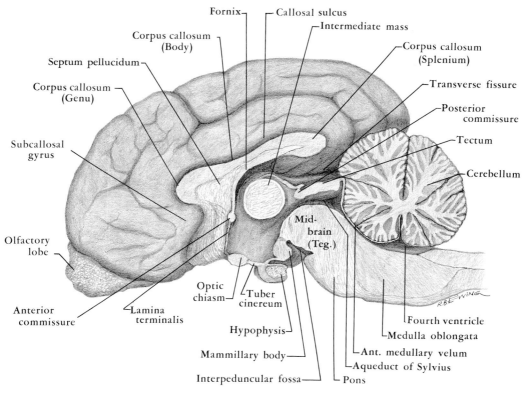

Fig. 2-7. Sagittal section of the brain of the dog.

length is in the midline, and laterally the pons tapers as it approaches the cerebellopontine angle. The transverse fibers may be followed grossly around the lateral side, and when the cerebellar hemisphere is lifted they may be seen to form the middle cerebellar peduncle (brachium pontis). The ventral midline contains a shallow indentation in direct continuation of the ventral median fissure of the medulla oblongata, and the basilar artery is lodged in the continuous groove.

The *trigeminal nerve* (V) emerges from the pons far laterally, in association with the brachium pontis. From the ventral view, the large trigeminal nerve is directly rostral to nerves VII and VIII, which are associated with the trapezoid body (Fig. 2-5). The trigeminal nerve is mostly sensory, and grossly it is possible to identify the large

sensory root (*portio major*) and the smaller motor root (*portio minor*), or masticator nerve, on the medial side of the whole nerve.

The *dorsal surface of the pons* is discussed above in part with the myelencephalon, because the floor of the fourth ventricle is composed of both the open portion of the medulla oblongata and the dorsal surface of the pons.

The *rostral (superior) cerebellar peduncles* (brachia conjunctiva) form the lateral walls of the fourth ventricle in the pontine area. These peduncles, which are covered by the cerebellum from which they pass rostrally as the most medial of the three cerebellar peduncles, enter the midbrain, where they soon lose their identity as they blend grossly with the mesencephalic tissue beneath the caudal colliculi. As mentioned previously, the rostral portion of the roof of the fourth ventricle is formed by the rostral (anterior) medullary velum, the thin white sheet that stretches between the two rostral (superior) cerebellar peduncles. Near its attachment on the midbrain above the aqueduct of Sylvius the anterior medullary velum contains the decussation of the trochlear nerves.

The Cerebellum

The cerebellum develops from the dorsal surface of the embryonic metencephalon. It is caudal to the mesencephalon and *in vivo* is separated from the caudal surface of the cerebral hemisphere by the osseous tentorium and the tentorium cerebelli within the transverse cerebral fissure. Obviously these structures are not present in the brain that has been removed from the skull and has had the dura mater removed. The cerebellum covers most of the fourth ventricle.

The cerebellum consists of three gross divisions: a small unpaired midline band, the *vermis,* named because of its resemblance to a worm, and on each side a large mass, the *cerebellar hemisphere.*

Similar to the cerebrum, the cerebellum has a peripheral cortex composed of gray matter. In addition, the cerebellar cortex is also similar to the cerebral cortex in that the cerebellar cortex consists of long slender convolutions, named *folia,* which are separated by somewhat parallel sulci, therefore corresponding to the gyri and sulci of the cerebrum. Also, like the cerebral hemispheres, the cerebellum has an internal white matter in which are embedded nuclei (gray matter). The white matter connects the cerebellar nuclei and cortex and the entire cerebellum with the brain stem by means of the three cerebellar peduncles (Fig. 2-6). Although in a transection through the three cerebellar peduncles on each dorsolateral side of the fourth ventricle they appear to be oriented in a lateral-medial plane, the nomenclature is based on the functional

rostrocaudal sequence as follows: *the rostral (superior) cerebellar peduncle (brachium conjunctivum),* attached to the midbrain, the *middle cerebellar peduncle (brachium pontis)* attached to the pons, and the *caudal (inferior) cerebellar peduncle (restiform body),* connected to the medulla oblongata. In a hemisected brain the sagittal view of the arrangement of the gray and white matter of the cerebellum resembles the appearance of a shrub or tree, and hence it is often referred to as the *arbor vitae* (Figs. 2-7 and 2-8 A).

Fissures and Lobes of the Cerebellum

Embryonically the first cerebellar fissure to appear is the *caudo(postero)lateral fissure,* which separates the flocculonodular lobe caudally from the body of the cerebellum (corpus cerebelli) rostrally. Further differentiation results in the flocculonodular lobe remaining small and the body of the cerebellum enlarging to form most of the cerebellar mass.

The next fissure to develop is the *primary fissure,* which separates the anterior and posterior lobes of the cerebellar body. In a hemisected cerebellum the lobules of the vermis are relatively easy to identify, as illustrated in Figure 2-8 A. As a point of interest the nine lobules named are exactly the same in dog (Fig. 2-8 A) and man (House and Pansky, 1967). In the sagittal view of the vermis, the lobules in a rostrocaudal sequence are: lingula, central lobule, culmen, declive, folium, tuber, pyramis, uvula, and nodulus. Notice that in the dog cerebellum the primary fissure is between the culmen and declive as in man, but in the dog the fissure is more perpendicular to the brain stem. Perhaps this is due to the fact that the lingula, which rests on the anterior medullary velum, is larger in the dog, and has pushed the primary fissure caudally to the dorsoventral orientation. According to Larsell (1952), among mammalian species the size of the cerebellar lingula relatively parallels that of the tail.

Meyer (1964) lists the classical nine lobules in the vermis as given above, but illustrates an additional lobule, *lobulus ascendens*, inserted between the central lobule and the culmen, as based on the reports of Dexler (1932) and Ackerknecht (1943). There are many additional names of subdivisions that would be considered in a more detailed study of the cerebellum. For the purposes of this text, however, it seems injudicious to include various uncommonly used synonyms and additional details. Figure 2-8 B, however, refers to a lobule, the simplex lobule (lobulus simplex), and its lateral extension, which contributes to a goodly portion of the dorsal hemisphere. This term is

also used in human nomenclature (Ranson and Clark, 1959) and refers to the most rostral portion of the posterior lobe of the cerebellum at the caudal border of the primary fissure. Based on the labels in Figure 2-8 A for the nine lobules of the vermis, the simplex lobule would be comparable to the rostral portion of the declive.

Grossly, the cerebellar hemispheres are divisible into the paraflocculus, the paramedian lobule, the ansiform lobule, and the lateral extension of the simplex lobule (Fig. 2-8 B).

The *paraflocculus* lies at the lateral base of the cerebellum as a doubled tier of short folia extending from the rostrolateral point

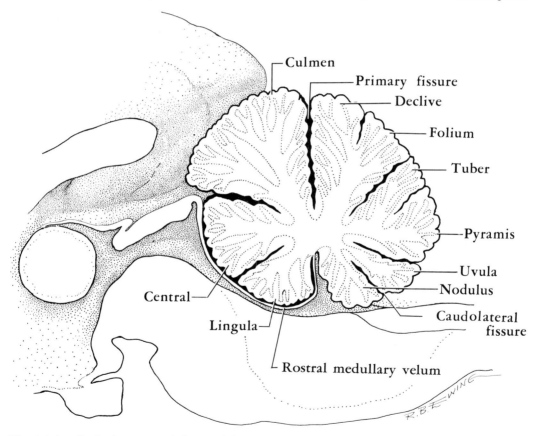

Fig. 2-8 A. Sagittal section of the caudal portion of the brain to show the nine lobules of the cerebellar vermis in relation to surrounding structures. The nomenclature used here for the dog agrees with human terminology, but it should be realized that there are other commonly accepted methods of identification which differ from the classical 'nine-lobule plan' given here.

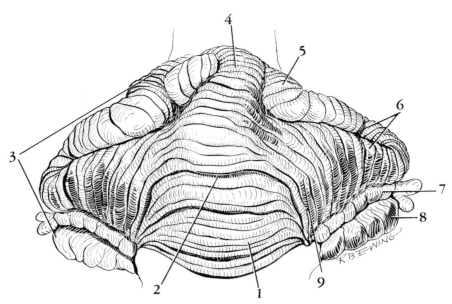

Fig. 2-8 B. Dorsorostral view of the cerebellum. 1, vermis portion of rostral (anterior) lobe; 2, primary fissure; 3, right hemisphere; 4, vermis portion of caudal (posterior) lobe; 5, paramedian lobule; 6, ansiform lobule; 7, dorsal paraflocculus; 8, ventral paraflocculus; 9, lateral portion of simplex lobule. (After Meyer, 1964.)

caudally around the base of the hemisphere to the caudal side where the dorsal paraflocculus meets the paramedian lobule and the ventral paraflocculus is lost to view immediately caudal to the dorsal cochlear nucleus of the eighth cranial nerve.

At the lateral apex, the ventral paraflocculus possesses a few long folia, mushroom-shaped in appearance, which project laterally into the cerebellar fossa of the petrosal part of the temporal bone.

At the lateral apex of the hemisphere both tiers, the dorsal and the ventral paraflocculus, pass dorsal to the *flocculus,* which is the lateral portion of the flocculonodular lobe. This lies immediately rostral to the cochlear nuclei and extends to the attachment of the trigeminal nerve.

From the caudal aspect of the cerebellum the *paramedian lobule* lies next to the vermis on each side as a band of transverse folia. At approximately half the distance to the primary fissure the paramedian lobule turns laterally to blend with the *ansiform lobule,* which continues laterally above the caudal

portion of the dorsal paraflocculus. Rostral to the ansiform lobule are the folia of the lateral portion of the *simplex lobule,* which extends to the primary fissure. Rostral to the primary fissure the *anterior lobe* extends to the caudal edge of the inferior colliculi of the midbrain.

The cerebellar peduncles have been discussed with their relationship to the fourth ventricle. The functional aspects are considered in Chapter 14. It seems appropriate also to consider the cerebellar nuclei in Chapter 14 because they can best be seen submacroscopically in cross sections of the metencephalon.

The Mesencephalon

The Midbrain

The midbrain, or mesencephalon is the segment of the brain stem between the diencephalon rostrally and the metencephalon caudally. Recall that, as discussed in the first part of this Chapter, the mesencephalon is the only one of the three original embryonic divisions that does not subdivide during growth and differentiation; both the prosencephalon and the rhombencephalon do. To mention briefly two of the gross anatomic features of the mesencephalon that reflect a lesser degree of modification of the original neural tube, the following may be considered: (1) the mesencephalon retains, to a large extent, the original tubular form, with no hemispheric outgrowths; (2) the mesencephalic segment of the ventricular system, the cerebral aqueduct, is retained as a small simple cylinder, with no expansions.

Grossly, the adult mesencephalon has a rather consistent shape throughout its short length. In cross section there are four areas in a dorso-ventral sequence (Fig. 2-9): *tectum*,

containing the corpora quadrigemina; *tegmentum*, containing the reticular core with its embedded nuclei; *substantia nigra*, a darkly pigmented layer best observed in old dogs; and the *crura cerebri* (*cerebral peduncles*), which form the 'legs' of the midbrain. The space between the cerebral peduncles is the *interpeduncular space*.

TECTUM

The tectum of the midbrain is not visible from a surface view from any angle of the whole brain, because of the covering of the cerebral hemispheres and the cerebellum. In a whole brain with the dura mater removed, one can separate the cerebral and cerebellar hemispheres and follow the

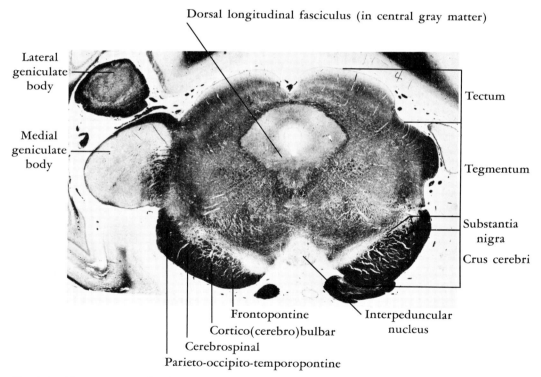

Dorsal longitudinal fasciculus (in central gray matter)

Lateral geniculate body

Medial geniculate body

Tectum

Tegmentum

Substantia nigra

Crus cerebri

Frontopontine
Cortico(cerebro)bulbar
Cerebrospinal
Parieto-occipito-temporopontine

Interpeduncular nucleus

Fig. 2-9. Photomicrograph of cross section through the rostral colliculus of the midbrain. The four areas are labeled from dorsal to ventral: tectum, tegmentum, substantia nigra, and crus cerebri (cerebral peduncle). The organization of the corticifugal fibers in the crus cerebri is in agreement with Singer (1962). Additional structures are labeled for orientation.

transverse cerebral fissure between the two hemispheres and observe that the delicate leptomeninges (pia mater and arachnoid) from the rostral surface of the cerebellum extend on the caudal colliculi of the mid-brain tectum. Recall that when dura mater is present the tentorium cerebelli is within the transverse cerebral fissure and therefore extends toward the tectum. The most conspicuous structures of the tectum are two pairs of rounded elevations, the *rostral (superior) colliculi* and the *caudal (inferior) colliculi.* Collectively, the four prominences are referred to as the *corpora quadrigemina,* or quadrigeminal bodies. The two rostral colliculi are connected by a relatively inconspicuous commissure (commissure of the rostral colliculi). The two caudal colliculi are also connected in a similar fashion, by the commissure of the caudal colliculi, which is much more distinct as a transverse band of fibers directly caudal to the rostral colliculi above the cerebral aqueduct (Fig. 2-6).

Each colliculus is connected to a geniculate body by an arm, or brachium, of the rostral or the caudal colliculus, respectively (Plates 20 and 21). The brachium of the rostral colliculus joins it to the lateral geniculate body, and the brachium of the caudal colliculus joins it to the medial geniculate body. Again, grossly, the brachia of the caudal colliculi and the medial geniculate bodies are more distinct than the respective connections of the rostral colliculi. Recall that the geniculate bodies are technically components of the metathalamus, a subdivision of the diencephalon. It is tempting to discuss here the functional aspects of these structures, but the primary objective of this Chapter is to consider the gross anatomy. Because the lateral geniculate bodies and the rostral colliculi are concerned functionally with vision and visual reflexes, the functional aspects are covered in Chapter 18. Since the medial geniculate bodies and the caudal colliculi are concerned functionally with hearing and audi-

tory reflexes, respectively, the neuroanatomic details are covered in Chapter 19.

CRANIAL NERVES OF THE
MESENCEPHALON

Two cranial nerves (III and IV) are attached to the mesencephalon. As a point of interest, both nerves go to the same area (extrinsic eye muscles), but one (III) emerges from the ventral side, and the other (IV) emerges from the dorsal side, of the mesencephalon.

The *oculomotor nerve (cranial nerve III)* (Fig. 2-5) emerges from the ventral surface of the cerebral peduncle as a strong-looking nerve. Close observation reveals that the thin fascicles (rootlets) emerge deeply from the midline in the center of the interpeduncular space and then pass lateroventrally around the medial surface of the peduncle to fuse as the single, relatively thick oculomotor nerve. The rootlets consolidate at the spot where the posterior cerebral artery within the circle of Willis (circulus arteriosus cerebri) makes a sharp bend laterally around the rostral base of the oculomotor nerve. Caudal to the third cranial nerve the anterior (superior) cerebellar artery arises from the posterior cerebral artery to pass laterally and dorsally toward the rostral surface of the cerebellum.

The *trochlear nerve (cranial nerve IV)* (Fig. 2-5) is the only nerve that emerges from the dorsal side of the brain stem. This very thin nerve may be seen as it emerges laterally from the attachment of the anterior medullary velum to the caudal surface of the mesencephalon ventral to the caudal colliculi. The trochlear nerves decussate within the anterior medullary velum to pass around the lateral surface of the cerebral peduncles within the transverse cerebral fissure between the cerebral and cerebellar hemispheres. On the ventral side of the brain the trochlear nerve emerges from the transverse cerebral fissure bordered by the rostrolateral edge of the pons, the rostro-medial tip of the ventral paraflocculus of the cerebellum, and the cerebral hemisphere. From here the nerve passes rostrally adjacent to the oculomotor nerve, and both follow a similar course to the orbit, as discussed in Chapter 17.

The Diencephalon

The diencephalon (Figs. 2-5, 2-6, 2-7, and 2-10) is the most rostral division of the brain stem. As illustrated in the figures, the diencephalon is between the telencephalon and the mesencephalon. Grossly, the bordering telencephalic structures are very intimately associated with the diencephalon. It should be remembered that the diencephalon and telencephalon are subdivisions of the embryonic prosencephalon, or forebrain.

Boundaries of the Diencephalon

Only the general boundaries that can be observed by gross dissection are considered here. The *dorsal, or superior, boundary* of the diencephalon is formed by the fornix and velum interpositum. *Caudally,* the diencephalon blends with the tegmentum of the midbrain without any definite line of demarcation. *Laterally,* the optic tract forms the rostral boundary, and the internal capsule borders the caudal area before its descending fibers enter the cerebral peduncles of the midbrain. *Rostrally,* the lamina terminalis delimits the diencephalon. There is no part of the brain that serves as a *ventral boundary,* because the hypothalamus is directly on the ventral surface of the brain. At the *medial boundary* is the third ventricle.

Divisions of the Diencephalon

The anatomic subdivisions of the diencephalon are as follows: (1) the epithalamus, which is the most dorsal portion, and in the midline; (2) the thalamus, or dorsal thalamus, the largest portion; (3) the hypothalamus, the basal area; (4) the metathalamus, composed of the geniculate bodies; and (5) the subthalamus, an area lateral to the hypothalamus.

EPITHALAMUS

The epithalamus (Figs. 2-7, 2-10) is best observed in a hemisected brain, because all of its components are in the midline and form the most dorsal thin area of the diencephalon. The following structures of the epithalamus should be identified in gross dissection: The *epiphysis,* or *pineal body (gland),* which is very small in the dog, is a caudal projection from the extreme dorsocaudal border of the diencephalon. The pineal body rests rostral to the groove between the two rostral colliculi of the midbrain tectum. This area is referred to as the pretectal area. Immediately rostral to the attachment of the pineal body are the *habenulae,* which are two small nuclear masses bordering the midline that are best seen in transverse sections (Plate 22). Gross transverse (cross) sections through the nuclei will reveal the habenulopeduncular tract which passes from the habenular nuclei to the interpeduncular space of the midbrain. The *stria medullaris thalami* is a thin bundle of myelinated fibers on the dorsal border of the diencephalon in the midline.

THALAMUS

The thalamus, or dorsal thalamus (Figs. 2-7, 2-10), is a large gray mass between the midline and the internal capsule on each side. In the midline the thalamus is directly ventral to the epithalamus as it forms the wall of the third ventricle. Laterally, however, the thalamus bulges so that it extends dorsally to the centrally placed epithalamus. This is best seen in cross section (Plates 22 and 23).

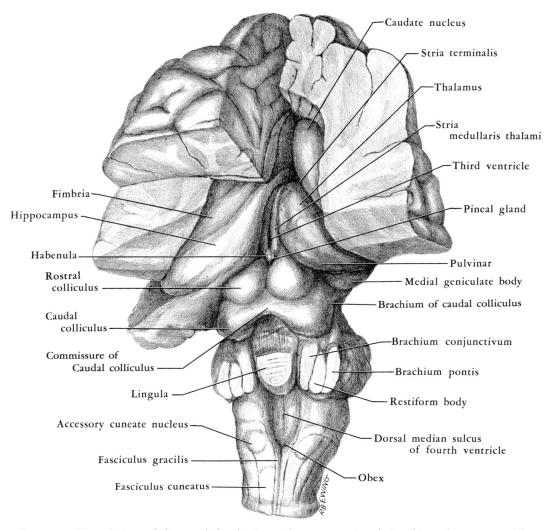

Caudate nucleus

Stria terminalis

Thalamus

Stria
medullaris thalami

Third ventricle

Pineal gland

Fimbria

Hippocampus

Habenula

Rostral
colliculus

Caudal
colliculus

Commissure of
Caudal colliculus

Lingula

Accessory cuneate nucleus

Fasciculus gracilis

Fasciculus cuneatus

Pulvinar

Medial geniculate body

Brachium of caudal colliculus

Brachium conjunctivum

Brachium pontis

Restiform body

Dorsal median sulcus
of fourth ventricle

Obex

Fig. 2-10. Dorsal view of dissected dog brain to show anatomic relationships of structures. The cerebellum has been removed. The right cerebral hemisphere is cut horizontally and the caudal portion removed. The rostral portion of the left hemisphere is intact, with most of the caudal portion removed. The structures viewed on the medial portion of the right cerebral hemisphere are ventral (deep) to those shown on the left.

The cavity of the diencephalon is the third ventricle, which is in the midline as a narrow cavity oriented in a dorsoventral plane bordered on each side by the thalamus and hypothalamus. In the center of the ventricle there is a large round connecting structure between the two thalami. This gray area which binds the two thalami is known as the *interthalamic connection* (*massa intermedia*, or *interthalamic adhesion* [*adhesio interthalamica*]). As a point of interest, this interthalamic connection is present in only approximately 80 per cent of human brains (Truex and Carpenter, 1969). The details of the thalamic nuclei are considered in Chapter 15. Grossly, the anterior nuclear group is identifiable as a rounded prominence, the *anterior tubercle of the thalamus,* at the rostro-dorsal edge of the thalamus.

HYPOTHALAMUS

On the basal surface of the brain the hypothalamus is composed of the following structures in a rostrocaudal sequence: optic chiasm, infundibulum (and neurohypophysis), tuber cinereum, and mammillary bodies. The *infundibulum* is the stalk of the hypophysis (pituitary gland), and the *neurohypophysis* is the posterior lobe of the pituitary, so named because it is composed of nerve tissue and not of glandular tissue, as is the adenohypophysis (the anterior lobe). In a brain in which the pituitary gland is not attached, usually the severance is through the infundibulum. Close observation of the infundibulum reveals a hollow center, which is the extension of the third ventricle into this structure. Since the neurohypophysis is connected to the infundibulum, which in turn attaches to the tuber cinereum, it is logical to consider only the three basic hypothalamic areas with reference to the optic chiasm, tuber cinereum, and mammillary bodies. It is customary to use these three areas, namely, the supraoptic, tuber, and mammillary areas, as points of reference for discussion of the organization of the hypothalamic nuclei, as in Chapter 15. The *mammillary bodies* are two rounded eminences between the cerebral peduncles fused at the midline immediately rostral to the posterior perforated substance in the interpeduncular space (Fig. 2-5).

METATHALAMUS

The metathalamus is composed of the two geniculate bodies, slightly elevated areas located at the dorsocaudolateral angle of the thalamus. The *lateral geniculate body* is less distinct, but may be grossly identified by following the optic tracts laterocaudally from the optic chiasm (Fig. 2-11). The *medial geniculate body* appears as a rounded mass ventral to the lateral geniculate body and lateral to the rostral colliculus. Because of the close gross relationship and the functional connections between the geniculate bodies and the colliculi of the midbrain tectum, one is tempted to erroneously consider the geniculate bodies as components of the mesencephalon rather than of the metathalamus (diencephalon), to which they properly belong.

SUBTHALAMUS

The subthalamus is best seen in transverse sections through the diencephalon. This is a transition zone between the tegmentum of the midbrain caudally and the thalamus rostrally. It is bounded by the hypothalamus medially, the internal capsule laterally, and the thalamus dorsally. The subdivisions of the subthalamus are not discernible by gross dissection; they are considered in Chapter 15.

The Telencephalon

The telencephalon is the rostral division of the original embryonic prosencephalon (forebrain). In the adult brain the *cerebral hemispheres* (*cerebrum*) constitute the telencephalon. The outermost gray surface area is the *cerebral cortex,* and the deeper substance of the cerebrum, composed of both white and gray matter, is the *cerebral medullary substance.* The bilateral symmetry of the telencephalon is apparent, in that there are two cerebral hemispheres separated in the dorsal midline by the *longitudinal fissure.* This fissure extends deeply to reach the corpus callosum (Fig. 2-7). When the dura mater is present, the falx cerebri is within the fissure, and the dorsal sagittal sinus within the dura follows the longitudinal fissure at its dorsal surface.

Cerebral Cortex

The surface of each hemisphere of the adult contains a pattern of small elongated

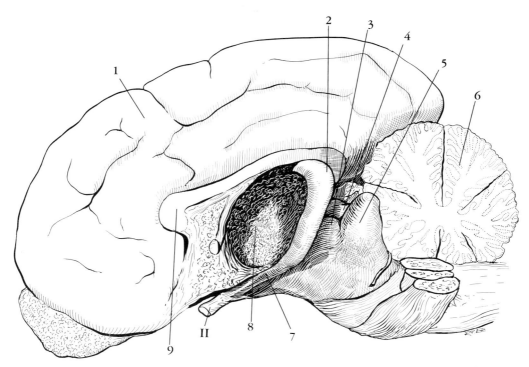

Fig. 2-11. Lateral view of brain with left cerebral and cerebellar hemispheres removed to show the optic tract in relation to the geniculate bodies and surrounding structures. 1, medial surface of right cerebral hemisphere; 2, lateral geniculate body; 3, medial geniculate body; 4, rostral colliculus; 5, caudal colliculus; 6, vermis of cerebellum (cut); 7, optic tract; 8, cut surface between left cerebral hemisphere and brain stem; 9, genu of corpus callosum; II, optic nerve.

elevations, the *gyri*, which are separated by shallow grooves, the *sulci*. Some sulci are unusually deep and are referred to as *fissures*; however, the terms 'sulcus' and 'fissure' are commonly used interchangeably.

The pattern of gyri and sulci is species specific and therefore the names of most of the gyri and sulci are not easily transferable from the brain of one animal to that of another; certainly they are not interchangeable between man and other animals, particularly subprimates. It has been pointed out previously that the general pattern of organization of the mammalian brain stem is consistent to the extent that specific names, relationships, and functions of nuclei and pathways may be readily compared among various higher mammals; for example, the brain stems of dog and

man are quite similar anatomically. In contrast, the cerebral cortices of dog and man are not similar enough to permit the interchange of names and relations of most gyri and sulci.

Brains that possess many gyri and sulci are referred to as 'convoluted,' whereas those brains with a smooth cerebral cortex, as in rat and similar mammals, are classified as 'lissencephalic.' The embryonic dog brain is lissencephalic, and the convolution pattern is simple at birth. Not until after the first month of life is the adult gyral and sulcal pattern well established (Meyer, 1964).

Since the outermost cerebral cortex covers the hemispheres like a cloak, or mantle, it is referred to as the *pallium*. Based on the phylogenetic age, the olfactory cortex at the base of the brain, which is the older cortex, is referred to as the archipallium (archicortex), or paleopallium (paleocortex).

Lobes of the Cerebral Hemispheres (Fig. 2-12)

Based on the bones of the calvaria under which the specific areas lie, the cortex is conveniently subdivided into four lobes, with names identical to those of the covering bones. The *frontal lobe* is the rostral portion, the *occipital lobe* is the caudal area, and the middle lateral portion is subdivided into the *temporal lobe* at the ventrolateral surface, and the *parietal lobe* above it at the dorsolateral surface between the frontal and occipital lobes. The four lobes named above are portions of the *neopallium* (*neocortex*); the olfactory portion of the hemispheres is not included, because this older portion of the brain is considered generally as a separate entity, named the rhinencephalon. This 'ol-

factory brain' is visible in part on the ventral surface of the brain (Fig. 2-5) as it is separated from the neocortex by the anterior and posterior rhinal fissures. As stated previously, the rhinal fissure is the first one to appear in the embryonic brain and is present in many so-called lissencephalic brains of lower mammals. The rhinal fissure (sulcus) is very constant as it extends the entire length of the cerebrum on the ventrolateral surface of the brain. At approximately its middle the rhinal fissure is joined obliquely by the *Sylvian* (*lateral*) *fissure,* which is short in the dog as it extends dorsally and slightly caudally. The junction of the Sylvian and rhinal fissures divides the latter into the so-called rostral (anterior) and caudal (posterior) rhinal fissures. In comparison with the primate brain, the brain of the dog has a poorly developed temporal lobe which does not extend as great a distance ventrorostrally. This results in the Sylvian fissure being more vertical

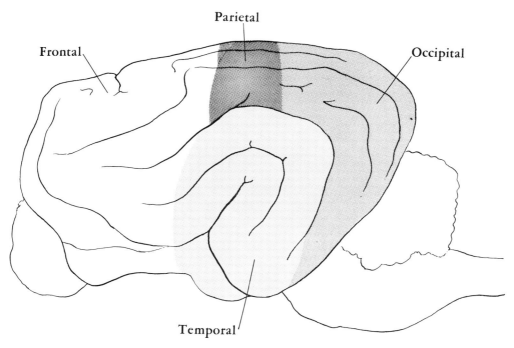

Fig. 2-12. Diagrammatic lateral view of the dog brain illustrating the relative positions of the four cerebral lobes. (Based on McGrath, 1960.)

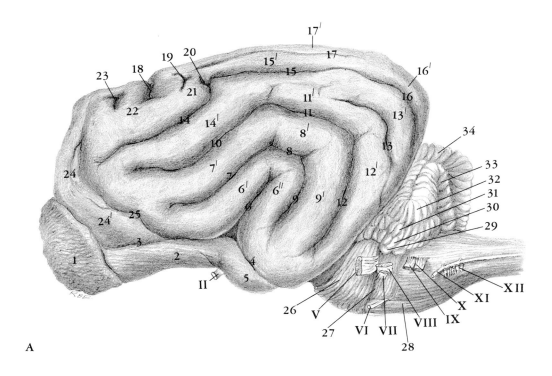

A

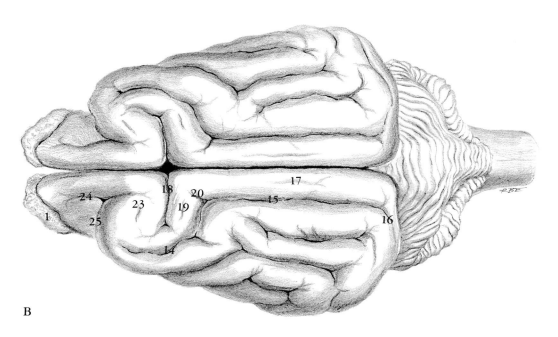

B

Fig. 2-13. See legend on facing page.

and not as deep as it is in primates (including man).

Figure 2-13 shows the major gyri and sulci of the cerebrum. The functional significance of cortical areas is considered in Chapter 16.

Cerebral Medullary Substance

The deeper substance of the cerebral hemispheres is composed of both white and gray matter. Basically the white matter is composed of myelinated fibers and the gray matter consists of cell bodies and mostly nonmyelinated fibers. In addition to these major differences between the white and the gray matter, the specific components discussed in Chapter 3 also apply to the cerebral medullary substance.

White Matter

There are three types of cerebral fibers, based on distribution: projection fibers, association fibers, and commissural fibers.

PROJECTION FIBERS

Projection fibers either originate or terminate in the cerebral cortex. Functionally, these fibers are either sensory (ascending) or motor (descending) fibers which compose the white substance of the cerebral hemispheres. Subcortically the fibers form the *corona radiata* before they converge deeply into an organized band of fibers, the *internal capsule* (Plates 23 and 24), which is discussed in more detail in Chapter 16. Grossly, in a horizontal cut the *anterior and posterior limbs* of the internal capsule can be identified, as well as the *genu* of the internal capsule, the angle formed by the union of the two limbs. The anterior limb passes through the *corpus striatum* (between the head of the caudate nucleus and the putamen). The posterior limb is bordered medially by the thalamus and laterally by the globus pallidus of the lenticular nucleus.

ASSOCIATION FIBERS

The cerebral association fibers connect areas of the cerebrum located on the same side of the midline, that is, within the same cerebral hemisphere. These fibers may be classified as either *intracortical* or *subcortical.* The intracortical fibers are confined to the cortex; the subcortical fibers pass deeply to and within the medullary substance. Association fibers are also classified as either short (intralobar) or long (interlobar), distinctions which are self-explanatory. The

Fig. 2-13. Views of the brain of the dog. A. Lateral view. B. Dorsal view, for identification of gyri and sulci on the dorsal surface of the cerebrum, which may be unclear on the lateral surface as shown in A. The same numbers are used for identification of the sulci in both views. 1, olfactory bulb; 2, olfactory tract; 3, rostral rhinal sulcus; 4, caudal rhinal sulcus; 5, piriform area; 6, (pseudo)sylvian fissure; 6', rostral sylvian gyrus; 6'', caudal sylvian gyrus; 7, rostral ectosylvian sulcus; 7', rostral ectosylvian gyrus; 8, middle ectosylvian sulcus; 8', middle ectosylvian gyrus; 9, caudal ectosylvian sulcus; 9', caudal ectosylvian gyrus; 10, rostral suprasylvian sulcus; 11, middle suprasylvian sulcus; 11', middle suprasylvian gyrus; 12, caudal suprasylvian sulcus; 12', caudal suprasylvian gyrus; 13, ectomarginal (ectosagittal) sulcus; 13', caudal ectomarginal (ectosagittal) gyrus; 14, coronal sulcus; 14', coronal gyrus; 15, marginal (sagittal) sulcus; 15', marginal (sagittal) gyrus; 16, caudal marginal (sagittal) sulcus; 16', caudal marginal (sagittal) gyrus; 17, endomarginal (endosagittal) sulcus; 17', endomarginal (endosagittal) gyrus; 18, cruciate sulcus; 19, postcruciate sulcus; 20, ansate sulcus; 21, postcruciate gyrus; 22, precruciate gyrus; 23, precruciate sulcus; 24, prorean sulcus; 24', prorean gyrus; 25, presylvian sulcus; 26, pons; 27, trapezoid body; 28, pyramid; 29, flocculus; 30, ventral paraflocculus; 31, dorsal paraflocculus; 32, ansiform lobule; 33, paramedian lobule; 34, vermis; II, optic nerve; V, trigeminal nerve; VI, abducens nerve; VII, facial nerve; VIII, vestibulocochlear nerve; IX, glossopharyngeal nerve; X, vagus nerve; XI, accessory nerve; XII, hypoglossal nerve.

association fibers are extremely well developed and profuse in man, and therefore many distinct long association fiber fasciculi can be demonstrated in the human brain (Truex and Kellner, 1948), but are either entirely absent or very nebulous in the brain of the dog. There are definite interlobar association bundles in dog brain, and for further details the reader is referred to Meyer (1964). The same author states that a search of the literature revealed that among the interlobar fiber systems the *cingulum* (Fig. 2-7) is the only long association bundle the identity and function of which have not been questioned.

COMMISSURAL FIBERS (FIG. 2–7)

Grossly, the commissural fibers of the cerebrum compose bundles or *commissures* identified as: the corpus callosum, the anterior commissure, and the hippocampal commissure. The commissural fibers connect corresponding areas of the two hemispheres; therefore they cross the midline. The greatest commissure is the *corpus callosum*. The rostral portion of the corpus callosum is referred to as the *genu*, the middle area as the *body*, and the caudal area as the *splenium*. The body of the corpus callosum forms the roof of the lateral ventricles. The anterior (rostral) horns of the ventricles are separated by the *septum pellucidum* (Fig. 2-7), a thin sheet of tissue extending in the midline between the corpus callosum and the fornix. The fibers of the corpus callosum extend laterally and are intersected by projection fibers of the internal capsule.

The *anterior commissure* (Fig. 2-7) is a round bundle of commissural fibers within the *lamina terminalis* at the rostral midline disappearance of the fornix, where it bifurcates to form the precommissural and postcommissural bundles of the fornix. The rostral portion of the anterior commissure connects the two olfactory bulbs, and the caudal portion connects the piriform and

olfactory areas on the ventral side of the hemispheres (Fig. 2-5).

The *posterior commissure* technically is a portion of the epithalamus and therefore is not considered here.

The *hippocampal commissure* at the ventral side of the splenium of the corpus callosum is composed of commissural fibers that connect the hippocampi of the two sides. Since this is the general area of fusion between the hippocampus and fornix and there is no line of demarcation between the two, the commissure is sometimes referred to as the *commissure of the fornix.*

Basal Ganglia (Nuclei)

The term 'basal ganglia' is established in the literature so that it is commonly used, although by definition the gray areas of the cerebral medullary substance should be named 'nuclei.' Four gray bodies constitute the basal ganglia: (1) the caudate nucleus, (2) the lenticular nucleus, (3) the claustrum, and (4) the amygdaloid body.

CAUDATE NUCLEUS

The caudate nucleus (Fig. 2-10) has a large head, which tapers to a body and a thin tail. The large bulbous head forms the lateral wall of the anterior horn of the lateral ventricle (Plate 25). The anterior commissure passes through the ventral edge of the head of the caudate nucleus.

LENTICULAR NUCLEUS

The lenticular (lentiform) nucleus is the gray triangular mass, as observed in horizontal section, that fits into the wedge between the anterior and the posterior limb of the internal capsule as they diverge from the genu. The lenticular nucleus is composed of an outer (lateral) cap, the *putamen* (Plate 25), which grossly in texture looks like the head of the caudate nucleus, with which it is connected to form the corpus striatum as discussed previously. In most horizontal sections, one does not see the connecting striations between the head of the caudate nucleus and the putamen, as they appear in a restricted area of the anterior limit of the internal capsule.

The triangular mass that is the smaller medial portion of the lenticular nucleus is the *globus pallidus.* At the medial apex is the genu of the internal capsule. The globus pallidus is separated from the putamen by a very thin white medullary lamina.

CLAUSTRUM

The claustrum is a layer of gray matter lateral to the lenticular nucleus, from which it is separated by the *external capsule.* On the lateral side of the claustrum is the thin line of white matter, the *extreme capsule,* which separates the claustrum from the gray matter of the cerebral cortex.

AMYGDALOID BODY

The amygdaloid body (amygdala) (Plate 25) is a relatively large nuclear mass deep within the piriform area at the rostral tip of the ventral horn of the lateral ventricle. In some gross dissections the amygdala is seen to be composed of a number of subnuclear groups.

The Rhinencephalon

The rhinencephalon, or 'olfactory brain,' has been alluded to several times previously, without crystallization of its identity and relationships. It has been stated that the rhinencephalon is better developed (grossly) in lower forms (*e.g.,* dog) than in man. This area of the brain is very complicated, and through the process of phylogeny the human brain shows evidence that there has been a decrease in olfaction *per se,* but the functions of the area have increased in complexity so that many lesions of the rhinencephalon in man produce derange-

ment of many human qualities (*e.g.,* personality), in addition to merely olfaction (see Chapter 16).

Grossly, the rhinencephalon is separated from the neocortex by the rostral and caudal rhinal fissures. Authors disagree on the components of the rhinencephalon, but the major gross structures discussed in Chapter 16 are: (1) olfactory bulbs, (2) olfactory tracts, (3) piriform area, and (4) hippocampus. These structures are illustrated in Figures 2-5, 2-10, and 2-15.

Rather than acquiring knowledge of the anatomic details of each component from separate discussions, the reader should find the figures referred to above helpful in locating the structures and the following brief remarks sufficient for a gross dissection of the brain. Further anatomic and functional considerations are covered in Chapter 16.

The *olfactory bulbs* in the dog are very large when compared with those in man (Fig. 2-14). Occasionally the olfactory bulb in the dog may be seen to contain a cavity that connects with the rostral portion of the lateral ventricle of the same side. The bulbs lie in the fossae bordered rostrally by the cribriform plate of the ethmoid bone. The fascicles of the olfactory nerve pass from the nasal cavity through the cribriform plate to enter the olfactory bulbs (see Fig. 16-4).

The *olfactory tract* extends caudally from the bulb and bifurcates to form the *medial* and *lateral olfactory striae* (Fig. 2-5). The area between the diverging striae has a perforated appearance, owing to the numerous small holes made by the small blood vessels which were pulled away during routine removal of the brain. This area logically is named the *rostral (anterior) perforated substance.* Figure 2-5 also shows the *lateral olfactory gyrus,* which is the cortex between the lateral olfactory stria and the rostral rhinal sulcus. The lateral olfactory stria and gyrus can be followed caudally and laterally to the anterior perforated substance, where they enter the piriform area. The amygdaloid complex is deep within this area.

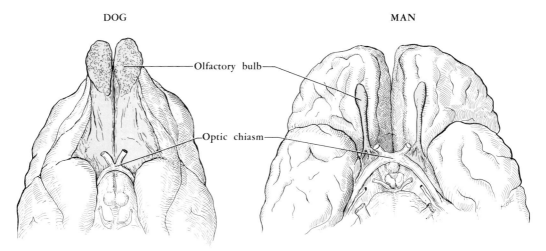

DOG   MAN

—Olfactory bulb—

—Optic chiasm—

Fig. 2-14. Olfactory bulbs and tracts in dog and man, showing comparative sizes.

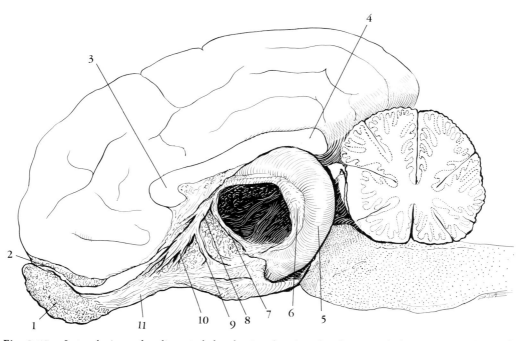

Fig. 2-15. Lateral view of a dissected dog brain, showing the rhinencephalic structures. Note the relationship of hippocampus and fornix. 1, left olfactory bulb; 2, right olfactory bulb; 3, genu of corpus callosum; 4, splenium of corpus callosum; 5, hippocampus; 6, fimbria of hippocampus; 7, column of fornix; 8, middle portion of rostral commissure; 9, caudal portion of rostral commissure; 10, rostral portion of rostral commissure; 11, olfactory tract.

Axons of the medial olfactory stria pass to the medial surface of the cerebral hemisphere in the subcallosal area and paraterminal gyrus ventral to the genu of the corpus callosum (Fig. 2-7).

The *hippocampus* (*Ammon's horn*) (Figs. 2-10 and 2-15) is a C-shaped structure beginning in the subcortical piriform area close to the amygdala and arching caudodorsally to pass beneath the splenium of the corpus callosum. The hippocampus extends rostrally in the floor of the caudal horn of the lateral ventricle and becomes the fornix. Beneath the splenium of the corpus callosum the two hippocampi are joined by the *commissure of the hippocampus;* this, as mentioned previously, is also called the *commissure of the fornix,* because of the fornix being considered a rostral extension of the hippocampus.

Veterinarians have a special interest in the hippocampus, because gross sections of this structure are taken from most animals suspected of having rabies. Proper technique of preparation of tissue specimens and staining of microscopic sections of the hippocampus are used to demonstrate Negri bodies, intracytoplasmic inclusion bodies that are considered pathognomonic for rabies. Although these inclusion bodies have a natural predilection for the hippocampus, other areas of the brain from an animal with rabies will contain them. In some species the semilunar ganglion or the cerebellum is preferred for sectioning to demonstrate Negri bodies.

In a gross dissection of the brain the hippocampus may be exposed by lateral and caudal decortication. A simpler dissection of a hemisected brain is performed by inserting a scalpel blade horizontally between the splenium of the corpus callosum and the hippocampus (fornix) and cutting caudally through the ventricle and hemisphere. The hippocampus and *fimbria,* a thin white lip on the lateral side of the rounded hippocampus in the floor of the rostral portion of the caudal horn of the lateral ventricle (Fig. 2-10), should be identified. From this region the hippocampus and fimbria may be exposed as they follow the inferior (temporal) horn of the lateral ventricle into the temporal lobe.

Bibliography

Ackerknecht, E.: Die Zentralorgane Rückenmark und Gehirn. In *Handbuch der vergleichenden Anatomie der Haustiere* (Ellenberger-Baum). Berlin, Springer, 1943. Cited by Meyer (1964).

Brodal, A.: *Neurological Anatomy in Relation to Clinical Medicine,* 2nd ed. New York, Oxford University Press, 1969.

Davies, J.: *Human Developmental Anatomy.* New York, Ronald Press Co., 1963.

Dexler, H.: Die Entwicklung und der feinere Aufbau des zentralen Nervensystems. In *Handbuch der vergleichenden Anatomie der Haustiere* (Ellenberger-Baum). Berlin, Springer, 1932. Cited by Meyer (1964).

House, E. L., and Pansky, B.: *A Functional Approach to Neuroanatomy,* 2nd ed. New York, McGraw-Hill Book Co., Blakiston Division, 1967.

Houston, M. L.: The early brain development of the dog. J. Comp. Neurol. *134:* 371–384, 1968.

Larsell, O.: The morphogenesis and adult pattern of the lobules and fissures of the cerebellum of the white rat. J. Comp. Neurol. *97:* 281–356, 1952.

McClure, R. C.: The Cranial Nerves. Chapter 10 in *Anatomy of the Dog* (Miller, M. E., Christensen, G. C., and Evans, H. E.). Philadelphia, W. B. Saunders Co., 1964. Pp. 544–571.

McGrath, J.: *Neurologic Examination of the Dog.* Philadelphia, Lea & Febiger, 1960.

Meyer, H.: The Brain. Chapter 8 in *Anatomy of the Dog* (Miller, M. E., Christensen, G. C., and Evans, H. E.). Philadelphia, W. B. Saunders Co., 1964. Pp. 480–532.

Patten, B. M.: *Embryology of the Pig,* 3rd ed. Philadelphia, The Blakiston Co., 1948.

Patten, B. M.: *Human Embryology,* 3rd ed. New York, McGraw-Hill Book Co., Blakiston Division, 1968.

Peele, T. L.: *The Neuroanatomic Basis for Clinical*

Neurology, 2nd ed. New York, McGraw-Hill Book Co., Blakiston Division, 1961.

Ranson, S. W., and Clark, S. L.: *The Anatomy of the Nervous System—Its Development and Function,* 10th ed. Philadelphia, W. B. Saunders Co., 1959.

Singer, M.: *The Brain of the Dog in Section.* Philadelphia, W. B. Saunders Co., 1962.

Thomas, J. B.: *Introduction to Human Embryology.* Philadelphia, Lea & Febiger, 1968.

Truex, R. C., and Carpenter, M. B.: *Human Neuroanatomy,* 6th ed. Baltimore, Williams & Wilkins Co., 1969.

Truex, R. C., and Kellner, C. E.: *Detailed Atlas of the Head and Neck.* New York, Oxford University Press, 1948.

chapter 3

Cells of the Nervous System

The nervous system is composed of two main categories of cells: neurons and several other kinds of cells (glial cells, interstitial cells), collectively referred to as neuroglia.

In the past, much more attention has been given to the neuron than to neuroglia. This was true primarily because the main function of the nervous system in adapting the organism to its environment depends most directly on the neuron. Neuroglia was spoken of as the supportive substance which held the other cells of the central nervous system together (*neuroglia*—Gk., nerve + glue). Recently, however, research has shown that the glial cells have many functions other than the principal one of support. As one cardinal function, they play a great role in transporting nutrients and metabolic wastes to and from the neurons, thereby enabling the latter to perform their functions. The different types of neuroglia are discussed later in this Chapter, following a consideration of neurons.

Neurons

Basic Terminology

A *neuron* (Fig. 3-1) is defined as a nerve cell body plus all of its processes. The *cell*

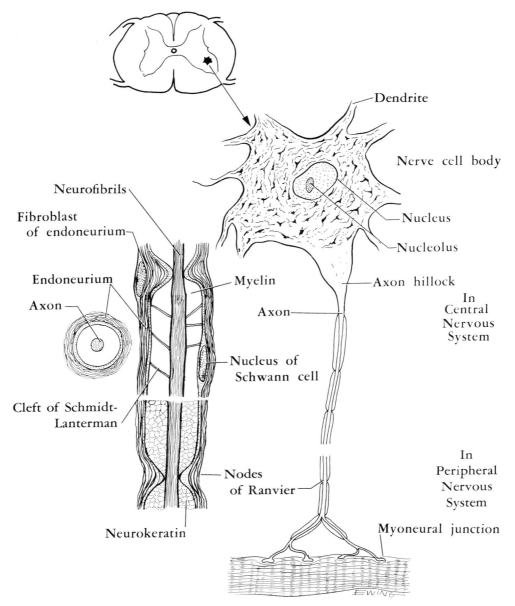

Fig. 3-1. The histologic characteristics of a 'typical' multipolar neuron. On the right is a diagrammatic drawing of an entire lower motor neuron. On the left is a high magnification of a section of the axon and its sheaths as found in the peripheral nervous system. The upper portion of the longitudinal view shows myelin with the clefts of Schmidt-Lanterman. The lower longitudinal section shows the neurokeratin network as one would observe it following dissolution of the fatty portion of the myelin.

body (*perikaryon*) is the portion of the neuron which contains the nucleus and a varying amount of surrounding cytoplasm. The processes are of two types: *dendrites*, which may vary in number, size, and shape; and the *axon*, which is always single but may vary in diameter, length, and the number of collaterals branching from it.

The classic conception of a neuron and its components is based on cell body location as a point of reference, as illustrated in Figure 3-1. Dendrites are said to transmit impulses toward the cell body, whereas axons are said to conduct impulses away from the cell body. This is true for the motor neuron, which generally is used as an example of a 'typical' neuron. Using this basic principle, one has difficulty in defining the dendrite and axon of a sensory neuron, such as one in a spinal nerve which has its cell body in a dorsal root ganglion (see Fig. 3-8).

According to the traditional concept of cell body location as the focal point, it is commonly taught that the peripheral process of a sensory ganglion cell body functionally is a 'dendrite,' because it conveys impulses toward the cell body. Anatomically, however, this 'dendrite' is identical to the axon of a motor neuron. To avoid this confusion, which is still perpetuated by many present-day texts, Bodian (1962) proposed definitions of axon and dendrite in a generalized neuron with *impulse origin* as the basic criterion rather than location of the cell body (Fig. 3-2).

Briefly, he considers the *'dendritic zone'* as the receptor membrane of a neuron consisting of tapering cytoplasmic extensions (dendrites) which either receive synaptic endings from other neurons or are differentiated to convert environmental stimuli into local response-generating activity.

The perikaryon (cell body) is the focal point of embryonic outgrowth of the neuronal processes, axon regeneration, and perhaps enzymatic synthesis in the differentiated neuron.

The axon is a single elongated cytoplasmic extension differentiated to conduct nervous impulses away from the dendritic zone. The axon is uniform in caliber and ensheathed by neuroglia in the central nervous system or by neurilemmal cells (of Schwann) in the peripheral nervous system. Many collaterals (branches) may arise from a single axon (Fig. 3-1). *Telodendria* are the differentiated axonal terminals, which are related to synaptic transmission or neurosecretory activity.

Figure 3-1 illustrates the typical microscopic characteristics of a longitudinal section and a cross section of a heavily myelinated axon in the peripheral nervous system. The terms *axon* and *axis cylinder* are synonymous; the term *fiber* also is used synonymously with axon. This means that technically the myelin and neurilemma are not a portion of the axon, but are sheaths enveloping it. Therefore, the *axolemma* is the true cellular membrane, and should not be confused with the neurilemma.

The Neuron Doctrine

In 1891 Waldeyer formulated the neuron doctrine, which is a direct application of the cellular theory to the neuron. Briefly, the main tenets of this doctrine state that the neuron is:

1. **The genetic unit:** Each neuron is derived from a single cell, called the *neuroblast.*

2. **The structural unit:** Each neuron is a cell and therefore is an enclosed mass of protoplasm. When portions of two or more neurons come together at a synapse, there is no continuity of protoplasm from one neuron to the other(s). Each neuron within pathways of the central nervous system may be compared to a link in a chain.

3. **The functional unit:** Although it is agreed that the neuron is the structural unit of the nervous system, there is some disagreement concerning it as a functional unit. On the basis that *in vivo* one single neuron cannot carry out the integrative function

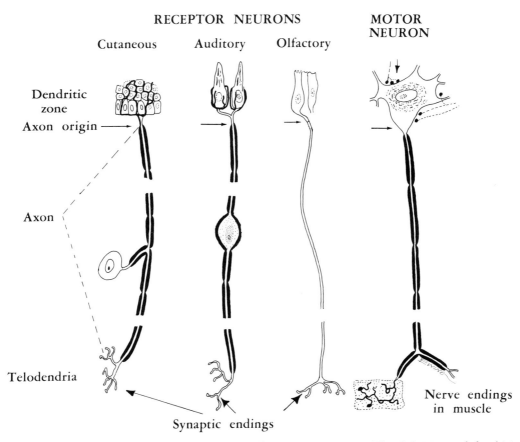

RECEPTOR NEURONS   MOTOR NEURON

Cutaneous   Auditory   Olfactory

Dendritic zone

Axon origin

Axon

Telodendria

Synaptic endings

Nerve endings in muscle

Fig. 3-2. Diagram of three sensory neurons and one motor neuron. The definitions of dendritic zone and axon are clarified in the concept based on the site of impulse origin rather than location of the cell body. Note that the cell body may be located either in the dendritic zone or in the segment of the axon. (After Bodian, 1962.)

necessary for any neuronal activity, physiologists in general consider the reflex arc to be the functional unit.

4. **The trophic unit:** The function of nutrition is attributed more directly to the nucleus within the cell body of the neuron. The nucleus maintains and governs the general nutrition and health of the entire cell. The neuron responds to injury as a whole. When the axon is injured, the morphologic and functional damage is reflected by the cell body. When the cell body containing the nucleus is destroyed, the entire neuron will degenerate and will not regenerate. The details of degeneration and

regeneration are considered in Chapter 8.

5. **The conduction unit:** The neuron is the only element directly concerned with conduction of the nerve impulse.

Histologic Characteristics of Neurons

Nucleus

The main morphologic component of the cell body is the nucleus. The nucleus within the cell body has the following general characteristics:

(*a*) It is usually centrally located within the cell body.

(*b*) It is characteristically large and spherical, measuring 3 to 18 micra in diameter.

(*c*) Generally the nucleoplasm is vesicular in appearance; it usually stains poorly with ordinary stains, whereas the nuclear membrane stains heavily with hematoxylin.

(*d*) Chromatin granules are very finely dispersed throughout.

(*e*) There is at least one large nucleolus, and occasionally there is more than one. The nucleolus is so large and generally stains so darkly, while the nucleus generally stains so lightly, that the beginning student may easily confuse the nucleolus with the nucleus and the nucleoplasm with cytoplasm.

(*f*) In female cats a small (approximately 1 micron in diameter) intranuclear body of sex chromatin is visible; this body is not as conspicuous in neurons of the male animal. In 1949 Barr and Bertram first observed this body attached to the surface of the nucleolus in nerve cells of female cats and therefore named it a 'nucleolar satellite.' Barr and his associates soon found that the position of the nucleolar satellite is not constant in other kinds of cells. As illustrated in Figure 3-3, even in different types of neuron cell bodies in the female cat they frequently found the chromatin body in two additional positions: (1) free in the nucleoplasm, and (2) apposed to and somewhat flattened against the inner side of the nuclear membrane (Barr *et al.*, 1950). Consequently, the term nucleolar satellite has been dropped and the body is now generally referred to as a 'Barr body.'

Organelles

Neuroplasm is a general term used in reference to the cytoplasm within a neuron. The neuroplasm within the axon is referred to as *axoplasm*. The neuroplasm of the cell body (perikaryon) contains filamentous, membranous, and granular organelles which tend to surround the nucleus. Some organelles are also located in the neuroplasm of the axon and dendrites.

Neurofibrils (Fig. 3-1)

Neurofibrils are small fibrils found within the dendrites, the cell body, and the axon. Within the axon they are confined to the central portion. Within the processes they tend to be oriented in a true longitudinal

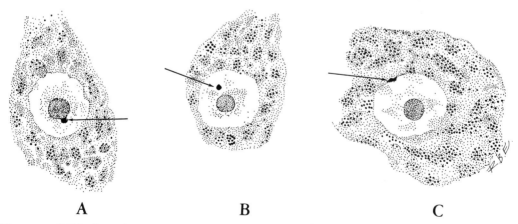

A	B	C

Fig. 3-3. Cell bodies of neurons of female cat showing the three locations of the intranuclear satellite bodies (arrows) labeled as sex chromatin: adjacent to the nucleolus as a true nucleolar satellite (A); free in the nucleoplasm (B); and apposed to and somewhat flattened against the inner nuclear membrane (C). (After Barr *et al.*, 1950.)

fashion and occur somewhat in bundles. Within the perikaryon the organization is more diffuse. With the aid of an electron microscope it can be seen that these fibrils are in turn made up of small filaments measuring about 100Å (Bloom and Fawcett, 1968).

Neurofibrils are best demonstrated in microscopic sections by the use of special silver stains, or in tissue culture with methylene blue or any aniline dye.

The functions of neurofibrils are not clear. It has been hypothesized that they play a role in support of the neuron. At one time it was thought that the conduction of the nerve impulse was a direct function of neurofibrils. Now, however, the conduction of the nerve impulse is considered a phenomenon of the cell membrane.

Nissl Bodies

Nissl bodies (chromophil substance, tigroid bodies, chromidial substance) are made up of a basophilic material and are characteristic of nerve cells. The Nissl bodies are found within the neuroplasm of the dendrite and the cell body, except at the specific site of origin of the axon. The absence of Nissl bodies at this location designates the *axon hillock,* which consequently appears as a light area. Nissl bodies are not found within the axoplasm. At one time these bodies were thought to be artifacts of fixation. However, it has been shown by means of tissue culture that they do exist in living cells. The largest Nissl bodies are characteristically found within the somatic efferent (motor) cell bodies. The size and distribution of Nissl bodies tends to be correlated with the functional activity of the cell. They are very high in RNA (ribonucleic acid), with RNA granules varying in size between 10 and 30 millimicra and occurring in clusters (Truex and Carpenter, 1969). These are the counterparts of the granular endoplasmic reticulum in other cells.

It is thought that the Nissl bodies are concerned with protein synthesis and the replacement of neuroplasm. Injury to any part of the neuron may be reflected in a change in the distribution and number of the Nissl bodies within the cell body. During axonal damage there may be a complete dissolution of many Nissl bodies (chromatolysis). It has been shown that these bodies decrease in number as the result of physiological changes, for example, in old age, overexertion, disease, and metabolic disturbances.

Golgi Apparatus

A Golgi apparatus is present within the cell body of all neurons. It varies in form and extent but commonly may be seen as an irregular wavy threadlike network surrounding the nucleus. It is thought to be the agranular endoplasmic reticulum.

The functions of the Golgi apparatus within neurons are somewhat obscure. However, it is thought that it does serve as a source of new synaptic vesicles. According to some authorities, the state of the Golgi apparatus reflects the general health of the entire neuron, and during neuronal injury it may be more sensitive and undergo dissolution more quickly than the Nissl bodies.

It is of interest to note that the Italian histologist Golgi (1844–1926) first discovered the 'Golgi apparatus' within the neuron.

Mitochondria

There seems to be nothing unusual about the mitochondria of the neuron. They are located in the dendrite, cell body, and axon, and may be granular, rod shaped, or filamentous. As in other cells throughout the body, mitochondria serve in glycolysis, cell respiration, biosynthesis, and general cellular metabolism.

Centrosomes

Centrosomes are usually absent in an adult neuron but may be demonstrated in the embryonic neuroblast. This is as expected, since the centrosomes are functionally associated with cell division, and adult neurons have lost the ability to divide.

Inclusions

Pigments

MELANIN

This black pigment is found in various tissues of the body in different concentrations. Within the nervous system it occurs in two main areas: (1) in the substantia nigra within the midbrain and subthalamus, and (2) in the nucleus pigmentosus (ceruleus) within the pons. Melanin is not present in the fetal or in the neonatal nervous system. In the human it appears at the end of the first year and tends to increase in concentration in the above-mentioned sites until the age of puberty. Its true significance or function within the central nervous system is not understood. Melanin also appears abundantly within the cranial meninges of sheep. Based on casual observations, I have not noticed it in the meninges of other animals, certainly not to the extent that it is found in sheep.

According to Bloom and Fawcett (1968) melanocytes are present in the pia mater of the ventral surface of the medulla in most mammals.

It is generally agreed that there is a positive correlation between meningeal melanosis and skin pigmentation; neural melanosis, on the other hand, is completely unrelated to skin pigmentation. The chemical pathways for neural and meningeal melanogenesis are different.

LIPOCHROME

Lipochrome, or lipofuscin, a light-colored, yellowish pigment, appears in selected areas within both the peripheral and the central nervous system. It tends to increase with age in a specific location within the perikaryon. In the human it is absent at birth but appears within the spinal ganglionic cell bodies at about the age of 6 years, within the spinal cord a few years later, and within the cortex after 20 years of age (Truex and Carpenter, 1969). Not all cell bodies within the specific areas cited contain this pigment, and its true significance is not known.

Fat

As a normal constituent of myelin within the white matter of the central nervous system and the myelin of the peripheral nervous system, fat is combined with protein as a lipoprotein. The breakdown or the dissolution of the lipoid portion of the myelin is indicative of certain pathologic conditions (*e.g.,* demyelinating disease).

Glycogen

Glycogen generally is not found in the nervous system of the adult. Within the embryo, however, glycogen may be found in the ependymal cells and choroid plexuses, as well as the neurons.

Iron-Containing Granules

Iron-containing granules are relatively restricted in location within the brain. They are found within the cell bodies of neurons in the globus pallidus, red nucleus, and substantia nigra. The number of granules increases with age, but their true significance is not known.

Axon (Figs. 3-1, 3-2, and 3-4)

As stated previously in this Chapter, the axon is a single cytoplasmic extension differentiated to conduct nerve impulses away from the dendritic zone of the neuron. A multipolar neuron has numerous dendrites,

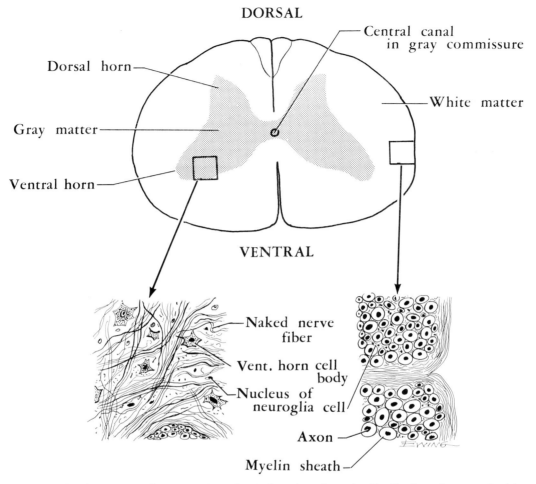

DORSAL

Dorsal horn

Central canal
in gray commissure

Dorsal horn

Gray matter

White matter

Ventral horn

VENTRAL

Naked nerve
fiber

Vent. horn cell
body

Nucleus of
neuroglia cell

Axon

Myelin sheath

Fig. 3-4. (Top) Diagram of cross section of spinal cord to show the distribution of gray and white matter. (Bottom) Diagrams of high-power magnification of sections of gray matter (left) and white matter (right). Note the various degrees of myelination of fibers and the absence of cell bodies in the white matter. (Modified from Ham, 1965.)

but only one axon. It should be recalled that axis cylinder is synonymous with axon, and that neurofibrils and mitochondria are the organelles found in the axoplasm. The axolemma is the limiting membrane of the axon, separating it from the myelin and the neurilemmal sheath in the peripheral nervous system.

Myelin

A great majority of the nerve fibers are myelinated (medullated). Figure 3-1 illus-

trates a typical axon with the heaviest myelin sheath. These largest fibers (approximately 20 micra in diameter) are not the only myelinated axons, since a typical peripheral nerve or a section of white matter of the central nervous system contains axons with myelin sheaths of different thicknesses (Fig. 3-4; see also Fig. 6-4, p. 95). The postsynaptic (postganglionic) axons of the autonomic neurons have no myelin sheaths and therefore are referred to as *nonmyelinated* or *nonmedullated fibers*, or

Remak fibers. The rate of nerve impulse conduction tends to be directly proportional to the thickness of the myelin sheath (Table 3-1). The myelin sheath is not continuous, but is interrupted at intervals by constrictions, the *nodes of Ranvier.* Distances between nodes vary from 50 to 1,000 micra, with the longer internodes present in the fibers with larger diameters (House and Pansky, 1967). In addition, the internodal segments are interrupted slightly by the thin oblique *clefts of Schmidt-Lanterman.*

Chemically, myelin is a lipoprotein consisting of thin concentric lamellae of protein alternating with layers of lipids. Developmentally, in the peripheral nervous system the myelin sheath represents concentric infolding of double layers of the neurilemmal (Schwann) cell (Fig. 3-5); in the central nervous system the oligodendroglial cell (a type of neuroglia discussed later) forms the myelin (Bunge, 1968). In nonmyelinated fibers of the peripheral nervous system one Schwann cell generally is associated with many axons (Fig. 3-6).

It should be realized that in the fresh state the gross appearance of myelin is glistening white. This accounts for the appearance of the white matter (of the central nervous system) as differentiated from the gray matter, which has no heavily myelinated axons (Fig. 3-4).

In the peripheral nervous system, three connective tissue coverings exterior to the neurilemma are prominent in a cross section of a nerve trunk (see Chapter 6). Note that the *endoneurium* is the innermost connective tissue sheath which envelops each fiber, the *perineurium* surrounds bundles of fibers, and the *epineurium* envelops the entire nerve trunk (see Fig. 6-4, p. 95).

Classification of Neurons

In this brief discussion only the morphological and physiological criteria for classification of neurons will be considered.

Morphology

Size

Neurons vary in size probably more than any other single type of cell within the body. The diameter of the cell body may vary from approximately 4 micra (in the granular layer of the cerebellum) to 150 micra for the cell bodies of the lower motor (somatic efferent) neurons within the ventral horn of the spinal cord. The length of the processes also varies tremendously. The dendrites on some neurons are so small that they barely can be seen with the ordinary light microscope. Other neurons have large, widespread arborizations extending for a considerable distance from the cell body (*e.g.,* Purkinje cells of the cerebellum).

The length of the axon is also used as a criterion of classification. Those neurons with extremely long axons are referred to as *Golgi Type I* neurons. The largest Type I axons may be measured in feet or meters. In the human, one neuron contributing to the sciatic nerve, as an example, may have an axon over 3 feet long, extending from the spinal cord down to the foot. In the giraffe a Golgi Type I axon may actually extend 15 feet or more. Protoplasm extends for this distance within a single cell. Those neurons that have extremely short axons, measured in micra or millimeters, are referred to as *Golgi Type II.* Within the central nervous system, these cells are confined to the gray matter.

Shape

Neurons may be classified according to the number of processes extending from the cell body. Three general types may be considered: bipolar, unipolar, and multipolar.

BIPOLAR CELLS (FIG. 3-7 A)

Bipolar cells are the most primitive and are found most commonly in the embryo.

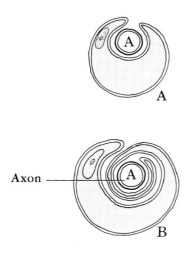

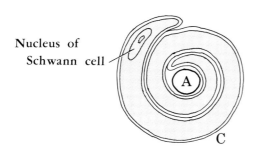

Axon

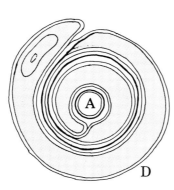

Nucleus of
Schwann cell

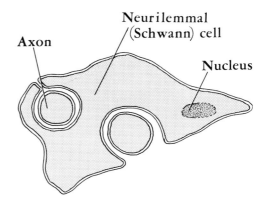

Axon

Neurilemmal
(Schwann) cell

Nucleus

Fig. 3-6. Diagram showing association of one neurilemmal cell to more than one nonmyelinated fiber.

In the adult, however, they are located in specific restricted areas, for example, in sensory neuroepithelia and within the ganglia of the vestibulocochlear (eighth cranial) nerve.

UNIPOLAR CELLS (FIG. 3-7 C)

This type of cell is derived from the original bipolar cell in a manner illustrated in Figure 3-7. It generally is found in the peripheral nervous system, specifically within the sensory dorsal root ganglia of the spinal nerves and within the cranial nerve sensory

Fig. 3-5. Diagram showing development of the myelin sheath enveloping an axon. Note that in the peripheral nervous system the myelin sheath evolves from continuous concentric infoldings of double layers of a neurilemmal (Schwann) cell, which accounts for the name "jelly-roll theory of myelin formation." (After Truex and Carpenter, 1969.)

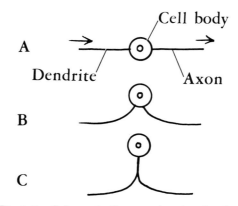

Fig. 3-7. Schematic diagram showing development of a unipolar cell from the primitive bipolar form. Note that the single 'pole' contains both the dendrite and the axon.

ganglia. It should be noted that only on a morphological basis are these considered unipolar; physiologically the single 'pole' is formed by fusion of the axon and dendrite. Because of the single-pole formation, with the cell body on top and the processes diverging at almost right angles, the unipolar cell is frequently referred to as a 'T' cell.

MULTIPOLAR CELLS

The multipolar cell is the most common type of neuron found within the central nervous system of higher mammals. Within the peripheral nervous system this type of cell body is located in the autonomic (motor) ganglia. The polarity of the neuron is designated according to the number of its dendrites, since, no matter how many dendrites a neuron possesses, it has only one axon. The most outstanding physiological advantage of a multipolar neuron is that the many dendrites permit more axodendritic synapses. Physiologic details of

different types of synapses are discussed in Chapter 7.

Physiology

According to function, neurons may be classified with regard to the direction the nerve impulse travels along its processes. Using this criterion and the neuraxis (spinal cord and brain) as a point of reference, as illustrated in Figure 3-8, there are three types of neurons: (1) afferent, or sensory; (2) connector, or internuncial; and (3) efferent, or motor.

AFFERENT (SENSORY) NEURONS
(FIG. 3-8, 2)

The sensory neurons of the peripheral nervous system transmit impulses from the periphery toward the central nervous system. The cell bodies of these spinal nerve neurons are located in the dorsal root ganglia and are classified as unipolar (Fig. 3-7).

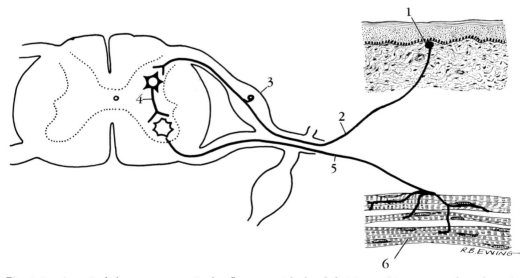

Fig. 3-8. A typical three-neuron spinal reflex arc, with the definitions of its neurons based on the direction of impulse conduction (see text). The somatic efferent (motor) neuron, which innervates the skeletal muscle, is also referred to as the lower motor neuron. *1*, receptor (in skin); *2*, sensory (afferent) neuron; *3*, dorsal root ganglion; *4*, internuncial (connector) neuron; *5*, motor (efferent) neuron; *6*, effector (skeletal muscle).

An afferent, or sensory, neuron terminates within the gray matter of the neuraxis. The specific location of its termination depends on the definite functional type of neuron as discussed in detail in Chapter 7 for spinal nerves and Chapter 17 for cranial nerves.

Within the spinal cord the sensory fibers are within ascending (sensory) tracts (bundles or fasciculi) of the white matter. 'Ascending' refers to the fibers which transmit impulses from spinal cord levels toward the brain.

CONNECTOR (INTERNUNCIAL) NEURONS (FIG. 3-8, 4)

In the typical three-neuron reflex arc the internuncial neurons connect the afferent with the efferent (motor) neurons. The cell bodies of these neurons are of the multipolar type and are located in the gray matter of the spinal cord or brain stem. Generally there are many axon collaterals (branches), which permit synapses with many neurons other than the single main efferent neuron illustrated in the figure.

In Chapter 7 a discussion of the reflex arc designates the connector neuron as the one that permits great versatility in spinal reflexes, depending on the specific types of afferent and efferent neurons connected. The importance of the connector neuron in this situation is emphasized by the fact that the physiological classification of spinal reflexes is based on the specific type of the connector neuron.

As shown in Figure 3-8, the entire internuncial neuron may be located within the gray matter. In addition, the axons of some connector neurons enter the white matter to pass a great distance prior to re-entering the gray matter for synapse with other neurons.

In summary, the cell bodies of all connector neurons and the terminals of their processes are located in gray matter.

It is the connector (internuncial) neuron that actually 'integrates' the neuronal activity within the central nervous system; it connects the afferent neuron with the proper efferent neuron(s) or with the next neuron in the chain so that the proper effect will result from a specific stimulus.

EFFERENT (MOTOR) NEURONS (FIG. 3-8, 5)

The multipolar cell bodies of the efferent (motor) neurons lie within the gray matter of the neuraxis, and their axons extend into the peripheral nerves via the ventral roots of the spinal nerves or directly into certain cranial nerves to terminate in the effectors (see Chapters 7 and 17 for details).

Within the central nervous system the brain also is used as a point of reference for defining long motor (efferent) axons in pathways (tracts). Therefore, the ascending cord pathways (tracts) are the sensory pathways, and the descending tracts (from brain to spinal cord) are classified as motor.

Rate of Impulse Conduction Correlated with Fiber Diameter

Under the general heading of physiology there are basic criteria of classifying neurons other than those given above. One common classification is based on the relation of fiber diameter and the rate of the transmission of impulses by that particular fiber. A general 'rule of thumb' is that the conduction rate in meters per second is six times the total diameter of the fiber in micra. This ratio applies directly to the nervous system of the cat, and perhaps will vary slightly with the species.

Since the thickness of the myelin sheath is the factor that varies the most and therefore largely determines the relative diameter of fibers, it appears that myelin is important for faster rates of impulse conduction. Because myelin functions as an insulator, only at the breaks in myelin, as at the nodes of Ranvier, can current flow out. Therefore, the concentration of flow becomes very great

TABLE 3-1. Properties of Mammalian Nerve Fibers
The top figures show the values most commonly observed; the figures in parentheses show the range of values recorded by various investigators.

Fiber	Diameter (μ)	Conduction Speed m/sec	Function
A (myelinated)	10–20 (1–22)	100 (5–120)	GSE motor GSA proprioceptive
B (myelinated)	5–9 (3–9)	10–15 (3–20)	GSA exteroceptive touch pressure GVE preganglionic
C (nonmyelinated)	1.3–2 (0.3–4)	0.6–2 (0.6–2.3)	GSA exteroceptive pain GVE postganglionic

at the nodes, as the self-propelling impulse is propagated by literally 'jumping' from node to node in a saltatory fashion. Since this is a self-stimulating mechanism, physiologists refer to this phenomenon as 'propagation without decrement.'

Nerve fibers may be classified into A, B, or C groups, according to their diameter and physiologic properties (Table 3-1). *A fibers,* the most heavily myelinated fibers to skeletal muscle (somatic efferent) and the sensory fibers from muscle-tendon receptors (general somatic afferent), are the largest in diameter (10–20 micra). These fibers have a conduction speed reaching a maximum of 100–120 meters per second. The *B fibers* (5–9 micra in diameter) are lightly myelinated, for example, the preganglionic autonomic (general visceral efferent) fibers. Their conduction rate is much slower. The *C fibers* (1.3–2 micra) are the nonmyelinated fibers, for example, the postganglionic autonomic (general visceral efferent), and the small nonmyelinated sensory neurons of the peripheral nerves and dorsal roots, for example, the general somatic afferent, or exteroceptive pain. These smallest fibers

conduct impulses most slowly. Although the anatomic characteristics alone may not distinguish marginal A, B, or C fibers, the physiological characteristics may positively identify specific fibers. As an example, B fibers may be distinguished from A fibers by the absence of a negative after potential (Ruch *et al.,* 1965). For further details the reader should consult a neurophysiology text.

Ganglia

A ganglion is defined as an aggregation of neuron cell bodies located outside of the central nervous system. There are two types of ganglia: sensory and motor.

Sensory Ganglia

Examples of sensory ganglia are: dorsal root ganglia of the spinal nerves, and the sensory ganglia of cranial nerves. Only cranial nerves V, VII, VIII, IX, and X have sensory ganglia.

The histologic features of the typical dorsal root ganglion are:

(*a*) Unipolar cell bodies. The single process emerging from the cell body, formed by fusion of the two processes (see Fig. 3-7), may coil about itself near the cell body before extending peripherally to divide. Because of the obvious resemblance, this coiled configuration is referred to as a *glomerulus.*

(*b*) The cell bodies have smooth surfaces and large, centrally located nuclei.

(*c*) Two types of cell bodies are present (Fig. 3-9): (1) large *clear* and (2) small *obscure* cell bodies. The former are light staining with vesicular nuclei and fine Nissl bodies; the latter are smaller in diameter and stain more deeply. The processes of the clear cell bodies are heavily myelinated, whereas those of the obscure cell bodies are very

thinly myelinated or nonmyelinated. The Nissl bodies within the dendrites of both types are extremely difficult to observe under the ordinary light microscope.

(*d*) Two separate layers of cells surround the cell bodies (Fig. 3-9): (1) *Satellite cells* with rounded nuclei enclose the glomerular formation as an inner layer. These cells are thought to be ectodermal in origin and homologous to the Schwann cells of the peripheral processes. (2) *Capsule cells* with fusiform, darkly stained nuclei form an outer layer covering the inner layer of satellite cells. The capsule cells are of mesodermal (connective-tissue) origin and extend to become continuous with the endoneurium of the peripheral process of the neuron.

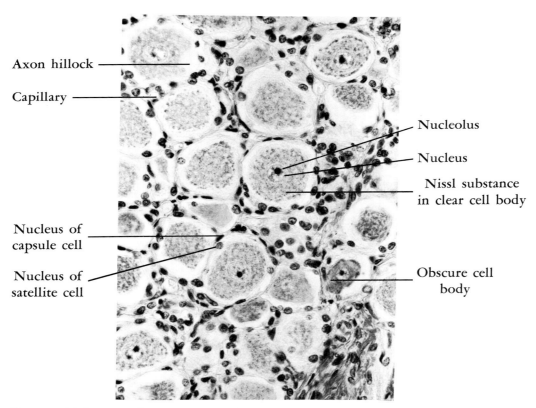

Fig. 3-9. High-power photomicrograph of a section through a dorsal root ganglion. In addition to the specific structures labeled, note the compact arrangement of the cell bodies and their centrally located nuclei.

(*e*) The cell bodies are arranged compactly. Therefore there is relatively little intercellular tissue.

(*f*) The entire ganglion is grossly covered with a tough connective-tissue capsule composed mostly of collagenous fibers. This capsule of the entire ganglion should not be confused with the layer of capsule cells surrounding the cell bodies of the individual neurons.

The detailed characteristics of the dorsal root ganglia cited above apply in general to the sensory ganglia of the cranial nerves, except for those of the vestibulocochlear nerve, which characteristically are composed of bipolar cell bodies without the surrounding layer of capsule cells.

Motor (Autonomic) Ganglia

The principal histologic characteristics of autonomic ganglia listed below should be compared with those of sensory ganglia.

(*a*) The cell bodies are multipolar, therefore their surface is not smooth.

(*b*) The nuclei are eccentrically located within the cell bodies.

(*c*) The satellite and capsular cells either cover the cell bodies incompletely or may be entirely absent.

(*d*) The smaller, irregularly shaped cell bodies are more loosely arranged, that is, there is more intercellular substance than is found in sensory ganglia.

(*e*) Grossly, a smooth connective-tissue capsule does not cover the entire ganglion, as is found in sensory ganglia. Adipose tissue frequently is observed inside and around the ganglia, which adds to the overall histologic picture of much less organization, when compared with sensory ganglia.

Neuroglia

In addition to the various types of neurons mentioned above, the central nervous system contains another large group of cells, the interstitial, or stromal, cells, which sometimes are referred to collectively as the neuroglia, or incorrectly called the 'connective tissue cells.' Based on their embryonic origin, it is erroneous to consider them true connective-tissue cells because all neuroglia, with the exception of one type of cell, are formed from ectoderm, whereas the connective-tissue cells are derived from mesoderm. It is true that glial cells are functionally similar to connective-tissue cells in that they are supportive. In the past this was considered their sole function; recent research, however, has revealed that they serve many other functions. Some glial cells are phagocytic, some are credited with the ability to form myelin for central nervous system neurons, some form cerebrospinal fluid, some contribute to metabolite transfer between blood vessels and neurons, and, as discussed in Chapter 5, some have perivascular feet which contribute to the anatomical concept of the blood-brain barrier. Disorders of the neuroglia are among the commonest and best understood of those affecting the central nervous system.

Casual observation of glial cells reveals their general similarities to neurons. However, their processes are not differentiated into dendrites and axons, there are no true synapses with other cells, and they do not participate directly in the propagation of nerve impulses. Healthy neuroglia are essential, however, for the proper maintenance of neurons and therefore for proper functioning of the central nervous system.

Brain and spinal-cord sections stained with general stains will show only the nuclei of the glial cells and not the processes. With practice, one may learn to identify quite accurately various glial cells stained with non-specific stains by the relative sizes, shapes, and locations of the nuclei. However special glial stains, mostly employing gold and silver salts, are necessary to demonstrate the outlines of the entire cells, including the processes. The various types of glial cells are illustrated in Figure 3-10, and each type is briefly discussed below.

Types of Glial Cells

PROTOPLASMIC ASTROCYTES (FIG. 3-10 A)

Protoplasmic astrocytes are confined in general to the gray matter of the central nervous system. Their processes tend to be very wavy, branching profusely, which accounts for these cells often being referred to as 'mossy cells.' The cytoplasmic processes can often be seen terminating on blood vessels by means of footplates, or *perivascular feet.* Recall that this association of the astrocytes with blood vessels forms the major anatomic feature of the blood-brain barrier.

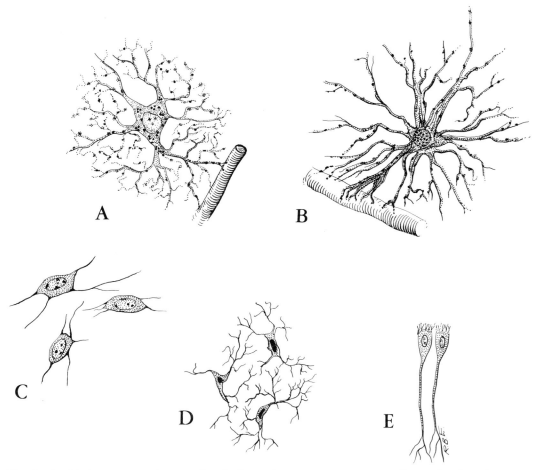

Fig. 3-10. Various types of neuroglia. A. Protoplasmic astrocyte, generally confined to the gray matter of the central nervous system. Note the special terminals of the process (perivascular feet), to form contact with blood vessels. B. Fibrous astrocyte, generally confined to the white matter. These glial cells also possess perivascular feet. C. Oligodendroglia, located in both white and gray matter (see text for comparison with neurilemmal cells of the peripheral nervous system). D. Microglia (mesoglia), located in both gray and white matter, thought to be the glial cells showing the greatest phagocytic activity. E. Ependymal cells, the embryonic-type cells which line the ventricular system of the brain and the central canal of the spinal cord.

FIBROUS ASTROCYTES (FIG. 3-10 B)

Fibrous astrocytes are spider-shaped cells which are confined in general to the white matter of the central nervous system. The cytoplasmic processes are longer than those found in the protoplasmic astrocytes and tend to be much straighter, with less branching; for this reason they are sometimes referred to as 'spider cells.' Processes of the fibrous astrocytes also terminate in perivascular feet.

As a result of injury to brain or spinal cord, astrocytes tend to multiply and enlarge. Protoplasmic astrocytes acquire fibrils, and the fibrils of fibrous astrocytes become more prominent. The result of this proliferation is called *gliosis,* and the culmination of the process is a *glial scar.*

OLIGODENDROGLIA (FIG. 3-10 C)

Oligodendroglia are found in both the gray and the white matter of the central nervous system. As the prefix indicates, oligodendroglial cells have fewer and smaller processes than the astrocytes, and the processes do not terminate in perivascular feet. Their nuclei are smaller in diameter than those of astrocytes. Within the gray matter of the central nervous system many oligodendrocytes tend to locate themselves close to the cell bodies of the neurons. Oligodendrocytes found in this specific location are referred to as *perineuronal satellites.*

Special studies in the distribution of oligodendrocytes by Cammermeyer (1960b) revealed that in the dog brain there are extremely numerous perineuronal satellites in the deep layers of the cortex, but relatively few oligodendrocytes in the subcortical white matter and the optic tract. Within the white matter oligodendrocytes tend to line up in rows between the myelinated fibers. This location gives a picture analogous to that of the neurilemmal cells of the peripheral nervous system. In addition to this analogy with the neurilemmal cells of

Schwann, the oligodendrocytes give rise to the myelin in the central nervous system, and the Schwann cells lay down the myelin in the peripheral nervous system.

Within both gray and white matter many of these glial cells may be found surrounding blood vessels, but they possess no perivascular feet. In this specific location they are referred to as *perivascular satellites.* Cammermeyer (1960a) reported that actually oligodendrocytes occupy only one position: everywhere they are in close association with blood vessels. Even the perineuronal satellites are in close proximity to blood vessels. Based primarily on this anatomic relationship, Cammermeyer suggested the possibility of oligodendrocytes controlling blood flow through capillaries within the central nervous system.

It is thought that the oligodendrocytes have some phagocytic ability and help digest the degenerating neurons following damage to the central nervous system. However, such neuronophagia is considered more properly a characteristic of the microglial cells.

MICROGLIA (FIG. 3-10 D)

The cells spoken of collectively as the microglia are the only ones that are derived from mesoderm. Therefore, microglia are sometimes referred to as *mesoglia.* As the prefix *micro-* indicates, these are small cells, with only a few spiny processes. The nuclei of these cells tend to be spindle-shaped or bizarre in shape with polymorphic characteristics. The processes contain no fibrils and do not terminate in perivascular feet. The cells are very mobile and occur in both white and gray matter, seemingly more within the latter. They act as a counterpart of the macrophages of other parts of the body, and with this phagocytic function they may be considered true members of the reticuloendothelial system. When microglia are actively participating in neuronophagic activities they are referred to as *gitter cells,* primarily because in appearance they resemble a grid, or lattice work.

EPENDYMAL CELLS (FIG. 3-10 E)

Ependymal cells, elongated cells lining the original neural canal within the embryonic neural tube, are also considered glial cells. In the adult these cells have retained their embryonic shape and position, lining the central canal of the spinal cord and the ventricles of the brain. In the area of the choroid plexuses, the ependymal cells are of great physiologic importance. In the walls of the lateral ventricles and roofs of the third and fourth ventricles, the single layer of ependymal cells combines with the pia mater to form the choroid membrane (*tela choroidea*). This membrane is then invaded by capillary tufts, the *choroid plexus.* The ependymal cells in these respective areas become modified to form a simple cuboidal epithelium.

The combination of choroid membrane plus the plexus invaginates restrictive areas of the ventricles. In these locations the combined structures become very 'fluffy' in appearance. Strict terminology with regard to these structures varies among different authorities. Technically, the tela choroidea, or choroid membrane, is definitely distinguishable from the choroid plexus. However, in general usage, one refers to the entire arrangement consisting essentially of the dilated vascular channels, the connective tissue of the pia mater, and the simple cuboidal epithelium, as the choroid plexus. The simple cuboidal epithelial cells in this location contain microvilli, which increase the cell surface and in that way enhance the secretory and possibly absorptive functions of the cells (Truex and Carpenter, 1969).

The role of the choroid plexus in the production of cerebrospinal fluid (CSF) is discussed in Chapter 5. The reader is referred to the same chapter for consideration of the dynamic functions of glial cells in the blood-brain and brain-CSF barriers. The neuroglia should be considered not merely as supportive cells which hold the tissue of the central nervous system together, but also as active regulators of the chemical milieu of nerve cells and therefore as important controllers of neuronal metabolism.

Pathologic Reactions of Glial Cells

Reactions of glial cells, particularly of astrocytes and microglia, along with vascular changes, are clear indicators of disease or insult of the central nervous system. Exceptions are acute disorders that are rapidly fatal (within minutes or hours), with no time for reactive glial changes to develop, and extreme chronic diseases, with only a relatively acellular fibrous gliosis remaining, situations in which no pathologic change may be visible in the usual histologic preparations (Adams and Sidman, 1968).

Most of the intracranial tumors (neoplasms) have their origin from either the embryonic or the adult form of neuroglia. The nomenclature and specific identification of various glial tumors is beyond the scope of this book. A general, non-specific term referring to a tumor of glial cells is *glioma.* Generally the pathologist may identify the tumor because of its histologic characteristics, based on the specific type of neuroglial cell which served as the parent cell. As an example, those gliomas arising from astrocytes are labeled *astrocytomas,* and gliomas of oligodendroglial origin are named *oligodendrogliomas.*

Neurohistologic Technique

The information given here is not intended to enable the reader to prepare neurohistologic sections. It is hoped, however, that he will acquire a general appreciation of the objectives of a few histologic procedures. In addition to numerous individual reports in the literature of special neurotechniques, most of the standard texts and manuals of microtechnique have a section on special procedures for staining nerve tissue (Conn *et al.*, 1960; Davenport, 1960; Emmel and Cowdry, 1964; Galigher and Kozloff, 1964; Gatenby and Beams, 1950; Humason, 1967; Luna, 1968; McClung, 1964).

The discussion below covers the basic principles of neurohistologic technique. Each step has many modifications, some of which are necessary to achieve a desired purpose; many modifications, however, have resulted from individual preferences and have been found successful in a particular laboratory.

1. **Removal of the brain and fixation.** In order to prevent postmortem degeneration and related artifacts it is essential that the brain be fixed with a suitable fluid as soon after death as possible. The brain may either be removed from the animal shortly after death and submerged in a fixative, or be perfused with a fixative via the blood vessels with the brain *in situ*. One can usually fix the brain of an experimental animal (*e.g.*, cat or dog), immediately after death by the submersion technique, or may actually begin perfusion while the animal is deeply anesthetized. For obvious reasons, it is generally necessary to wait several hours after death to fix a human brain. This is one important reason why it is much more difficult to achieve slides of human brain sections that are completely satisfactory, that is, free of cellular distortion by shrinkage or swelling, and with good staining qualities.

A good fixative should: (*a*) kill bacteria and inhibit enzymatic activity so quickly that few structural changes can occur; (*b*) neither shrink nor distend the tissue; (*c*) render the tissue insoluble and prevent postmortem changes; (*d*) penetrate all parts of the tissue equally well; and (*e*) not interfere with subsequent staining procedures. The fixative most commonly used for nervous tissue is 10% buffered formalin, that is, a 10% solution of commercial formalin, which is 36–38% formaldehyde gas dissolved in water. One of the many methods for buffering is to saturate this solution with

calcium carbonate. Various combinations of formalin and other ingredients are preferred by some laboratories, and some investigators use a higher percentage of formalin. An example of fixation variation is the use of 10% formalin as given above, followed by a mordant such as potassium dichromate, which preserves the lipid portion of myelin. This serves as an excellent preparation for nerve tissue that is going to be stained with a myelin stain.

2. **Washing, dehydration, and clearing.** The general principle is to wash the fixative from the tissue and then remove the water from the tissue by a slow method of immersion in many changes of solution of gradually increased concentration of ethyl alcohol. The alcohol used for dehydration will not dissolve or mix with paraffin, therefore a 'clearing agent' is used because it is miscible with both. The clearing agent serves as an intermediary solution between the dehydrating agent and the paraffin used for infiltration and embedding. The most common reagents used for this purpose are the hydrocarbons: xylene, toluene, and benzene. Cedarwood oil is also used as a clearing agent. For the last immersion prior to infiltration, many laboratories use a mixture of the clearing agent and the paraffin used for infiltration.

3. **Infiltration, embedding, and cutting.** The purpose of infiltration and embedding (blocking) is to add support to the tissue so that it may be sliced with a microtome into thin sections, usually 10, 15, or 20 micra thick. In addition to supporting the tissue internally by infiltration, the embedding mass surrounding the tissue in the block serves as a means of attaching it solidly to the microtome for cutting. Melted paraffin or a modified paraffin (*e.g.*, Tissuemat* or Paraplast†) is commonly used for both infiltration and embedding. The brain sections

*Tissuemat—Fisher Scientific Co., Chicago, Illinois 60651.

†Paraplast—Aloe Scientific Co., Oak Brook, Illinois 60521.

used for the photographs in the Atlas of this book were infiltrated and embedded in Paraplast, a specific mixture of paraffin and several plastic polymers of regulated molecular weights (Humason, 1967). Some laboratories use a type of nitrocellulose for embedding, which requires a procedure different from the one that is discussed here for paraffin. The frequently heard phrase, 'celloidin technique,' is commonly used in the broad sense to mean embedding with any kind of nitrocellulose. After the tissue is embedded in the chosen matrix, the blocks are cut into thin sections with a microtome. The sections of tissue surrounded by the embedding matrix are affixed to glass slides, the embedding matrix is dissolved, and the tissue is ready for staining.

4. **Staining.** The main purposes in neurostaining are to enhance the morphologic properties of specific portions of the nerve cells or to color the entire cell body and processes for study of their shapes or outlines. Neurostaining is a special technique which ideally should be reserved only for the experienced histotechnician who has the time and interest to devote to the subject. It should be appreciated that, from a practical standpoint, one can make many general neurohistologic and neuropathologic observations by study of sections stained with the hematoxylin-eosin (H. and E.) combination. This stain probably is the most universally used general stain for most body tissues and is relatively easy to employ. Note that this is a 'general' stain, and therefore it will not stain specific nerve cell components to best advantage for study. There is no single stain that can be considered the 'best' for all neuronal or glial components. Therefore, special neurostains are grouped as: myelin, Nissl, nerve fiber or neurofibril, and neuroglial stains.

Not only are there a great number of special stains in each group, but there are specific stains that are preferred for individual types of cells, for example, glial cells. It is not true that one cannot see any neu-

roglia without the use of a special stain; the nuclei of neuroglia can be seen in a brain section that is stained with the general hematoxylin-eosin combined stain. If one wishes to see the detailed glial processes, however, a special glial stain must be used.

A few brief comments are given below regarding only a sample of special neurologic stains commonly used, so that the student may correlate the names of the stains with the specific anatomic structures for which they are employed. It should be remembered that there is no single staining agent that will give the best quality stain of all the anatomic components of a neuron.

Myelin stains. Some of the more popular stains in this category are: Weigert, Weil, osmic acid (osmium tetroxide), and Luxol fast blue. Since these stain myelin, they may be used for pathologic study of demyelination. Degenerated myelin will not take the stain, and therefore degenerating areas appear clear. The Marchi staining method is designed to stain degenerating myelin, but leaves normal myelin unstained and therefore clear. Thus the normal myelin stains and the Marchi stain present opposite histologic pictures. Myelin stains show an affinity for the lipid portion of myelin (a lipoprotein), which is mostly dissolved in ordinary dehydration and clearing procedures. The use of potassium dichromate after the fixative is a method of preserving myelin for better results with these stains. Osmic acid frequently is used for staining myelin in peripheral nerves. Students sometimes fail to realize that, because of affinity of fat for the stain, not only the fatty portion of myelin but also the adipose elements in the connective-tissue coverings of the nerve stain black. If one wishes to stain primarily the protein portion of myelin (neurokeratin), the lipid portion is dissolved and then a connective-tissue stain may be used to stain the finely trabecular network of the neurokeratin (Fig. 3-1).

Nissl stains. Cresyl violet, neutral red, and thionine are probably the most popular Nissl stains. These are of value in neuropathology, because injured neurons demonstrate chromatolysis as reflected by a lack of Nissl stain in their cell bodies (see Chapter 8).

Neurofibril stains. Neurofibrils have a strong affinity for silver (*argyrophilia*—silver + love for). This category includes staining procedures named after famous past neuroanatomists (*e.g.,* Golgi, Ramon y Cajal, Ranson, Bielschowsky), as well as some leading neuroanatomists of the present day (*e.g.,* Bodian, Nauta, and Gygax). The Nauta-Gygax technique selectively impregnates preterminal degenerating axons, using the frozen-tissue–section method. Modifications of this technique have been developed, in which paraffin sections may

be used. Some silver techniques are considered as 'bulk,' or 'block,' staining, because the staining is done prior to embedding. In the Golgi technique the use of silver solutions for neurostaining is similar to that in photographic techniques. The tissues are placed in solution of silver salts for several days, and then placed in a reducing solution. The silver particles are deposited within and on the surface of the cell body and its processes. The reduced silver particles are then removed from the non-neural elements, but the neurons hold them because of their greater argyrophilia. In areas like the cerebral cortex, this gives a delicately detailed black silhouette of the neurons on a light background (Fig. 3-11).

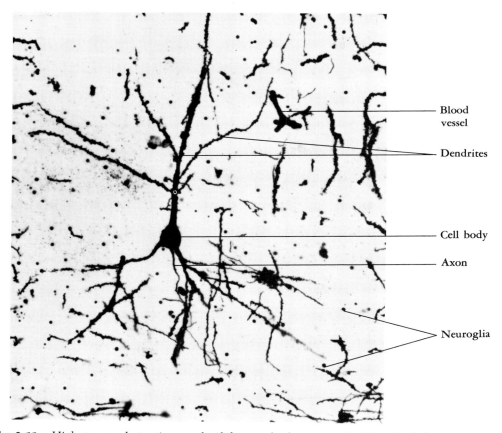

Blood vessel

Dendrites

Cell body

Axon

Neuroglia

Fig. 3-11. High-power photomicrograph of dog cerebral cortex stained by the Golgi-Cox method. A large pyramidal neuron is shown with its parts labeled. Note that this stain reveals no internal cellular details, but is specific for demonstrating the cellular outlines as silhouettes. Blood vessels also have an affinity for this stain. (See text for further discussion of staining techniques.)

Because the silver is deposited on and in the neurons in such procedures, the method technically is referred to as 'silver impregnation,' rather than staining, which is a molecular combination between dye and tissue. There are modifications of the original silver technique of Golgi. Figure 3-11 is a photomicrograph of a preparation stained by a Cox modification, which is therefore referred to as the Golgi-Cox method (Davenport, 1960).

Combinations of stains. Neurohistologic literature gives many combinations of stains. For a variety of reasons, the original methods have been modified by the use of various colored counterstains. The slide preparations used for the Atlas of this book are examples of such modifications. These slides were stained with a popular present-day combination of Luxol fast blue for staining myelin, and cresyl violet for staining Nissl substance. This combination was introduced by Klüver and Barrera (1953). A more recent variation is the combination of Luxol fast blue and neutral red (Lockard and Reers, 1962). The Nissl bodies stain more brilliantly with neutral red than with the cresyl violet.

Bibliography

Adams, R. D., and Sidman, R. L.: *Introduction to Neuropathology.* New York, McGraw-Hill Book Co., Blakiston Division, 1968.

Barr, M. L., Bertram, L. F., and Lindsay, H. A.: The morphology of the nerve cell nucleus, according to sex. Anat. Rec. *107:* 283–297, 1950.

Bloom, W., and Fawcett, D. W.: *A Textbook of Histology,* 9th ed. Philadelphia, W. B. Saunders Co., 1968.

Bodian, D.: The generalized vertebrate neuron. Science *137:* 323–326, 1962.

Bunge, R. P.: Glial cells and the central myelin sheath. Physiol. Rev. *48:* 197–251, 1968.

Cammermeyer, J.: Reappraisal of the perivascular distribution of oligodendrocytes. Am. J. Anat. *106:* 197–231, 1960a.

Cammermeyer, J.: The distribution of oligodendrocytes in cerebral gray and white matter of several mammals. Am. J. Anat. *107:* 107–127, 1960b.

Conn, H. J., Darrow, M. A., and Emmel, V. M.: *Staining Procedures Used by the Biological Stain Commission,* 2nd ed. Baltimore, Williams & Wilkins Co., 1960.

Davenport, H. A.: *Histological and Histochemical Technics.* Philadelphia, W. B. Saunders Co., 1960.

Emmel, V. M., and Cowdry, E. V.: *Laboratory Technique in Biology and Medicine,* 4th ed. Baltimore, Williams & Wilkins Co., 1964.

Galigher, C. E., and Kozloff, E. N.: *Essentials of Practical Microtechnique.* Philadelphia, Lea & Febiger, 1964.

Gatenby, J. B., and Beams, H. W., Eds.: *The Microtomist's Vade-Mecum (Bolles Lee),* 11th ed. Philadelphia, Blakiston, 1950.

Ham, A. W.: *Histology,* 5th ed. Philadelphia, J. B. Lippincott Co., 1965.

House, E. L., and Pansky, B.: *A Functional Approach to Neuroanatomy,* 2nd ed. New York, McGraw-Hill Book Co., Blakiston Division, 1967.

Humason, G. L.: *Animal Tissue Techniques,* 2nd ed. San Francisco, W. H. Freeman and Co., 1967.

Klüver, H., and Barrera, E.: A method for the combined staining of cells and fibers in the nervous system. J. Neuropath. Exp. Neurol. *12:* 400–403, 1953.

Lockard, I., and Reers, B. L.: Staining tissue of the central nervous system with Luxol fast blue and neutral red. Stain Techn. *37:* 13–16, 1962.

Luna, L. G., Ed.: *Manual of Histologic Staining Methods of the Armed Forces Institute of Pathology,* 3rd ed. New York, McGraw-Hill Book Co., Blakiston Division, 1968.

McClung, C. E.: *McClung's Handbook of Microscopical Technique,* 3rd ed. New York, Paul B. Hoeber, Inc., 1964.

Ruch, T. C., Patton, H. D., Woodbury, J. W., and Towe, A. L.: *Neurophysiology,* 2nd ed. Philadelphia, W. B. Saunders Co., 1965.

Truex, R. C., and Carpenter, M. B.: *Human Neuroanatomy,* 6th ed. Baltimore, Williams & Wilkins Co., 1969.

chapter 4

Blood Supply and Dural Venous Sinuses

The importance of an adequate blood supply to the central nervous system cannot be overemphasized. Damage to the human brain as a result of hypoxia is very well known. The tolerance to hypoxia among various loci in the brain and the total effect of hypoxia on the body vary greatly. If certain vital centers within the medulla are deprived of oxygen for only a few minutes, death will result. However, some small areas of the brain may be deprived of oxygen, causing death of the cells (neurons) within the area, without producing a perceptible effect on the body.

A great number of paralyses in man result from the so-called 'stroke,' which generally is a disturbance in the blood supply to a specific motor area within the brain. Blood supply to a specific area of the brain may be depleted in many ways. As an example, anoxia may result from a bursting aneurysm (a circumscribed dilatation of an artery), causing hemorrhage and thereby depriving distal areas supplied by that particular vessel of blood, or a blood vessel may be blocked (*e.g.,* by a thrombus), causing infarction of a localized area of the brain, resulting in neuronal death. Any such episode of circulatory disturbance producing a neurological deficit may be referred to as a *cerebral vascular accident* (CVA).

From reading the literature and comparing clinical cases in veterinary and human medicine, one may conclude that the dependency of the central nervous system on its blood supply seems to be much greater in man than it is in animals. Probably it is not an inherent difference in the tolerance of the nervous tissue *per se,* but since dogs and cats have a rich physiologically effective anastomotic cerebral blood supply the symptoms of a disturbance in the cerebral circulation are usually much less severe and extensive.

Cerebral vascular accidents probably occur in old dogs much more frequently than most clinicians realize. Possible symptoms seen in old dogs which could have resulted from a CVA are unilateral sagging of the lip and eyelid, as well as a drooping of the ear, and occasionally one sees a weakness of the front- and hindlimb on the same side (Hoerlein, 1965). Another factor is that in veterinary medicine the critical clinical cases often are not seen in the hospital; too frequently animals with such neurological deficits are euthanatized without a complete study.

The gray matter in the central nervous system is much more vascular than the white. This is as expected, since the gray matter contains the cell bodies which, among other functions, serve as the trophic centers for the entire neurons. It has been reported that one cubic millimeter of gray matter in certain areas of human brain contains over 1,000 linear millimeters of capillaries (Ranson and Clark, 1959).

Arterial Blood Supply

The major afferent blood supply to the brain arises from the two internal carotid arteries and the two vertebral arteries. The general anatomic pattern of the two paired afferent vessels is similar in all mammals, but the details of the vessels, the extent of anastomoses, and therefore the efficiency of physiologic compensatory mechanisms differ.

Circulus Arteriosus Cerebri (Circle of Willis)

The bilateral sources of blood supply are indirectly joined at the base of the brain by the circulus arteriosus cerebri (circle of Willis). This arterial structure is characteristic of mammals, although the exact pattern of the arterial circle varies among them. As illustrated in Figure 4-1, the circulus arteriosus cerebri is formed by the junction of the internal carotid arteries and the dichotomy of the basilar artery, which is formed caudally by fusion of the two vertebral arteries.

The carotid and vertebral arteries are located so that the internal carotids and their branches supply the middle and rostral brain areas, whereas branches from the vertebral and basilar arteries supply the caudal area of the cerebral cortex, the cerebellum, and the more caudal divisions of the brain stem. The two sources are joined on each side by the caudal (posterior) communicating arteries, to complete the circle.

Anatomically the arterial 'circle' actually is in the form of a hexagon or heptagon, depending on whether or not the rostral (anterior) communicating artery is present. Schematically the two common configurations for the circulus arteriosus cerebri in the dog are illustrated in Figure 4-2.

Functionally, the arterial circle (1) provides anastomotic channels for distribution of blood as needed because of different physiologic demands, and (2) serves as a stabilizer for the maintenance of constant blood pressure in the terminal arteries which arise from the circle. The efficiency of the circulus arteriosus cerebri varies among different species of mammals and among individuals. In adult man there is

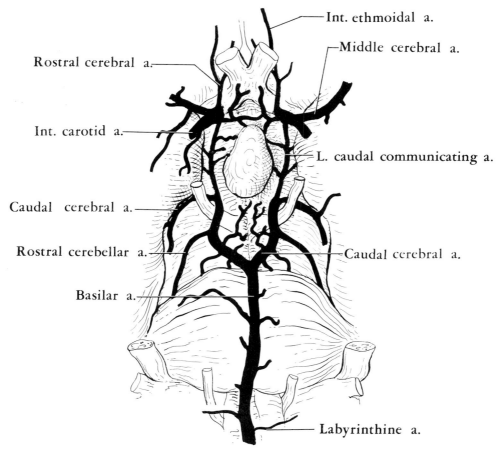

Int. ethmoidal a.

Middle cerebral a.

Rostral cerebral a.

Int. carotid a.

L. caudal communicating a.

Caudal cerebral a.

Rostral cerebellar a.

Caudal cerebral a.

Basilar a.

Labyrinthine a.

Fig. 4-1. The circulus arteriosus cerebri (circle of Willis), on the ventral surface of the brain.

evidence that the circle of Willis is not an efficient anastomotic channel, is not bilaterally symmetrical, and does not efficiently distribute blood between the carotid and the vertebral circulation (Hale, 1960).

Intracranial-Extracranial Anastomoses

In addition to the two main sources of blood to the brain, in the dog and cat there are great anastomoses between the intracranial and extracranial blood vessels. These extensive anastomoses undoubtedly furnish the major reason for infrequent drastic

effects of cerebral vascular accidents in these species.

In one study (Whisnant *et al.,* 1956) six of nine dogs survived a cervical one-stage ligation of all major afferent blood vessels to the brain. This included bilateral ligation of the vertebral, common carotid, internal and external carotid, occipital, and ascending pharyngeal arteries. The symptoms varied and included ataxia, confusion, and bilateral paresis, but in only two dogs did the syndrome include visual impairment and generalized convulsions. Symptoms which developed gradually disappeared within 7 to 21 days. This indicated that

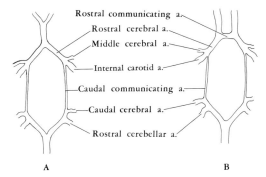

Rostral communicating a.
Rostral cerebral a.
Middle cerebral a.
Internal carotid a.
Caudal communicating a.
Caudal cerebral a.
Rostral cerebellar a.

A   B

Fig. 4-2. A schematic diagram showing a common variation in the basic pattern of the circulus arteriosus cerebri in the dog. Whether the hexagon (A) or the heptagon (B) pattern is present depends on the presence or absence of the rostral (anterior) communicating artery.

collateral circulation develops rapidly and effectively. The main channels for collateral circulation were reported to be branches of the omocervical and costocervical arteries to the muscular branches of the vertebral arteries at multiple sites. Evidence indicated that these channels ordinarily are not functional. There were no gross pathologic changes in the brains at necropsy, and very minor microscopic changes were noted in the brains of only two animals.

In a similar study, the common carotid and vertebral arteries on both sides were doubly ligated and cut between the ties in seven dogs. No special nursing care was needed, and the postoperative course from one to seven months (the time of writing) was uneventful in each case. It was concluded that the anterior (ventral) spinal artery alone is able to transport sufficient blood to the brain to sustain life in the dog (Bunce, 1960).

The most obvious and most important anastomosis between intracranial and extracranial circulation in the dog has been reported to be the anastomosis between the internal carotid and the internal maxillary artery (De La Torre *et al.,* 1959). They found this anastomosis in all dogs studied by both angiography and dissection.

A similar, but perhaps greater, anastomosis can be located in the cat and identified as a *rete mirabile.* This is a complex network of fine arterial vessels interlaced with vessels of the pterygoid plexus that completely surrounds the trunk of the internal maxillary artery (extracranially) from the foramen rotundum to the optic foramen (Fig. 4-6). Branches from the rete are distributed to the brain (substituting for the internal carotid artery in the cat), to the orbit, to the adjacent muscles of mastication, and to the nasal region.

The veins leaving the rete are connected with the dural venous sinuses, the deep and superficial facial veins, the ophthalmic veins, and the nasal veins (Gillilan and Markesbery, 1963). These authors found major arteriovenous (A-V) communications in: cat, dog, rat, guinea pig, rabbit, and to a lesser extent in monkey. In all these animals the A-V communications were located in the retro-orbital region.

Internal Carotid Artery

Internal Carotid Artery in the Dog

The divisions of the common carotid artery in the cervical region are more complex in the dog than in man. Frequently in the dog one finds a trifurcation of the common carotid artery due to participation of the occipital artery in addition to the external and internal carotid arteries. The ascending pharyngeal artery arises in close association with the division. At the origin of the internal carotid there is a bulbous enlargement, the *carotid sinus,* which functions in reflex regulation of blood pressure (carotid sinus reflex). The internal carotid artery courses cephalically and medially to reach the base of the skull and enters the caudal (posterior) carotid foramen within the petrobasilar fissure.

Within the skull of the dog the artery traverses the carotid canal, from which it exits at the rostral end by emerging ventrally through the external carotid canal. It then continues rostrally and pierces the outer dura to enter the cavernous sinus, which it traverses before perforating the inner dura and arachnoid as it enters the subarachnoid space immediately caudal to the optic tract and lateral to the infundibulum. At this location the artery contributes to the circulus arteriosus cerebri, where it terminates by trifurcating into the rostral (anterior) cerebral, middle cerebral, and caudal (posterior) communicating arteries (Figs. 4-1 and 4-2).

Termination of Internal Carotid Artery at Circulus Arteriosus Cerebri in the Dog

In the description that follows no attempt is made to describe in detail the distribution of all the arteries in relation to specific loci of the brain. Only that of the main vessels and of their major branches is described. In addition to the distributions discussed it should be understood that all of these arteries give rise to many branches throughout their courses. It should be remembered that the terminal trifurcation of the internal carotid includes the following three main vessels.

ROSTRAL (ANTERIOR) CEREBRAL ARTERY (FIG. 4-1)

The rostral (anterior) cerebral artery runs rostromediad dorsal to the optic nerve near the chiasm toward the base of the rostral longitudinal fissure. At this location it either unites directly with its mate from the opposite side or is connected with it by the rostral (anterior) communicating artery (Fig. 4-2). The rostral cerebral artery divides into: (1) a rostral extension which passes on the ventral side of the brain near the bottom of the rostral longitudinal fissure toward the olfactory bulb, and (2) a dorsal branch which passes dorsally toward the genu of the corpus callosum.

Two branches arise from this rostral portion of the rostral cerebral artery. (1) The internal ethmoidal artery continues toward the cribriform plate of the ethmoid bone, where its branches anastomose with external ethmoidal branches in a retelike formation. According to Miller et al. (1964) most of the branches of the internal ethmoidal artery perforate the cribriform plate to supply the ethmoturbinate bones and the nasal septum. (2) The internal ophthalmic artery follows the optic nerve through the optic canal to enter the caudal portion of the orbit, where it anastomoses with branches of the external ophthalmic artery. From this anastomosis arises the arterial supply to the eye, via the ciliary artery and the central artery of the retina.

The second main branch of the rostral (anterior) cerebral artery referred to above can be followed best from the medial surface in the sagittal view as shown in Figure 4-3. This branch courses dorsally to curve around the genu of the corpus callosum and extend caudally within the callosal sulcus almost to the splenium of the corpus callosum, where its terminal branches anastomose with those of the caudal cerebral artery on the medial surface. Throughout its course it gives rise to many branches which supply the medial cortical surface, and dorsally its terminal rami emerge from the dorsal longitudinal fissure to extend laterally for a short distance to supply gyri bordering the dorsal longitudinal fissure on the rostral two-thirds of the dorsal surface of the brain (Fig. 4-4).

MIDDLE CEREBRAL ARTERY (FIGS. 4-4 AND 4-5)

The middle cerebral artery supplies the largest area of the brain surface. At its origin from the arterial circle it appears as a large direct continuation of the internal carotid artery. Near its origin, as it lies on the anterior perforated substance, it gives rise to the choroidal artery, which penetrates the brain to supply the adjacent areas and the choroid plexus of the lateral ventricle. It proceeds dorsally to the piriform area to

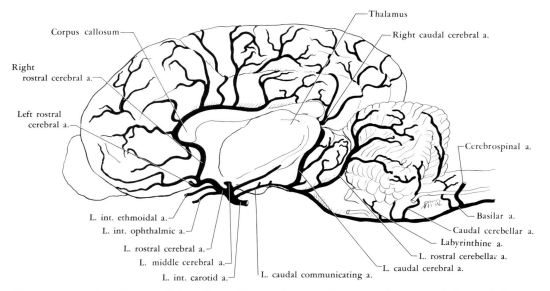

Fig. 4-3. Arteries of the medial surface of the cerebrum and the lateral surface of the cerebellum. (After Miller *et al.,* 1964.)

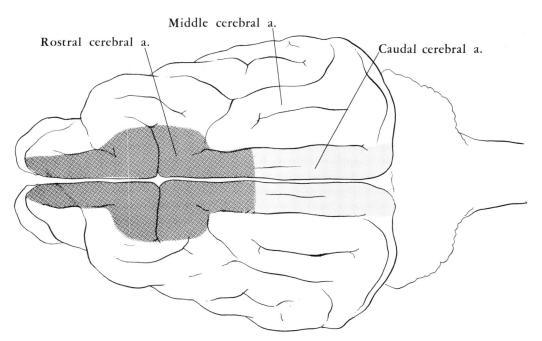

Fig. 4-4. Dorsal view of the dog brain showing the distribution of the three main cerebral arteries on the dorsal surface. (After Miller *et al.,* 1964.)

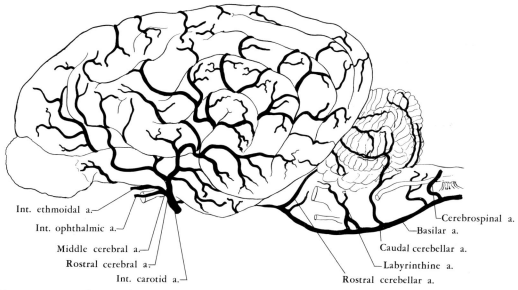

Int. ethmoidal a.
Int. ophthalmic a.
Middle cerebral a.
Rostral cerebral a.
Int. carotid a.

Cerebrospinal a.
Basilar a.
Caudal cerebellar a.
Labyrinthine a.
Rostral cerebellar a.

Fig. 4-5. Lateral view of the dog brain to show the areas of cerebral cortex supplied by the middle cerebral artery. Major arteries to the cerebellum and brain stem also are shown. (After Miller *et al.*, 1964.)

enter the Sylvian fissure and distribute numerous ramifying branches which extend over the entire lateral surface of the cerebral hemisphere. The cortical branches anastomose with those of the rostral and caudal cerebral arteries.

In man the lenticulostriate arteries, arising from the middle cerebral artery to supply the lenticular nucleus and internal capsule area, are extremely important. Disturbance of the blood supply to the internal capsule is commonly the cause of stroke, which characteristically results in hemiplegia. In dog the corticospinal tract is less important than in man. Therefore, a disturbance of the blood supply to the internal capsule would not necessarily result in hemiplegia. Also, rupture of the vessels supplying this area is believed to occur less frequently in the dog than it does in man.

CAUDAL (POSTERIOR) COMMUNICATING ARTERY

The caudal (posterior) communicating artery generally is considered as connecting

the two afferent sources to the circulus arteriosus cerebri. Specifically, it is a connection between the internal carotid artery and the caudal cerebral artery (Figs. 4-1 and 4-2). De La Torre *et al.* (1959) consider this interpretation as applicable to man, but not to the dog, primarily because the rostral terminal (bifurcation) of the dog's basilar artery is farther caudad. This results in a relatively greater distance from the end of the basilar artery to the infundibulum. Evidence that they consider that, in the dog, the caudal communicating artery, which is larger than in man, is the important artery for the caudal portion of the arterial circle is their statement that the two caudal communicating arteries fuse to form the basilar. Therefore, they consider the caudal cerebral and the rostral (superior) cerebellar arteries as branches of the caudal communicating artery.

As illustrated in Figures 1-1 and 4-2, based on the anatomic construction of the arterial circle, one can logically agree with both interpretations. The important conclusion is that the basic anatomic plans of the arterial circle are similar in man and dog, although in the dog the circle is elongated rostrocaudally, because of the longer caudal communicating arteries.

The major portion of the pituitary gland is supplied by the numerous rostral (anterior) hypophyseal arteries, which arise from the caudal communicating arteries.

Internal Carotid Artery in the Cat

Unlike the patent internal carotid artery that is present in adult man and dog, the internal carotid artery of the cat is patent embryonically, but in the adult it is a vestigial ligament. The distal portion of this artery is patent as it approaches the brain and anastomoses with the ascending pharyngeal artery (Walker, 1967).

The following points of summary of the carotid circulation in the cat are according to Davis and Story (1943) and illustrated in Figure 4-6. Instead of the normal internal carotid afferent supply, the brain receives its blood from the external carotid artery via three channels: (1) a large anastomotic vessel connecting the internal maxillary artery via the external rete mirabile through the orbital fissure; (2) an additional anastomotic vessel, part of the original middle meningeal artery, between the internal maxillary artery and the internal rete mirabile which extends to the circulus arteriosus cerebri; and (3) the trunk of the ascending pharyngeal artery, which anastomoses with and has completely taken over the distal segment of the internal carotid artery.

Vertebral Artery

The vertebral artery arises from the subclavian artery to penetrate deeply between the sixth and seventh cervical vertebrae and traverses the transverse foramina of the upper six cervical vertebrae. During its course the vertebral artery gives off many rami to the cervical muscles. As noted above, the multiple anastomoses of the ver-

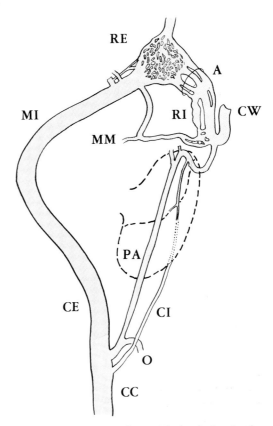

Fig. 4-6. Diagram of carotid circulation in the cat. Note the vestigial nonfunctional proximal segment of the internal carotid artery. The distal segment is functional due to an anastomosis with the ascending pharyngeal artery, a branch of the external carotid artery. In addition, the cat brain receives its afferent supply from two other external carotid sources: (1) a large anastomotic artery connecting the external rete mirabile and the circle of Willis, and (2) an anastomotic vessel, part of the original middle meningeal artery, between the internal maxillary artery and the internal rete mirabile, which extends to the circle of Willis. CC, common carotid artery; CE, external carotid artery; MI, internal maxillary artery; RE, external rete; A, anastomotic artery; RI, internal rete; O, occipital artery; CI, internal carotid artery; PA, ascending pharyngeal artery; MM, middle meningeal artery; CW, circle of Willis. The outline of the tympanic bulla is shown by broken lines. (After Davis and Story, 1943.)

tebral arteries with the omocervical and costocervical arteries are extensive enough to sustain life in a majority of dogs following bilateral ligation of both the internal carotid and the vertebral artery at their sites of origin (Whisnant *et al.*, 1956).

During its course within the cervical transverse canal, the vertebral artery gives off the first seven spinal rami which enter the vertebral canal at each of the first seven intervertebral foramina. Within the canal these rami divide and anastomose longitudinally on the ventral side, to contribute to the formation of the ventral spinal artery which extends caudally at the base of the ventral median fissure of the spinal cord. The dorsal branches of the spinal rami follow the dorsal roots of the spinal nerves to reach the spinal cord, where they disperse to supply the dorsal and lateral portions of the cord without formation of a definitive dorsal or lateral spinal artery.

At the wing of the atlas the vertebral artery anastomoses with branches of the occipital artery. The large dorsal branch of the vertebral artery, after having merged with the occipital branches, enters the transverse foramen of the atlas. It continues cephalad and passes through the intervertebral foramen of the atlas, pierces the dura and arachnoid to enter the vertebral canal on the ventral surface of the spinal cord at about the level of the first cervical nerve. At this site it is referred to as the *cerebrospinal artery*, which divides into: (1) a descending (spinal) branch which meets its mate and contributes to the ventral spinal artery; and (2) a cerebral branch which courses rostrally to meet its counterpart from the opposite side to form the *basilar artery* (Fig. 4-1). In the dog this large artery passes rostrally along the ventral midline of the brain stem as the largest source of blood to the circulus arteriosus cerebri and therefore to the brain (Miller *et al.*, 1964).

In its rostral course on the ventral surface of the brain stem the basilar artery gives rise to the following bilateral branches from caudal to rostral—caudal (posterior) cerebellar, labyrinthine (acoustic), and pontine branches—before bifurcating at the pontine-mesencephalic junction. The most common interpretation of the basilar bifurcation is to consider the basilar artery as dividing into the two caudal cerebral arteries with the rostral (superior) cerebellar arteries arising from them.

CAUDAL (POSTERIOR) CEREBRAL ARTERY (FIGS. 4-1, 4-3, AND 4-4)

The caudal (posterior) cerebral artery is formed by the terminal bifurcation of the basilar artery and contributes to the caudal third of the arterial circle. This artery is easily identified as it leaves the circle because it passes immediately rostral to the oculomotor nerve after it is joined by the caudal extension of the caudal communicating artery. The caudal cerebral artery extends laterally over the ventral surface of the cerebral peduncle to enter the transverse fissure between the cerebellum and the cerebrum. It continues dorsally around the brain stem caudal to the splenium of the corpus callosum at the base of the longitudinal fissure. Here it ramifies extensively over the medial surface of the caudal portion of the cerebrum, where it anastomoses with the rostral cerebral artery. It reaches the dorsal surface of the brain on the lateral wall of the longitudinal fissure and extends over the top to supply the most dorsal gyri in the caudal third, where it anastomoses with the cortical terminals of the anterior and middle cerebral arteries.

ROSTRAL CEREBELLAR ARTERY (FIGS. 4-1 AND 4-3)

The rostral cerebellar artery arises from the caudal cerebral artery immediately caudal to the emergence of the oculomotor nerve. It enters the bottom of the transverse fissure lateral to the cerebral peduncle-pons junction to extend dorsally and ramify throughout the rostral and lateral surfaces of the cerebellum. Along its way it gives off branches to the brain stem.

Major Veins of the Brain

A general description of the major veins of the brain will suffice for the purposes of this text. A more complete discussion of the dural sinuses, into which the veins drain, is given. The veins, like the arteries, lie in the pia mater. Histologically, they are similar to the dural sinuses in that they do not contain valves nor do they possess a muscular coat.

DORSAL CEREBRAL VEINS (FIG. 4-7)

The dorsal cerebral veins extend over most of the cerebral surface. They are inconstant in number (perhaps 4 to 7) and

tend to occur on both hemispheres, but not symmetrically. All of these veins enter directly into the dorsal sagittal sinus. They drain the cortical surface areas supplied by the anterior, middle, and posterior cerebral arteries. Numerous diploic veins from the calvaria enter these veins prior to their entrance into the dorsal sagittal sinus.

CAUDAL (POSTERIOR) CEREBRAL VEIN (FIG. 4-7)

The caudal (posterior) cerebral vein drains most of the cortex of the temporal lobe and usually empties into the caudal portion of the dorsal petrosal sinus.

GREAT CEREBRAL VEIN (OF GALEN) (FIG. 4-7)

The great cerebral vein is formed by a caudal confluence of the vein of the corpus

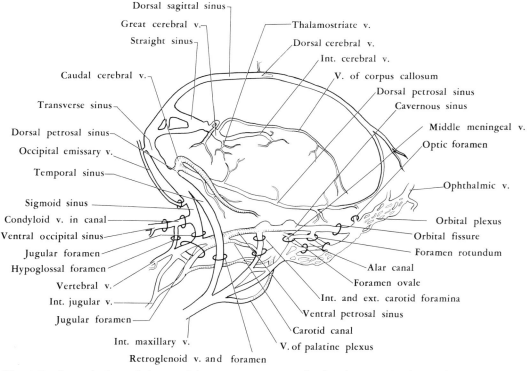

Fig. 4-7. Lateral view of the cranial venous sinuses and related veins. (Redrawn from Reinhard *et al.*, 1962.)

callosum, the internal cerebral vein, and the thalamostriate vein. The great cerebral vein serves as the principal exit of venous blood from the deeper portions of the cerebrum. The vein terminates caudally by extending directly into the straight sinus.

INTERNAL CEREBRAL VEIN (FIG. 4-7)

The internal cerebral vein receives tributaries from the area ventral to the corpus callosum and the dorsal area of the midbrain. It enters the rostral end of the great cerebral vein.

THALAMOSTRIATE VEIN (FIG. 4-7)

The thalamostriate vein, as the name indicates, drains the thalamus and the corpus striatum. It enters into the formation of the great cerebral vein at the ventrorostral side.

Cranial Dural Sinuses

As discussed in Chapter 5, the cranial dura mater actually is composed of two layers: the outer periosteal and the inner meningeal layer. Over most of the brain, the two are fused into one thick layer, but in certain isolated linear areas, the two layers are separated by passages known as the venous sinuses. The blood from the veins of the brain and calvaria drains into the venous sinuses as it courses back toward the heart. Histologically the venous dural sinuses lack a tunica media and have no valves within their lumina. The sinuses are connected in a continuous venous channel, and some of those at the base of the brain are interconnected, as will be explained.

DORSAL (SUPERIOR) SAGITTAL SINUS (FIG. 4-7)

The dorsal (superior) sagittal sinus originates rostrally by a fusion of the rhinal veins in the area of the olfactory bulbs near the middle of the cribriform plate of the ethmoid bone. It continues dorso-caudally in the midline directly above the median longitudinal fissure at the dorsum of the falx cerebri between the two hemispheres. Caudally at the occipital pole, or the area of the internal occipital protuberance, the superior sagittal sinus receives the straight sinus immediately before entering the left and right transverse sinuses. This junction of sinuses at the occipital pole is referred to as the *confluence of sinuses,* which in the dog is within the occipital bone at the *foramen impar.*

STRAIGHT SINUS (FIG. 4-7)

The straight sinus is a very short sinus beginning at the posterior margin of the falx cerebri as a direct continuation of the great cerebral vein (of Galen) and extending for only about 5 mm. This sinus lies rostrodorsal to the osseous tentorium. As indicated above, it may join the dorsal sagittal sinus as it approaches the foramen impar, or it may enter the confluence of sinuses independently at the same location.

TRANSVERSE SINUS (FIG. 4-7)

Each of the paired transverse sinuses originates by receiving the straight and dorsal sagittal sinuses and merges with its fellow in the midline at the confluence of the sinuses within the occipital bone. From here each runs laterally within its respective transverse canal, then in its transverse groove to follow the border of the tentorium cerebelli to terminate laterally by dividing into the temporal and sigmoid sinuses. Along their course the transverse sinuses receive emissary veins from extracranial areas.

TEMPORAL SINUS (FIG. 4-7)

The temporal sinus is a direct rostroventral continuation of the transverse sinus and extends between the petrous and squamous portions of the temporal bone. It continues distally to make its exit from the skull via the retroglenoid foramen to become the retroglenoid vein, which empties into the internal maxillary vein as one of its largest tributaries.

SIGMOID SINUS (FIG. 4-7)

The sigmoid sinus is somewhat S-shaped and extends ventrally as the other continuation of the transverse sinus. In its short course medially around the caudal end of the petrous temporal bone to the jugular foramen, it receives veins from the cerebellum and medulla. It joins the rostral end of the condyloid vein, within the condyloid canal, and the ventral petrosal sinus, within the petrobasilar canal at the base of the petrobasilar bone. The condyloid vein courses caudally as the *ventral occipital sinus,* which then becomes the *vertebral sinus* as it passes through the foramen magnum.

Also at the jugular foramen the sigmoid sinus, after receiving the ventral petrosal sinus, leaves the skull to form the origin of the internal jugular and vertebral veins. Since the internal jugular vein usually is small and relatively insignificant in the dog (as compared with that in man) the main venous drainage from the brain might be expected to be as Pilcher (1930) demonstrated, namely, dorsal sagittal, transverse, sigmoid, and vertebral sinus system. The anatomic and functional aspects of the longitudinal vertebral venous sinuses of the dog have been studied extensively by Worthman (1956).

DORSAL PETROSAL SINUS (FIG. 4-7)

The dorsal petrosal sinus is formed rostrally at the apex of the petrosal crest of the petrous temporal bone from veins draining the caudal midbrain and rostral areas of the cerebellum. According to Reinhard *et al.* (1962) the tributaries of the dorsal petrosal sinus of the dog are larger and ramify more extensively than any other intracranial veins. The sinus courses within the tentorium cerebelli along the lateral surface of the pyramid (petrous temporal bone). At this location, a groove in the bone indicates the course of the sinus. The dorsal petrosal sinus terminates caudally by entering the distal end of the transverse sinus. Near its termination the sinus receives a relatively large posterior cerebral vein.

VENTRAL PETROSAL SINUS (FIG. 4-7)

The paired ventral petrosal sinus connects the cavernous sinus rostrally with the ventral end of the sigmoid sinus caudally. The ventral petrosal sinus courses within the petrobasilar canal along the ventromedial base of the pyramid between the petrous temporal and the basioccipital bone.

CAVERNOUS SINUS (FIG. 4-7)

The large paired cavernous sinus is the largest of the venous dural sinuses. Including their transverse connections, these two sinuses occupy most of the central basal area of the middle cranial fossa. They extend from the orbital fissures rostrally to the ventral petrosal sinus at the rostral orifice of the petrobasilar canal. The two sinuses are connected medially by two transverse connections: the *rostral intercavernous sinus* in front of the dorsum sellae, immediately behind the infundibular stalk of the pituitary gland, and the *caudal intercavernous sinus* behind the dorsum sellae.

Throughout its extent the large cavernous sinus connects with numerous extracranial veins and plexuses. The details of these connections may be summarized by stating that rostrally the sinus is connected with orbital plexuses via the orbital fissure. Laterally there are many emissary veins traversing various foramina to connect with the internal maxillary vein. In addition, the cavernous sinus receives the middle meningeal vein from the dura mater.

The cranial nerves that make their exit from the skull via the orbital fissure are found within the wall of the cavernous sinus. These are: the oculomotor (III), the trochlear (IV), the ophthalmic division of the trigeminal nerve (V), and the abducens (VI). As the internal carotid artery approaches its termination in the circulus arteriosus cerebri, it passes through the cavernous sinus. In addition there are smaller arteries that course through the sinus, for example, the anterior portion of the middle meningeal artery and the anastomotic ramus of the external ophthalmic artery (Miller *et al.*, 1964).

Bibliography

Bunce, D. F. M., II: The arterial supply to the brain of the dog. Anat. Rec. *136:* 172–173, 1960 (abstract).

Davis, D. D., and Story, H. E.: Carotid circulation in the domestic cat. Field Museum of Nat. Hist., Zoological Series *28:* 1–47, 1943.

De La Torre, E., Netsky, M. G., and Meschan, I.: Intracranial and extracranial circulations in the dog. Anatomic and angiographic studies. Am. J. Anat. *105:* 343–381, 1959.

Gillilan, L. A., and Markesbery, W. R.: Arteriovenous shunts in the blood supply to the brains of some common laboratory animals—with special attention to the rete mirabile conjugatum in the cat. J. Comp. Neurol. *121:* 305–311, 1963.

Hale, A. R.: Circle of Willis—Functional concepts, old and new. Am. Heart J. *60:* 491–494, 1960.

Hoerlein, B. F.: *Canine Neurology—Diagnosis and Treatment.* Philadelphia, W. B. Saunders Co., 1965.

Miller, M. E., Christensen, G. C., and Evans, H. E.: The Heart and Arteries, *and* The Venous System. Chapters 4 and 5 in *Anatomy of the Dog.* Philadelphia, W. B. Saunders Co., 1964. Pp. 267–388, 389–429.

Pilcher, C.: A note on the occipito-vertebral sinus of the dog. Anat. Rec. *44:* 363–367, 1930.

Ranson, S. W., and Clark, S. L.: *The Anatomy of the Nervous System—Its Development and Function,* 10th ed. Philadelphia, W. B. Saunders Co., 1959.

Reinhard, K. R., Miller, M. E., and Evans, H. E.: The cranioribral veins and sinuses of the dog. Am. J. Anat. *111:* 67–87, 1962.

Walker, W. F., Jr.: *A Study of the Cat.* Philadelphia, W. B. Saunders Co., 1967.

Whisnant, J. P., Millikan, C. H., Wakim, K. G., and Sayre, G. P.: Collateral circulation to the brain of the dog following bilateral ligation of the carotid and vertebral arteries. Am. J. Physiol. *186:* 275–277, 1956.

Worthman, R. P.: The longitudinal vertebral venous sinuses of the dog. I. Anatomy. Am. J. Vet. Res. *17:* 341–348, 1956; II. Functional aspects. *idem. 17:* 349–363, 1956.

chapter 5

Meninges and Cerebrospinal Fluid

Meninges

The three meninges are sheet-like connective-tissue coverings of the central nervous system. In this anatomic relationship to the spinal cord and brain they have three main functions: protection, support, and nourishment. These membranes are named from the outermost inwardly: *dura mater, arachnoid,* and *pia mater.* Frequently the term leptomeninges (singular, leptomeninx) or pia-arachnoid is used to designate the combination of the pia mater and arachnoid. A synonym for dura mater is pachymeninx.

Dura Mater

The dura mater, histologically composed mostly of white collagenous fibers, is the toughest of the three meninges. The relationship of the dura mater to the adjacent bone in the area of the brain, as compared with that of the spinal cord, differs in the following respects. (1) In the cerebral region, the dura mater is composed of two layers. In most areas covering a major surface of the brain these two layers are fused, separated only in the restricted areas of the venous sinuses, as discussed in Chapter 4.

Spinal dura does not possess venous sinuses. (2) In the cerebral region the outer dural layer is actually the periosteum (or endosteum) of the skull bones. In contradistinction, a large extradural space separates the dura mater of the spinal cord from the bony wall of the vertebral canal, which has its own periosteum. Within this spinal extradural space there is an abundance of fatty areolar tissue with many blood vessels. The subdural space in both the cerebral and spinal regions is actually a potential space that does not contain cerebrospinal fluid.

In the cranial region, from the innermost (meningeal) layer of the dura mater there are two main projections approximately at right angles to each other, which help to support the brain within the cranial vault. The first projection, which extends ventrally within the median longitudinal fissure from the dorsal sagittal sinus toward the corpus callosum of the brain, is the *falx cerebri* (Fig. 5-1). Caudally beyond the corpus callosum this septum fuses with the second septum, the *tentorium cerebelli,* named because of its tentlike shape. This connective-tissue septum is oriented in a diagonal plane within the *transverse fissure* between the cerebral and cerebellar hemispheres. This septum is an extension of the osseous tentorium of the skull which is located in the same relative position and serves as a base for attachment of the connective-tissue tentorium cerebelli. The inner concave tentorial border forms the *tentorial notch* directly dorsal to the midbrain. Laterally the tentorium cerebelli attaches to the dorsal ridge of the petrous temporal bone, where it encloses a portion of the dorsal petrosal venous sinus.

The *diaphragma sellae* is a rostral continuation of the tentorium cerebelli over the posterior clinoid processes on the base of the skull to the anterior clinoid processes. As it passes over (dorsal to) the hypophysis (pituitary gland) it has a central foramen through which passes the infundibulum. In the process of removing a brain, it is necessary to keep this relationship in mind, because the diaphragma sellae must be incised if one wishes to have the hypophysis remain attached to the hypothalamus.

In consideration of the spinal dura mater the following points should be realized: (1) At the intervertebral foramen, the outer dural sheath extends to the bony vertebra. (2) The inner part of the dural sheath con-

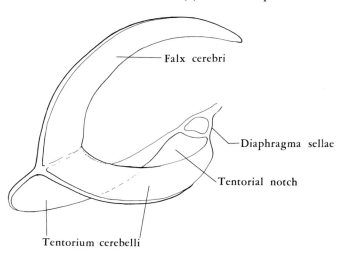

Fig. 5-1. Diagram illustrating the two principal inward projections of the dura mater: the falx cerebri and the tentorium cerebelli. *In vivo* the falx cerebri is located in the dorsal longitudinal (interhemispheric) fissure, and the tentorium cerebelli is within the transverse fissure between the cerebral and cerebellar hemispheres.

tinues through the intervertebral foramen and continues with the epineurium of the spinal nerve. (3) In the cervical region the first five cervical nerves in the dog have only one sheath surrounding both the sensory and motor roots, whereas those nerves caudal to the fifth cervical level have separate coverings of the dura (Fig. 5-2) investing each root (McClure, 1964). (4) Caudally the dura terminates by intimately surrounding the filum terminale (the extension of the spinal pia mater) and the arachnoid at the level of the first sacral vertebra, and extends caudally to attach to the periosteum of the vertebral canal at the seventh or eighth coccygeal vertebra (McClure,

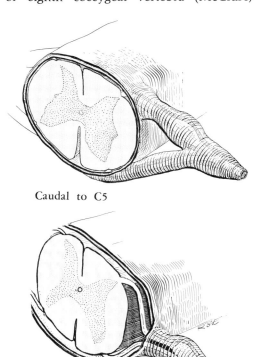

Caudal to C5

Cervical 1-5

Fig. 5-2. Diagrams illustrating relations of the spinal dura mater to the roots of the spinal nerves. Note that both dorsal and ventral roots of the first five cervical nerves in the dog are enclosed by a single dural sheath, whereas caudal to this level each root is enclosed separately.

1964). This thin supporting fibrous cord is known as the *coccygeal ligament.* (5) The spinal dura is continuous with the cranial dura at the foramen magnum.

The venous sinuses of the cranial dura mater are considered in Chapter 4.

Arachnoid

The arachnoid is much thinner than the dura mater. Histologically it is composed of delicate collagenous fibers with many elastic fibers interspersed. Unlike the dura mater, the arachnoid has a large underlying space, the *subarachnoid space,* which *in vivo* contains cerebrospinal fluid. This space possesses many arachnoid trabeculae, which extend from the arachnoid inwardly to the pia mater. As one studies the meninges in the cadaver, he does not find the large subarachnoid space, because the cerebrospinal fluid is not present and the arachnoid adheres directly to the pia mater (Fig. 5-3).

SUBARACHNOID CISTERNS

The subarachnoid cisterns are focal enlargements of the subarachnoid space which occur at specific cerebral locations. From the

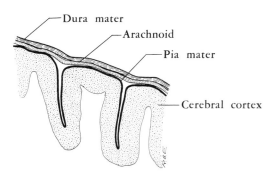

Dura mater
Arachnoid
Pia mater
Cerebral cortex

Fig. 5-3. Meningeal relations observed in the cadaver. Due to the absence of cerebrospinal fluid in the cadaver, the arachnoid adheres directly to the pia mater. The membranes may be separated at the surface of a sulcus, where the pia mater follows the walls of the sulcus deeply, but the arachnoid bridges the space of the sulcus.

clinical standpoint, only one is of major importance in the dog. This is located outside the fourth ventricle at the level of the foramen magnum, within the angle of the cerebellum and medulla oblongata. Based on these reference points, it is logically named the *cerebellomedullary cistern,* or the *cisterna magna.* In the dog this is the common site for aspiration of samples of cerebrospinal fluid rather than by the 'lumbar puncture' commonly performed in man. Two anatomic characteristics of the canine lumbar region make it an unfavorable site for puncture: (1) the termination of the spinal cord (conus medullaris) is at the level of the sixth or seventh lumbar vertebra, and (2) the subarachnoid space is small. Two of the relatively small cisterns the names of which explain their location are the *interpeduncular* and the *chiasmatic cistern.* In the dog the subarachnoid space does not envelop the pituitary gland. The chiasmatic cistern extends caudally only to the proximal end of the anterior lobe of the pituitary, and then the subarachnoid space passes around the pars tuberalis to join the rostral portion of the interpeduncular cistern (Schwartz, 1936).

Pia Mater

The innermost membrane, the pia mater, is the most delicate of the three, and has a high concentration of very fine elastic fibers. The pia mater is the meninx mainly concerned with nutrition of the central nervous system in that this is the vascular meninx. Actually the so-called 'blood vessels of the central nervous system' are technically located within the pia mater. In addition to blood vessels, the pia mater contains fibroblasts and macrophages. The delicate pia follows the surface of the brain and spinal cord, including the walls of the sulci. In the cerebral region of the cadaver one may separate the pia and arachnoid at the surface of a sulcus. There the arachnoid, which does not follow the walls of the

sulcus but merely bridges the indentation, may be lifted from the pia. The pia adheres to the walls of each sulcus, following the convolutions of the brain (Fig. 5-3).

The *denticulate ligament* (Fig. 5-4) is a lateral intermittent extension of the pia mater from the spinal cord to the arachnoid and dura between the dorsal and ventral roots of the spinal nerves. It extends throughout the length of the cord from the foramen magnum to the interspace between the roots of nerves L5 and L6 and helps secure the cord within the subarachnoid space (Fletcher and Kitchell, 1966). From a dorsal or ventral view of the spinal cord *in situ* the delicate denticulate ligament appears scalloped in a cephalocaudad sequence, having a free curved edge except where it is attached to the arachnoid and dura between sites of exit of the spinal nerves.

Caudal to the conus medullaris (L6–L7 in the dog) the pia mater extends caudally in the center of the subarachnoid space as a tenuous projection, the *filum terminale.* At the first sacral level it is joined by the arachnoid and the dura, to form the *coccygeal ligament,* as mentioned previously in discussion of the dura mater. The roots of the sacral and coccygeal nerves surround the filum terminale as they descend within the vertebral canal. In this fashion the roots produce a stringlike formation called the *cauda equina,* because of its likeness to a horse's tail (see Fig. 6-2).

Clinical Remarks about the Meninges

From the clinical standpoint, it is important to know that all three meninges follow the nerves of special sense peripherally to their respective sense organs. Therefore the subarachnoid space, and hence the cerebrospinal fluid, continue to the cribriform plate of the ethmoid bone by means of the olfactory fasciculi and to the eye via the optic nerve, and the cerebrospinal fluid can mix with the perilymph of the inner ear via the meninges accompanying the vestibulo-

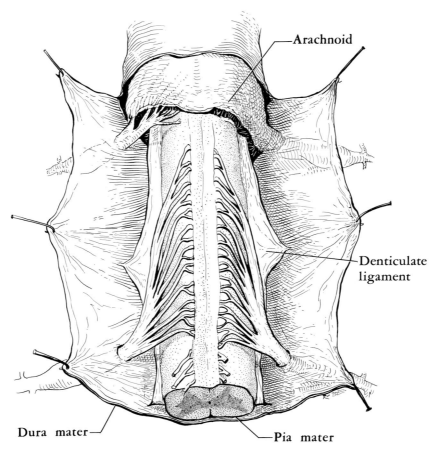

Fig. 5-4. Illustration of the denticulate ligament, a supportive lateral extension of the pia mater between the dorsal and ventral roots of the spinal nerves. Note the longitudinal spread of the dorsal rootlets of the spinal nerves as they enter the dorsolateral sulcus of the spinal cord.

cochlear nerve. The special anatomic relations of the meninges to these nerves provide a 'portal of entry' for such foreign invaders as bacteria and viruses. Consequently, these factors contribute to a favorable environment for a progression of disease conditions from neuritis and eye, ear, or nose infections, to meningitis, and ultimately to encephalitis. This pathway may be reversible; that is, certain types of encephalitides may lead to eye, ear, or nose infections.

Meningiomas (tumors of the meninges) are reported to be more common in cats, especially if 10 years old or older, than in dogs (McGrath, 1960). These tumors usually are benign and grow slowly, and therefore may not cause clinical symptoms. They produce brain damage usually by compression rather than by invasion. Diagnosis is difficult, but if they are discovered early and they are in a favorable location, meningiomas may be removed surgically with favorable results.

Cerebrospinal Fluid

The cerebrospinal fluid (CSF) is a clear, colorless fluid that fills the brain's ventricular system, the subarachnoid space of both the brain and the spinal cord, and the central canal of the spinal cord. The normal rate of production and flow of the cerebrospinal fluid within the ventricles and subarachnoid space are essential for maintaining a normal central nervous system. Flow within the central canal of the spinal cord is not considered to be important clinically. Actually, in mature animals, it is not uncommon to find the central canal partially obliterated with desquamated cells and cellular debris.

Although it is not the purpose of this text to consider the strictly clinical aspects of the nervous system or of the CSF, it should be realized that the changes in the fluid reflect changes in the health of the central nervous system. The cerebrospinal fluid is the only subdural substance that can be easily sampled. Obviously, in order to evaluate changes, it is necessary to know the normal properties of the cerebrospinal fluid.

Table 5-1 gives some normal values of cerebrospinal fluid in dog according to Fankhauser (1962). Normal values for CSF pressure in the dog vary among authors; for example, Schirmer (1958) reported a range between 22 and 144 mm of water, with an average of 73.2 mm. Because of such variations of normal, he suggested that determination of pressure values would appear to be of little clinical use. He found CSF pressure to fluctuate with muscular tension, straining, body position, respiration, and pulse. Hoerlein (1965) observed little change in CSF pressure in thoracolumbar compressions resulting from disk protrusion in dogs.

It is generally agreed that the cells normally found in CSF are small mononuclear cells similar to the small lymphocyte.

Formation of Cerebrospinal Fluid

The formation and flow of CSF plus its absorption back into the blood are highly controversial subjects. To avoid the many unnecessary controversies, for our purpose, we will agree with McGrath (1960), who shares probably the most popular belief: that the CSF is derived mainly from the blood in the choroid plexuses of the lateral ventricles by both filtration and secretory processes. Experimental evidence shows that, in the dog, bulk CSF is secreted at a mean normal rate of 0.035 $\pm$ 0.002 ml/min (Cserr, 1965). Other suggested sources of

TABLE 5-1. Normal Values of CSF in the Dog (values selected from Fankhauser, Table IX, p. 44, Innes and Saunders, Academic Press, 1962)

	Minimum	Maximum	Average
1. Quantity (ml)	0.9	16	6.5–7.0
2. Appearance		clear, colorless	
3. Pressure (mm H_2O)	24	172	86.5
4. Specific gravity	1.0033	1.0125	1.0056
5. Cells (per mm^3)	0	8	2
pups under 7 mo.	1	8	5
6. Sugar (mg%)	61	116	74
7. Total protein (mg%)	11	55	27.5
8. Albumen (mg%)	16.5	37.5	27
9. Globulin (mg%)	5.5	16.5	9.0
10. Protein quotient	.14	.75	.35

CSF are: ependyma, cerebral perivascular spaces, the general subarachnoid space and the vessels therein (Sweet *et al.*, 1956), and blood vessels of the pia mater and the brain substance (Bloom and Fawcett, 1968). Bering and Sato (1963) calculated that the normal dog (12–17 kg) produces 0.016 ml/min of CSF in the lateral and third ventricles, 0.011 ml/min in the fourth ventricle, and 0.020 ml/min in the subarachnoid space. Note that some of the sites listed above as sources are also given as sites for reabsorption of CSF as discussed later.

Flow of Cerebrospinal Fluid

As a starting place for flow of CSF, it is customary to begin with the *lateral ventricles.* From these ventricles the fluid normally flows through the interventricular foramina (of Monro) to reach the single midline *third ventricle.* The choroid plexus in the thin roof of the third ventricle contributes a small additional amount of fluid. From the third ventricle within the diencephalic division of the brain the fluid passes caudally through the small canal, the *cerebral aqueduct (of Sylvius)* within the mesencephalic division of the brain. The fluid then continues caudally to enter the *fourth ventricle,* the cavity of the rhombencephalon, where its choroid plexus adds more CSF. From here the fluid leaves the ventricular system to pass either (1) through two lateral apertures, the *foramina of Luschka,* to enter the cisterna magna, a dilatation of the subarachnoid space, or (2) directly into the central canal of the spinal cord. Some investigators (Fitzgerald, 1961) maintain that in addition to the lateral foramina of Luschka there is a median *foramen of Magendie* in the central portion of the posterior medullary velum. Other authors, however, report that the foramen of Magendie is absent in the dog (McGrath, 1960). According to Blake (1900) there is no foramen of Magendie in animals below the anthropoid apes. The extreme thinness of the posterior medullary velum covering the

caudal portion of the fourth ventricle is one factor that accounts for the difficulty in ascertaining positively whether this foramen is normally present or not. The fluid within the subarachnoid space may flow caudally into the spinal cord region as well as rostrally within the cranial region.

Exit of Cerebrospinal Fluid from the Subarachnoid Space

Classically it is stated that the cerebrospinal fluid leaves the subarachnoid space to return to the vascular system by passing through the *arachnoid villi* into the dorsal sagittal sinus.

Weed (1923) used dogs and cats in his classic investigation of the absorption of cerebrospinal fluid into the venous system. He observed arachnoid villi in all of the common laboratory mammals and in infants. He concluded that under normal conditions the pathway of absorption of the CSF is by way of the arachnoid villi into the dural venous sinuses. In addition to the villi in the dorsal sagittal sinus, he reported villi to be present in the sinuses at the base of the brain, especially in the cavernous sinus.

In a detailed study on the canine arachnoid villi Pollay and Welch (1962) described two structural types, both of which project directly into the dorsal sagittal sinus: (1) A meshwork of cytoplasmic prolongations of mesothelial cells and fibers. When open they are bulbous and project into the sinus; when collapsed they are flat. (2) Nodular balls of mesothelial cells within the subendothelial space of the sinus, which may be adjacent to extensions of the first type. Arachnoid villi of the second type never occur independently but always with villi of the first type interposed between them and the venous blood. Pollay and Welch liken the function of the villus mechanism in dog to that of a ball valve. In their perfusion experiments they observed open flow of fluid in only one direction, namely, from meninges to the dorsal sagittal sinus. They found the minimum opening flow pressure for the villus mechanism to be 30 mm of water.

McGrath (1960) denies the presence of distinct arachnoid villi in the dog, but admits to the probability that most of the CSF is absorbed into the 'subdural' venous sinuses.

As indicated above, not all authorities agree that the arachnoid villi are the sole means of exit for CSF from the subarachnoid space. A few of the many other possible routes of exit that have been suggested are: (1) spinal veins outside the cord (Elman, 1923); (2) perineural lymphatics associated with the spinal nerve roots (Woollam and Millen, 1953); (3) from the perivascular spaces directly into the smaller blood vessels (McGrath, 1960); (4) direct venous drainage from the spinal subarachnoid space (Howarth and Cooper, 1949), and (5) a generalized absorption of CSF throughout the lining of the subarachnoid space and the walls of the vessels within the space (Sweet et al., 1956).

The many controversies result from the varied interpretations of the results of different experimental approaches and pathologic conditions of the test animals. Due to lack of conclusive proof that these alternative routes indeed function to any great extent for drainage of CSF under normal conditions, the arachnoid villi are considered to be the main structures for this purpose. Their ideal anatomic location and structure, combined with the lack of proof of other functions, indirectly suggest that they are the main structures actively involved with CSF drainage.

Functions of Cerebrospinal Fluid

Cerebrospinal fluid has a variety of functions, as indicated by the following list.

1. The most obvious function generally thought of is the physical cushioning effect against accidental injury to the brain and spinal cord (*i.e.*, the anticoncussive function of CSF).

2. Closely related to the first-mentioned function, the fluid modulates the fluctuations of pressure within the skull.

3. The possibility that CSF may serve as a nutrient for nervous tissue has been suggested, but most authors agree that this is unlikely.

4. CSF has bactericidal and antitoxic properties and contains antibodies (Fankhauser, 1962).

5. Ions as well as organic molecules are known to be actively transported between CSF and blood, and there is evidence to suggest that the CSF contained in the ventricles and subarachnoid space may have the same composition as the interstitial fluid surrounding the neurons and glial elements in the brain.

It has been shown in dogs that the stability of cerebrospinal fluid potassium is the result of an active ion-transport process which pumps K^+ out of CSF (Cserr, 1965). Magnesium normally has a higher concentration in CSF than in plasma; Oppelt et al. (1963) showed in the dog that the higher concentration in CSF is maintained by an active transport process involved in moving Mg^{++} from blood to CSF. Ca^{++} concentration, which normally is lower in CSF than in plasma, also is thought to be regulated by an active transport process (Graziani et al., 1967). A carrier-mediated transport process for the passage of glucose from blood to CSF has been suggested (Fishman, 1964). It has been shown in goats that Diodrast, phenolsulfonphthalein, and p-aminohippuric acid are actively transported from CSF to blood (Pappenheimer et al., 1961).

6. CSF has been implicated in the control of respiration. A receptor which is sensitive to H^+ concentration in CSF and interstitial fluid in the brain has been located by some on the ventrolateral surface of the medulla (Mitchell et al., 1963) and by others beneath the surface of the medulla (Pappenheimer

et al., 1965). Respiratory responses to acidosis or alkalosis of respiratory or metabolic origin can be explained by changes in H^+ concentration of brain interstitial fluid. In chronic conditions of acidosis or alkalosis, the H^+ concentration of CSF equals that calculated for interstitial fluid (Fencl et al., 1966). Ionic regulation and transport appear to be active, but whether H^+ or HCO_3^- is being transported is not known.

7. The analysis of CSF as a clinical aid for indicating the general health of the central nervous system has been referred to earlier in this Chapter.

Barriers Related to Blood, Brain, and CSF

The physiologic compatibility between neurons of the central nervous system and the chemical milieu (internal environment) determines the functional efficiency and survival of the neurons. As indicated above in consideration of functions of the cerebrospinal fluid, the maintenance of a constant neuronal internal environment (homeostasis) is achieved by the active transport of ions and chemical substances among blood, cerebrospinal fluid, and nervous tissue. The differential rates of passage for various substances across the cerebral capillary walls to brain tissue have been studied extensively by such methods as microchemical assays, perfusion of various types of dyes, electron microscopy, and use of radioactive isotopes.

Blood-Brain Barrier

Arteries within the subarachnoid space of the central nervous system are invested with connective tissue of the leptomeninges. As a vessel penetrates into the nervous tissue, the pia and subjacent glial membrane blend with the vessel wall. The smaller arterial branches within the nervous tissue have only thin neuroglial membrane investments (astrocytic perivascular feet) which continue to the capillary level. It has been estimated from studies of rat brains with the electron microscope that such perivascular feet cover approximately 85 per cent of the total capillary surface (Maynard et al., 1957).

The capillary endothelium, its continuous and homogeneous basement membrane, and the numerous astrocytic perivascular feet are all that separate the plasma in the vessel from the interstitial fluid within the nervous tissue. These are the anatomic structures that have been equated by some investigators with the blood-brain barrier (Truex and Carpenter, 1969).

The same authors cite areas of the brain that presumably are devoid of a blood-brain barrier when tested with intravenous injections of vital dyes or radiotracers, namely, the pineal body, pituitary gland, area postrema, subfornical organ, supraoptic crest, and choroid plexus. All of these areas are highly vascular, and many are known to have or are suspected of having secretory function.

The function of the blood-brain barrier may be influenced by the composition of the blood and the metabolism of brain cells. The barrier may protect the brain from variations in composition of the blood and from the entry of toxic compounds or foreign substances. As the brain matures, the barrier becomes more selectively efficient, in that the passage of certain substances becomes impossible or more difficult. Intravenously injected trypan blue, an azo dye, and ferricyanide both penetrate freely into the brain of very young but not of mature laboratory animals (Chusid, 1970).

Brain-CSF Barrier

The ependymal cells and the subjacent glial elements constitute a potential barrier between the interstitial fluid of the brain and the CSF. The similarity in ionic composition of the CSF and of the interstitial fluid has been discussed previously. If the composition of the CSF reflects that of the interstitial fluid, perhaps a stable ionic milieu is maintained for the neurons and glial cells of the brain and spinal cord. The differential ionic concentrations such as those mentioned above under Functions of Cerebrospinal Fluid point to the possibility that active transport of constituents and regulation of molecular concentrations in brain interstitial fluid and cerebrospinal fluid constitute the physiologic aspects of the 'blood-brain' and 'brain-CSF' barriers.

Hydrocephalus

Hydrocephalus is a pathologic condition marked by an excessive accumulation of cerebrospinal fluid and subsequent dilatation of the ventricles (*internal hydrocephalus*) and thinning of the brain. If the subarachnoid space also is dilated, the condition is referred to as *external,* or *communicating, hydrocephalus.* The condition can be congenital, sometimes inherited as well, or acquired later in life as a secondary phenomenon (Innes and Saunders, 1962). According to McGrath (1960), internal hydrocephalus probably is the most common congenital lesion of the nervous system in the dog. Brachycephalic and toy breeds may possess varying degrees of hydrocephalus with very slight or no clinical symptoms.

There are many reports in the literature of single cases of hydrocephalus in animals, but for a comprehensive consideration of the subject the reader is referred to the well-documented monograph by Russell (1949). Although her report considers the pathology of hydrocephalus in man, it is generally agreed that the pertinent factors apply equally well to animals.

Causes of Hydrocephalus

Russell discusses a long list of causes of hydrocephalus, but states that in her extensive study the interposition of some obstruction in the cerebrospinal fluid pathway was responsible for at least 99 per cent of all cases of internal hydrocephalus. Among the causes which she considers in detail, including case histories, are: (1) Maldevelopments. Common causes of internal hydrocephalus in this category are stenosis and atresia of the aqueduct of Sylvius. Other developmental malformations associated with hydrocephalus are: spina bifida, meningomyelocele, Arnold-Chiari malformation, and deformities of the base of the skull. Other causes of hydrocephalus are: (2) gliosis of the aqueduct; (3) inflammations, including bacterial and non-bacterial infections, postmeningitic hydrocephalus, and granulomatous inflammations; (4) dural sinus thrombosis and thrombophlebitis; and (5) neoplasms so located as to obstruct flow of cerebrospinal fluid.

The most deleterious effect of the excess CSF in either the ventricles or the subarachnoid space is produced by the tremendous pressure which causes a thinning of the brain. The effects of this may lead to physical impairment, combined with mental retardation, or ultimately cause death.

Diagnosis and Surgical Treatment of Canine Hydrocephalus

Pneumoventriculography (radiography after the injection of air directly into the ventricles of the brain) and contrast ventriculography (radiography after injection of a relatively nontoxic radiopaque medium, such as ethyl iodophenylundecylate*) are procedures which have been used successfully for the diagnosis of hydrocephalus in dogs (Few, 1966). Few also described a successful surgical technique utilizing a ventriculocaval shunt valve to transfer excess CSF from the ventricles of the brain to the atrium of the heart. He reported that this technique is similar to that commonly used in man.

Bibliography

Bering, E. A., Jr., and Sato, O.: Hydrocephalus. Changes in formation and absorption of cerebrospinal fluid within the cerebral ventricles. J. Neurosurg. 20: 1050–1063, 1963.

Blake, J. A.: The roof and lateral recesses of the fourth ventricle, considered morphologi-

*Pantopaque—Lafayette Pharmacal Inc., Lafayette, Indiana.

cally and embryologically. J. Comp. Neurol. *10:* 79–108, 1900.

Bloom, W., and Fawcett, D. W.: *A Textbook of Histology,* 9th ed. Philadelphia, W. B. Saunders Co., 1968.

Chusid, J. G.: *Correlative Neuroanatomy & Functional Neurology,* 14th ed. Los Altos, Cal., Lange Medical Publications, 1970.

Cserr, H.: Potassium exchange between cerebrospinal fluid, plasma, and brain. Am. J. Physiol. *209:* 1219–1226, 1965.

Elman, R.: Spinal arachnoid granulations with especial reference to the cerebrospinal fluid. Bull. Johns Hopkins Hosp. *34:* 99–104, 1923.

Fankhauser, R.: The Cerebrospinal Fluid. Chapter 3 in *Comparative Neuropathology* (Innes, J. R. M., and Saunders, L. Z.). New York, Academic Press, 1962. Pp. 21–54.

Fencl, V., Miller, T. B., and Pappenheimer, J. R.: Studies on the respiratory response to disturbances of acid-base balance, with deductions concerning the ionic composition of cerebral interstitial fluid. Am. J. Physiol. *210:* 459–472, 1966.

Few, A. B.: The diagnosis and surgical treatment of canine hydrocephalus. J.A.V.M.A. *149:* 286–293, 1966.

Fishman, R. A.: Carrier transport of glucose between blood and cerebrospinal fluid. Am. J. Physiol. *206:* 836–844, 1964.

Fitzgerald, T. C.: Anatomy of cerebral ventricles of domestic animals. Vet. Med. *56:* 38–45, 1961.

Fletcher, T. F., and Kitchell, R. L.: Anatomical studies on the spinal cord segments of the dog. Am. J. Vet. Res. *27:* 1759–1767, 1966.

Graziani, L. J., Kaplan, R. K., Escriva, A., and Katzman, R.: Calcium flux into CSF during ventricular and ventriculocisternal perfusion. Am. J. Physiol. *213:* 629–636, 1967.

Hoerlein, B. F.: *Canine Neurology—Diagnosis and Treatment.* Philadelphia, W. B. Saunders Co., 1965.

Howarth, F., and Cooper, E. R. A.: Departure of substances from the spinal theca. Lancet *2:* 937–940, 1949.

Innes, J. R. M., and Saunders, L. Z.: *Comparative Neuropathology.* New York, Academic Press, 1962.

McClure, R. C.: The Spinal Cord and Meninges. Chapter 9 in *Anatomy of the Dog* (Miller, M. E., Christensen, G. C., and Evans, H. E.).

Philadelphia, W. B. Saunders Co., 1964. Pp. 533–543.

McGrath, J. T.: *Neurologic Examination of the Dog,* 2nd ed. Philadelphia, Lea & Febiger, 1960.

Maynard, E. A., Schultz, R. L., and Pease, D. C.: Electron microscopy of the vascular bed of rat cerebral cortex. Am. J. Anat. *100:* 409–433, 1957.

Mitchell, R. A., Loeschcke, H. H., Massion, W. H., and Severinghaus, J. W.: Respiratory responses mediated through superficial chemosensitive areas on the medulla. J. Appl. Physiol. *18:* 523–533, 1963.

Oppelt, W. W., MacIntyre, I., and Rall, D. P.: Magnesium exchange between blood and cerebrospinal fluid. Am. J. Physiol. *205:* 959–962, 1963.

Pappenheimer, J. R., Fencl, V., Heisey, S. R., and Held, D.: Role of cerebral fluids in control of respiration as studied in unanesthetized goats. Am. J. Physiol. *208:* 436–450, 1965.

Pappenheimer, J. R., Heisey, S. R., and Jordan, E. F.: Active transport of Diodrast and phenosulfonphthalein from cerebrospinal fluid to blood. Am. J. Physiol. *200:* 1–10, 1961.

Pollay, M., and Welch, K.: The function and structure of the canine arachnoid villi. J. Surg. Res. *2:* 307–311, 1962.

Russell, D. S.: *Observations on the Pathology of Hydrocephalus.* Med. Res. Council (Brit.) Spec. Rep. Ser. No. 265, 1949.

Schirmer, R. G.: Analysis of the cerebrospinal fluid. Its significance in diagnosis. Mod. Vet. Pract. *39:* 38–40, 1958.

Schwartz, H. G.: The meningeal relations of the hypophysis cerebri. Anat. Rec. *67:* 35–52, 1936.

Sweet, W. H., Brownell, G. L., Scholl, J. A., Bowsher, D. R., Benda, P., and Stickley, E. E.: The formation, flow, and absorption of cerebrospinal fluid. Res. Publ. Ass. Nerv. Ment. Dis. *34:* 101–159, 1956.

Truex, R. C., and Carpenter, M. B.: *Human Neuroanatomy,* 6th ed. Baltimore, Williams & Wilkins Co., 1969.

Weed, L. H.: The absorption of cerebrospinal fluid into the venous system. Am. J. Anat. *31:* 191–221, 1923.

Woollam, D. H. M., and Millen, J. W.: An anatomical approach to poliomyelitis. Lancet *1:* 364–367, 1953.

chapter 6

Anatomy of Spinal Nerves

Gross Anatomy

Although the spinal nerves are in the peripheral nervous system, their attachment to the spinal cord (a portion of the central nervous system) is essential to their functional integrity. The paired spinal nerves are attached to the spinal cord at their respective segments by dorsal and ventral roots. The *spinal nerve proper* is formed by a union of these two roots at approximately the level of the intervertebral foramen through which it leaves the vertebral canal (Fig. 6-1). Immediately proximal to the fusion of the two roots, the dorsal root has a localized enlargement, the *dorsal root (spinal) ganglion.*

Numbering Spinal Nerves

The nerves are numbered according to the number of the vertebral segment with which it is grossly associated. With the exception of the cervical nerves, the number of the spinal nerve corresponds to the number of the vertebra immediately cephalic to the intervertebral foramen through which the nerve leaves the vertebral canal. Recall that in most mammals there are eight pairs of cervical nerves, but only seven cervical vertebrae. In the dog, the first pair of cervical nerves pass through foramina which are not between vertebrae but perforate the craniodorsal segment of the dorsal

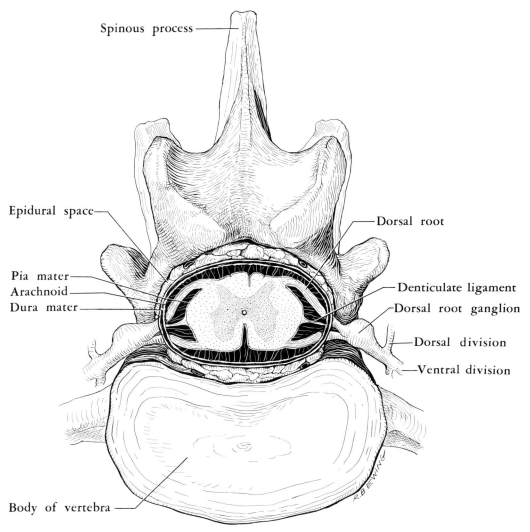

Spinous process

Epidural space

Pia mater
Arachnoid
Dura mater

Dorsal root

Denticulate ligament
Dorsal root ganglion

Dorsal division
Ventral division

Body of vertebra

Fig. 6-1. Transverse section of spinal cord within the vertebral canal to show the formation of a spinal nerve in relation to the vertebra and meninges. Note the adipose tissue and blood vessels in the epidural space.

arch of the atlas (the first cervical vertebra). The second pair of cervical nerves pass through the first true intervertebral foramina, between the atlas and the axis (the second cervical vertebra). The remainder of the nerves appear in succession, so that the first seven pairs of cervical nerves emerge above (cephalic to) the vertebra of the corresponding number, respectively, and the eighth pair emerges below (caudal to) the seventh cervical vertebra and above (cephalic to) the first thoracic vertebra. The remaining spinal nerves emerge through the intervertebral foramen which is immediately caudal to the vertebra of the same number (*e.g.,* the first thoracic nerve emerges caudal to the first thoracic vertebra). Therefore, throughout the thoracic, lumbar, and sacral regions there are the same number of nerves as of vertebrae, that is, 13 thoracic, 7 lumbar, and 3 sacral in the dog and cat. In these animals there is an average of only 5 coccygeal nerves, but usually there are 20 coccygeal vertebrae. Based on these figures, there are 36 pairs of spinal nerves in dog and cat.

The reader should be aware of the differences in the numbers of spinal nerves and vertebrae in man when compared with carnivores. The same general plan as described above for dog and cat exists in man, with a few exceptions; (1) the first cervical nerve passes between the occipital bone of the skull and the first cervical vertebra; (2) in man the 31 pairs of spinal nerves are distributed as 8 cervical (same as dog and cat), 12 thoracic, 5 lumbar, 5 sacral, and 1 coccygeal (Noback, 1967).

Spinal Cord Segments

In the preceding discussion it was emphasized that the spinal nerve is usually identified according to the number of the vertebra directly above (cephalic to) the exit of the nerve from the vertebral column. Because the adult spinal cord is shorter than the vertebral column, the spinal cord level of origin (attachment) for a specific spinal nerve is usually cephalic to the intervertebral foramen through which the nerve passes. As it obviously is not feasible to count spinal cord levels in a clinical situation, vertebral levels are logically used for designating the number of a spinal nerve. This discrepancy in the location of the segment of its origin from the spinal cord and the level of exit for a spinal nerve from the vertebral column results in progressive increase in the lengths of the spinal nerve roots in a cephalo-caudal sequence.

Keep in mind that the spinal cord segment to which the roots of a nerve are attached will have the same number as the nerve at the intervertebral foramen through which it emerges, based on the number of the appropriate vertebra, but the exact levels will be different. The student who

understands this principle should have no trouble in answering and explaining the classical question, "If we know that the spinal cord terminates at the level of the junction of the sixth and seventh lumbar vertebrae, how is it possible to have cross sections of the spinal cord correctly labeled sacral or coccygeal levels?" The explanation requires a knowledge of the two criteria used: the last coccygeal nerve roots attach at the coccygeal *spinal cord segment,* which is located at *vertebral segment* (level) junction of L6–L7. The following discussion should serve as an explanation.

Within the young mammalian embryo the bony vertebral column and the spinal cord are exactly the same length. Therefore, the spinal nerves pass through intervertebral foramina that are directly lateral to the spinal nerve attachments, so that the identically numbered spinal cord and vertebral segments are at the same level. As the embryo grows longer, the length of the spinal cord is not able to keep up with the increased length of the vertebral column. Therefore, the caudal tip of the spinal cord (*conus medullaris*) recedes, relatively speaking, to a higher (more cephalic) level of the vertebral column. As a result of this relative caudal migration of the bony vertebrae, the spinal roots within the vertebral canal are stretched caudally, particularly in the lumbosacral and coccygeal regions. Consequently, in the adult the upper (cervical) nerves possess relatively short roots and tend to retain their embryonic position, in that they emerge at approximately right angles to the spinal cord.

The roots of the spinal nerves innervating the hindlimb and caudal regions of the body, however, emerge from the cord at a progressively more acute angle with the midline in passing caudally. The roots of the more caudal nerves extend within the subarachnoid space for a considerable distance beyond the conus medullaris, which is at vertebral level L1 in man and L6–L7 in the dog. Finally, the roots emerge through the dura mater at the intervertebral foramen far caudal to their level of attachment to the spinal cord. At the intervertebral foramen the roots fuse to form the spinal nerve proper, as discussed previously. This general concept of nerve roots descending within the subarachnoid space is true for animals and man, but the situation in the dog is more complicated, as discussed in detail in Chapter 10, based on the report of Fletcher and Kitchell (1966).

As illustrated in Figure 6-2 the caudal extension of the spinal nerve roots accounts for the *cauda equina* (so named because of its gross resemblance to a horse's tail), which is a collective term used to include the filum terminale plus the dorsal and ventral roots of the more caudal spinal

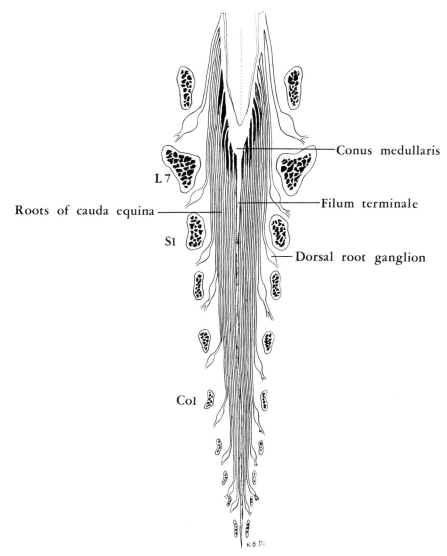

Fig. 6-2. Dorsal view of the cauda equina, composed of the filum terminale, plus the roots of the more caudal spinal nerves.

nerves which extend beyond the conus medullaris within the vertebral canal. It should be emphasized that the cauda equina is composed of the roots of the caudal spinal nerves and not the entire nerves themselves. Some textbooks erroneously refer to the component structures as 'spinal nerves.'

Anatomic Components (Subdivisions) of Spinal Nerves

Figure 6-3 schematically illustrates the anatomic components of a typical spinal nerve as exemplified by a thoracic nerve. Recall that the dorsal and ventral roots fuse to form the spinal nerve proper at approximately the level of the intervertebral foramen. The true spinal nerve segment is very short (a few millimeters), since almost immediately the nerve bifurcates into two major divisions (*primary rami,* or *branches*): (*a*) the *dorsal division,* which passes to the epaxial region, and (*b*) the longer *ventral division,* which serves as the intercostal 'nerve' or which together serve as the 'nerves' of the cervical, brachial, and lumbosacral plexuses. The *ramus communicans* attaches to the ventral side of the thoracic nerve at approximately the point of its bifurcation, as a connection with the paravertebral ganglion of the sympathetic chain.

Remember that the spinal 'nerves' as dissected peripherally in the limbs as extensions of the brachial and lumbosacral plexuses are technically ventral divisions of the spinal nerves. In the dog the spinal cord segments (which bear the same numbers as the nerves) from which the fibers forming

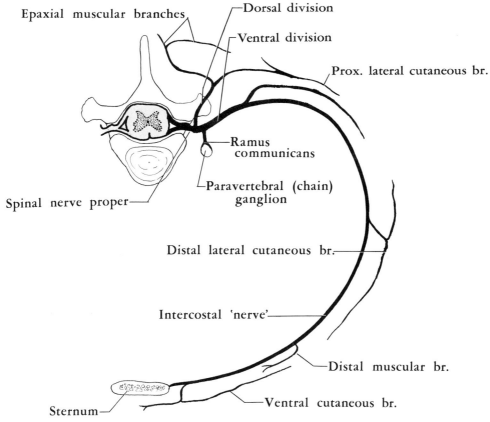

Fig. 6-3. Diagram showing the anatomic components of a typical spinal nerve.

the brachial plexus arise are the sixth, seventh, and eighth cervical plus the first two thoracic. In the same animal those contributing to the lumbosacral plexus are the last five lumbar and all three sacral segments (Miller *et al.,* 1964).

Although the microscopic details of neurons are considered in Chapter 3 and the functional aspects in Chapter 7, the remainder of this Chapter is devoted to a basic correlation of structure and function, with pertinent remarks on embryonic development as a basis for better understanding of the succeeding chapters.

General Microscopic Anatomy (Correlated with Function)

The typical isolated 'nerve,' as identified by dissection, consists of bundles (fascicles) of nerve fibers (dendrites and axons). Sizes of the bundles vary among nerves of different caliber, but the types of fibers are rather consistent. Obviously the nerves in a Great Dane are larger than the corresponding nerves in a Chihuahua. Histologically, within each bundle of nerve fibers there is a mixture of the smallest nonmyelinated C fibers, the thinly myelinated or medullated B fibers, and the largest, heavily myelinated A fibers. These structural characteristics of fibers are correlated with speed of conduction and functional components (see Table 3-1).

Figure 6-4 illustrates the appearance of a cross section of a nerve. In addition to the various fiber sizes, note the connective-tissue coverings enclosing different portions of the sectioned nerve. The innermost connective-tissue sheath surrounding individual neuron processes (axons and dendrites) immediately outside the neurilemma is named *endoneurium.* Bundles (fascicles) of nerve fibers are surrounded by a heavier connective-tissue covering, the *perineurium.* The entire 'nerve' surface is covered by the outermost connective-tissue sheath, the *epineurium.* The connective tissue gives

the entire nerve its characteristic firmness, cylindrical shape, and strength. In addition, the small blood vessels that supply the nerve fibers course within the connective-tissue sheaths.

The dorsal root (spinal) ganglion contains unipolar cell bodies of both visceral and somatic neurons. The processes of these cell bodies are the sensory fibers which convey impulses from viscera (general visceral afferent), from the dermal areas (general somatic afferent, exteroceptive), and from muscles, tendons, and joints (general somatic afferent, proprioceptive) (cf. Chapters 3 and 7). The ventral (motor) root has no ganglion and therefore contains only A fibers to skeletal muscle (general somatic efferent) and B fibers (general visceral efferent, preganglionic autonomic).

The fusion of the dorsal and ventral roots results in an intermingling of sensory and motor fibers. In considering the total extent of the nerve, the long central portion is mixed (both sensory and motor), whereas at each end of the nerve the sensory and motor components tend to be separated (Fig. 6-4). Therefore, a complete severance (*e.g.,* traumatic lesion) of a peripheral 'nerve' (*e.g.,* of the brachial or lumbosacral plexus) will generally cause both a motor deficiency (paresis or paralysis) and a sensory loss (hypesthesia or anesthesia). In addition, the loss of general visceral efferent function will produce autonomic changes, as discussed in Chapter 9. Innervations of muscle and tendon spindles are not considered in the above discussion.

Embryologic Basis for Somatic Motor and Sensory Distribution

A general explanation of the differentiation of the typical embryonic somite and its relation to the developing spinal nerve may aid in an understanding of the distribution of the somatic motor and sensory components as introduced above and discussed in detail in Chapter 7. Since the

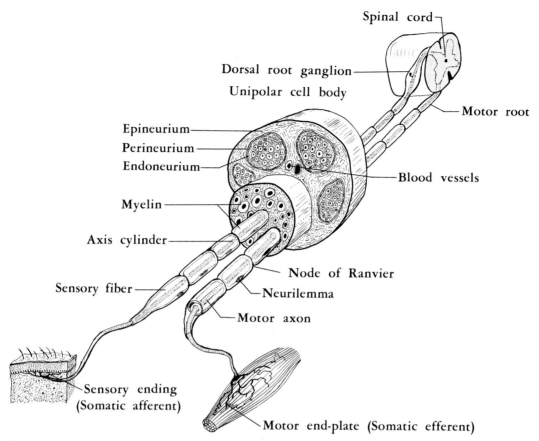

Fig. 6-4. Diagram showing the histologic characteristics of a nerve in cross section, correlated with the general gross relationships. (Modified from Ham, 1965.)

processes described below are similar for domestic animals and man, the following generalities are based in part on somite differentiation in the fifth week of human development, as reported by Thomas (1968).

Somites, a characteristic of all embryonal vertebrates (Romer, 1970), are paired segmented blocks of mesoderm arranged in a cephalocaudal sequence along the neural tube (Fig. 6-5). The same figure shows that three portions of each somite are distinguishable: (1) the dermatome is the lateral segment closest to the body surface; (2) the myotome is the mesenchymal (mesodermal) tissue deep to the dermatome; and (3) the sclerotome is the ventromedial layer of

mesodermal tissue. Each of the three portions migrates and differentiates according to the plan discussed in the following paragraphs and illustrated in Figure 6-5.

Dermatomes

The cells of each dermatome migrate peripherally to lie beneath the body surface ectoderm to form the dermis of the adult. The sensory portion of each spinal nerve accompanies this migration, so that each segment of dermatome is innervated by the exteroceptive general somatic afferent component of a single spinal nerve. This is of clinical importance in that the clinician can test in the adult the specific area of skin

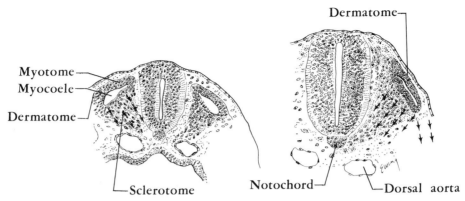

Fig. 6-5. Cross sections of the dorsal embryonic body wall to show the divisions of a somite. The differentiation and migratory paths are indicated. The section on the right is from a later stage of development than that on left.

innervated by the somatic sensory component of a single nerve, which is also designated a *dermatome.* Technically there is an overlapping of these areas (dermatomes) in the adult, so that in man it requires the total loss of exteroceptives (usually dorsal roots) of three consecutive spinal nerves to result in complete anesthesia of a single dermatome. Because of the somite origin and the short migration of the embryonic dermatome, the dermatomes in the adult tend to retain the segmental arrangement as evidence of the original embryonic metamerism. This is best demonstrated in the thoracic region, whereas the dermatomes in the limbs are modified. The dermatomes of the human body have been very thoroughly mapped, to designate the specific spinal nerve sensory areas, but our knowledge of the dermatomes in the dog is much less complete.

Myotomes

The cells of the myotome, which are the precursors of skeletal muscle, migrate in two directions: (*a*) dorsally between the neural tube and the dorsolateral body surface ectoderm to form the *epimere,* and (*b*) ventrally to contribute to the somatopleure and form the *hypomere.* The epimeres differ-

entiate into the epaxial muscles, which are the central dorsal muscles of the trunk and function as extensors of the vertebral column. The hypomeres differentiate into the hypaxial muscles, which are the ventral and lateral muscles of the trunk and function as flexors of the vertebral column.

The differentiating muscles of each myotome receive the motor innervation of the spinal nerve of the same segmental level. There is greater migration of the myotome than of the dermatome. The innervation of the epimere and hypomere from the same spinal nerve establishes the dorsal and ventral divisions (primary rami) of the spinal nerve as found in the adult and illustrated in Figure 6-3.

Modification of Segmental Innervation of Muscles

1. The subdivisions of the myotome and their migrations have been mentioned above. An embryonic muscle primordium receives its spinal nerve motor innervation from the original segmental level, which is usually the same as that from which the dermatome receives somatic sensory innervation. It should be noted that as the primordial muscle mass migrates it carries its innervation to its final destination, which

may be a considerable distance from its origin. An obvious example of this phenomenon is the diaphragm, which separates the abdominal and thoracic cavities in mammals. Originating in the cervical region of the fetus, the pre-diaphragmatic muscle mass receives somatic motor innervation from the cervical spinal cord levels according to the developmental plan described. The muscle mass migrates caudally to its final destination and carries its nerve supply with it. This explains the main innervation of the diaphragm by the phrenic nerve, which usually arises from the fifth, sixth, and seventh cervical spinal cord levels in the dog.

2. A second important factor that modifies the serially segmented motor innervation of skeletal muscle is that the limb muscles are derived from more than one myotome and therefore the nerves (ventral divisions) innervating the limb muscles are formed from contributions of more than one spinal cord segment. A recent study on the pelvic limb myotomes correlated with the lumbosacral plexus in the dog revealed that the intrinsic musculature of the pelvic limb was derived from five myotomes (L4 through S1). All pelvic limb muscles were derived from at least two myotomes, and most were derived from three (Fletcher, 1970). An important clinical application of this embryonic fact is that complete destruction of a single ventral root or proximal portion of a division will result in merely partial paralysis (paresis) of more than one muscle rather than total paralysis of a single muscle innervated by that segment.

Sclerotomes

The cells of this portion of each pair of somites migrate ventromedially to join each other and completely envelop the notochord, ventral to the developing spinal cord, to establish the future body of the vertebra (Fig. 6-5). Other cells of the sclerotome migrate dorsally to form the spinous and transverse processes of the vertebra. As illustrated in Figure 6-6 A, originally each segmental block of sclerotome is oriented in series so that the intersegmental arteries

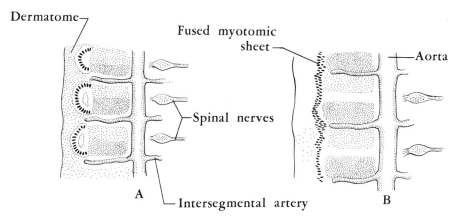

Fig. 6-6. Dorsal view of a frontal section of the embryonic body, showing two stages in the differentiation of somites in relation to intersegmental arteries and roots of the spinal nerves. A. Individual sclerotomes are shown to consist of rostral and caudal portions with the intersegmental arteries between the original sclerotomes and the spinal nerve roots positioned opposite the somites. B. Each sclerotome splits transversely and the cranial half of one sclerotome fuses with the caudal half of the preceding one to form a new mesodermic block (primordium of the vertebral body). The intersegmental arteries now cross the bodies of the vertebrae, and the roots of the spinal nerves lie between the vertebral bodies (*i.e.,* at the intervertebral foramina). (After Davies, 1963.)

are between the blocks, and the spinal nerve of the segment penetrates the center of the cellular mass. Each sclerotome splits transversely so that the cranial half of one sclerotome unites with the caudal half of the preceding one to form a new block of mesoderm, the precartilaginous primordium of the vertebral body. The importance of this migration pattern lies in its determination of the final adult relationship so that the spinal nerve is between the sclerotome blocks (vertebrae), thus establishing the position of the intervertebral foramina (Fig. 6-6 B). The intervertebral disks are formed from condensed transverse plates that were originally in the middle of the sclerotome and are ultimately located between the vertebral bodies. In the intervertebral disk regions the notochord undergoes mucoid degeneration to become the postnatal *nucleus pulposus* of the fibrocartilaginous intervertebral disk.

The clinical significance of these relationships in the dog are apparent in the well-known disk protrusions which commonly cause posterior paraplegia (paralysis of the hindlimbs). Hoerlein (1965) presents a detailed coverage of disorders of the intervertebral disks in dogs and cats, including surgical procedures and clinical management.

Bibliography

Davies, J.: *Human Developmental Anatomy*. New York, Ronald Press Co., 1963.

Fletcher, T. F.: Lumbosacral plexus and pelvic limb myotomes of the dog. Am. J. Vet. Res. *31:* 35–41, 1970.

Fletcher, T. F., and Kitchell, R. L.: Anatomical studies on the spinal cord segments of the dog. Am. J. Vet. Res. *27:* 1759–1767, 1966.

Ham, A. W.: *Histology*, 5th ed. Philadelphia, J. B. Lippincott Co., 1965.

Hoerlein, B. F.: *Canine Neurology—Diagnosis and Treatment*. Philadelphia, W. B. Saunders Co., 1965.

Miller, M. E., Christensen, G. C., and Evans, H. E.: The Spinal Nerves. Chapter 11 in *Anatomy of the Dog*. Philadelphia, W. B. Saunders Co., 1964. Pp. 572–625.

Noback, C. R.: *The Human Nervous System*. New York, McGraw-Hill Book Co., Blakiston Division, 1967.

Romer, A. S.: *The Vertebrate Body*, 4th ed., Philadelphia, W. B. Saunders Co., 1970.

Thomas, J. B.: *Introduction to Human Embryology*, Philadelphia, Lea & Febiger, 1968.

chapter 7

Physiology of Spinal Nerves

The structure and distribution of the spinal nerves as discussed in Chapter 6 are highly correlated with their functional characteristics. Because it is impractical to attempt to completely separate structure (anatomy) from function (physiology), Chapter 6 included mention of some physiologic aspects which will be further elucidated in the discussion in this Chapter.

In Chapter 6 it was established that the dorsal roots are sensory (afferent) and the ventral roots are motor (efferent). This was first shown by Bell at the beginning of the nineteenth century, and the statement has since been known as *Bell's law.* The following functional classification of spinal nerve fibers is the basis for understanding of the functional anatomy of the peripheral nervous system.

Functional Components of Spinal Nerves

Each specific functional component which refers to a particular kind of sensory or motor function pertains to the individual neuron, not to the nerve division (primary ramus) as identified in dissection. Although a peripheral nerve fascicle (Fig. 6-4, p. 95) conveys impulses in both directions, each individual neuron process within a fascicle *in vivo* normally transmits impulses in only one direction.

It is classically stated that the typical spinal nerve is composed of four functional types of neurons: (1) general somatic afferent, (2) general visceral afferent, (3) general visceral efferent, and (4) general somatic efferent. This means that somewhere within the 'typical' spinal nerve there are neuron processes of two types of sensory neurons which terminate in the dorsal horn of gray matter in the spinal cord, and two types of motor axons which originate from cell bodies within the intermediolateral and ventral gray horns of the spinal cord (Fig. 7-1).

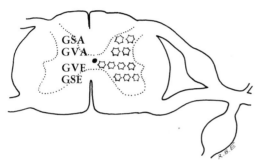

Fig. 7-1. Schematic representation of a cross section of spinal cord, indicating the relative positions in the gray matter of the cell bodies of the neurons of a typical spinal nerve, classified according to their function. The upper two rows of cell bodies (sensory) are known as columns of termination, and the lower two rows (motor), as columns of origin (see text).

General somatic afferent (GSA) fibers transmit sensory impulses that arise in the body wall and extremities. These include both *exteroceptive* impulses from the skin area and *proprioceptive* impulses from muscles, tendons, and joints.

General visceral afferent (GVA) fibers transmit sensory impulses that arise within the viscera.

General visceral efferent (GVE) fibers transmit autonomic impulses of both sympathetic and parasympathetic divisions to smooth muscle, cardiac muscle, and glands.

General somatic efferent (GSE) fibers transmit motor impulses to skeletal muscle.

Visceral vs Somatic Afferent Neurons

In general terms the distinction between the two types of *sensory neurons* is based on *their location in the body.* The GVA receptors are chiefly in the connective tissue of the viscera, whereas the GSA receptors are located in the body wall and limbs (Fig. 7-2). The further subdivisions of GSA may be considered in the same schema: *exteroceptive* receptors are within the dermal areas, and *proprioceptors* are found directly within the muscles, tendons, and joints of the body wall and the limbs.

Visceral vs Somatic Efferent Neurons

The criterion for the separation of somatic and visceral *efferent neurons* is the *histologic construction* of the target organs. GVE fibers from the central nervous system pass peripherally to innervate glandular epithelium, cardiac muscle, and smooth muscle, regardless of location. GSE spinal nerve fibers terminate in the skeletal muscle of the body wall and limbs. The visceral afferents and efferents are discussed and illustrated in Chapter 9. The above schema and discussion for spinal nerves have some exceptions and do not apply to cranial nerves.

The fundamental concept of four functional components in a typical nerve can be misleading in the sense of: (1) not defining a 'typical' spinal nerve, and (2) not conveying the fact that the spinal nerve proper, as pointed out in Chapter 6, is a very short segment. A so-called 'typical spinal nerve,' such as a thoracic or upper lumbar nerve, as illustrated in Figure 6-3 (p. 93), possesses all four functional components in the small spinal nerve proper.

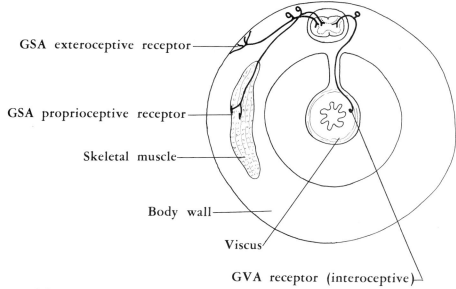

GSA exteroceptive receptor

GSA proprioceptive receptor

Skeletal muscle

Body wall

Viscus

GVA receptor (interoceptive)

Fig. 7-2. Schematic cross section of the body, illustrating the difference in locations for a majority of GSA (body wall) and GVA (visceral) nerve endings (receptors). Note the two types of GSA receptors: exteroceptive, located in the skin area, and proprioceptive, in skeletal muscles, tendons, and joint capsules. The GVA receptors and impulses are referred to as interoceptive.

As illustrated in Figure 7-1 and discussed in Chapter 10, the GSA and GVA cell bodies in the dorsal horn gray matter of the spinal cord are *not* those of the sensory neurons that have processes within the spinal nerve. It has been established that the dorsal root ganglion is the site of the unipolar cell bodies for those neurons. The designated visceral and somatic afferent strata of cell bodies contain the multipolar cell bodies of connector (internuncial) neurons with which the peripheral neurons of the dorsal root synapse. Since these synapses demarcate the termination of the peripheral neurons, and an aggregation of cell bodies within the central nervous system has been defined as a nucleus, it logically follows that the clusters of GSA and GVA cell bodies each constitute a *nucleus* of *termination,* or *sensory nucleus.*

Similar to the sensory nuclei of the dorsal horn gray matter, the clusters of multipolar cell bodies within the intermediolateral and the ventral gray horns constitute nuclei: those of the general visceral efferent (GVE or autonomic) are in the intermediolateral horn, and those of the general somatic efferent (GSE) are in the ventral horn. These *motor nuclei* differ from the sensory nuclei in that they are composed of cell bodies of neurons that have axons contributing the respective motor components to the peripheral nerve. Because these cell bodies are the origins of axons within the peripheral nerve, each GVE and GSE nucleus is logically named a *nucleus of origin,* or *motor nucleus.* Actually, in reference to the functional component representation in the spinal cord gray matter, it seems preferable to use the term 'column' rather than 'nucleus,' because of the continuous longitudinal layers of cell bodies of the specific functional components in the spinal cord. In the brain stem the 'columns' are interrupted and are therefore separated into isolated aggregations of cell bodies which by definition are 'nuclei' in the true sense of the word (see Fig. 11-3, p. 174).

Spinal Reflex Arcs

Definition and Components of Reflex Arc

The simple reflex arc is the basic compound unit of structure and function which includes the fundamental essentials of neuronal activity as they are found *in vivo*. The simplest form of reflex arc is composed of only two neurons: one sensory and one motor. Two-neuron (monosynaptic) reflex arcs, however, are very rare within the body. A classic example is that involved in the stretch, or myotatic, reflex, commonly demonstrated by the stifle (knee) jerk, which is elicited by tapping the tendon of the quadriceps femoris muscle immediately distal to the patella (Fig. 7-3). By tapping this tendon in the proper fashion, the muscle is stretched and the neuromuscular spindles are stimulated to initiate the afferent impulse. This travels within the afferent neurons in the femoral nerve to the spinal cord, where they synapse with the large multipolar cell bodies of the somatic efferent neurons (*lower motor neurons*). From here the impulse is transmitted peripherally through the fibers of the ventral root and within the femoral nerve to the specific motor end-plates of the quadriceps femoris muscle, resulting in its contraction and extension of the leg (crus).

Most spinal reflex arcs are composed of three neurons. The additional (*connector, or internuncial*) *neuron* is between the afferent and the efferent neurons and therefore is located within the gray matter of the spinal cord, as illustrated in Figure 7-4. Describing the complete route of the impulse, the segments of the reflex arc are as follows:

The *receptor* is the peripheral specialized nerve ending of the afferent neuron which responds to a specific change in the environment (*stimulus*). If the stimulus is of the necessary type and intensity (*i.e.*, if the receptor is capable of responding to the specific type of environmental change and the

threshold for that receptor is reached), a nerve *impulse* is generated in the afferent nerve terminal and propagated along the nerve process.

The *afferent* (*sensory*) *neuron* transmits impulses from the receptor toward the spinal cord. The cell body of this neuron is of the unipolar type and is located in the dorsal root ganglion. This neuron terminates within the gray matter of the spinal cord by synapsing with the connector neuron.

The *connector neuron* (*internuncial neuron, neuron of the second order*) actually defines the type of reflex arc. It is this neuron that gives versatility to the arc and determines the integrative ability of the central nervous system at the reflex level by connecting the proper afferent and efferent neurons. The multipolar cell body of this neuron is located in the dorsal horn of the gray matter and its axon passes to one of many places, depending on the neuronal activity involved. This axon terminates within gray matter by synapsing with the efferent neuron(s).

The *efferent* (*motor*) *neuron* has its large multipolar cell body located within the ventral horn gray matter of the spinal cord and its axon traverses the ventral root and a division of the spinal nerve peripherally to the effector.

Target Organ of Reflex Arc

The effector, or target organ, is the peripheral structure that is acted upon to produce the effective response. If the efferent axon is that of a somatic neuron, as illustrated in Figure 7-4, the effector is skeletal muscle; if, however, the efferent neuron is general visceral efferent (autonomic), the effector is either smooth muscle, cardiac muscle, or glandular epithelium. In reviewing the literature, much confusion may arise in the interpretation of whether the effector specifically refers to the target organ as defined above, or to the specialized terminal of the efferent fiber (*e.g.*, the motor end-

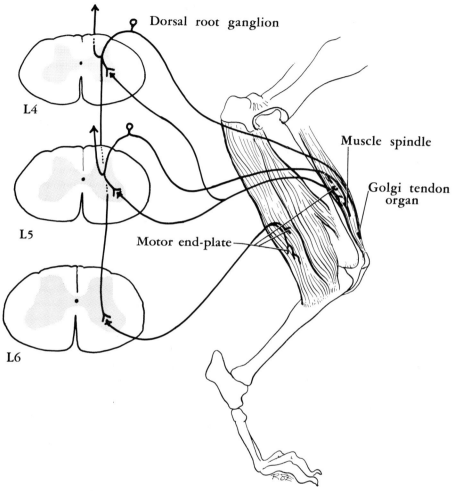

Fig. 7-3. Illustration of a two-neuron reflex arc, exemplified by that involved in the stretch, or myotatic, reflex. The femoral nerve contains both the afferent and efferent segments of the arc. This nerve is associated with spinal segments L4–6 in the dog. The principal receptors are the muscle spindles and Golgi tendon organs. These are stimulated when the muscle and tendon are stretched by tapping the tendon of the quadriceps femoris muscle immediately distal to the patella. Impulses are conveyed into the spinal cord via the femoral nerve dorsal roots and by means of a single synapse the efferent impulses are transmitted to motor end-plates in the same muscle by the same nerve, to complete the monosynaptic reflex. Motor fibers from spinal level L6 are shown innervating the caudal thigh muscles to convey impulses having an inhibitory effect on the antagonistic muscle mass during the reflex. The sciatic nerve (L5–7 and S1–S2), innervating the antagonists, shares some spinal levels with the femoral nerve for origins of motor neurons.

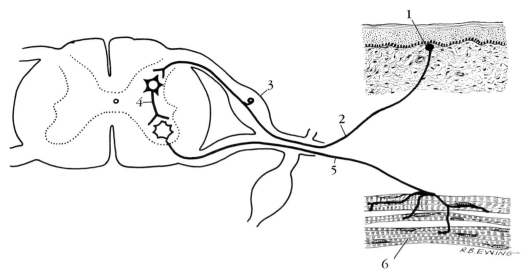

Fig. 7-4. A typical three-neuron spinal reflex arc, with the definitions of its neurons based on the direction of impulse conduction. The somatic efferent (motor) neuron, which innervates the skeletal muscle, is also referred to as the lower motor neuron. *1*, receptor; *2*, afferent (GSA) neuron; *3*, dorsal root ganglion; *4*, connector (internuncial) neuron; *5*, efferent (GSE) neuron; *6*, effector (target organ).

plate for skeletal muscle innervation). Gardner (1968) judiciously clarifies the nebulous terminology by introducing the term *neuro-effector junction* for the peripheral specialized ending of the somatic and visceral efferent fibers at the effectors.

The GSE motor neuron as illustrated in Figure 7-4 is designated as the *lower motor neuron* (motoneuron). These neurons form the basis of the *final common path* concept because of their role as the sole central nervous system outlet for control of skeletal muscle. In this context, numerous axons from various places converge on the lower motor neuron for synapses and thus influence it (see Figs. 10-5, p. 155; 16-2, p. 236). This focusing of many axons on the synaptic pool of the cell body of the lower motor neuron is referred to as *convergence.*

The significance of the foregoing concepts with reference to the lower motor neuron is important in understanding dysfunctions that may result from injury. From the clinical standpoint, such injuries are referred to

as *lower motor neuron lesions.* Insults to this neuron need not be due only to trauma; viral infections, such as poliomyelitis in man, may cause a lower motor neuron effect. Not only are there numerous types of insult to the lower motor neuron, but there are four different main sites in the efferent part of the reflex arc that may be damaged, to result in abnormal muscular action. As illustrated in Figure 7-5, these four sites are: (1) the cell bodies within the ventral horn gray matter of the spinal cord, (2) the axons within the ventral root or primary rami of the peripheral nerve, (3) the neuro-effector junction (*i.e.,* the motor end-plates of the somatic efferent neurons), and (4) the effector (*i.e.,* the skeletal muscle). Often it is extremely difficult to differentiate between lesions of the nerve innervating the target organ and lesions of the neuro-effector junction or the effector. It must be understood that a defect in any of the parts of a reflex arc (afferent or efferent, or both) will either diminish or abolish the response.

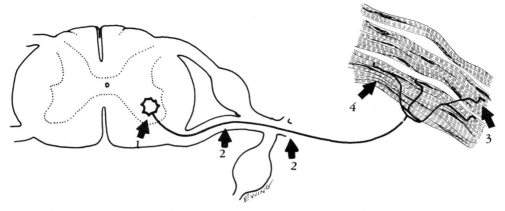

Fig. 7-5. Four principal sites in the efferent portion of a reflex arc which, if damaged, may result in abnormal muscular activity: (1) the cell bodies within the ventral horn gray matter of the spinal cord; (2) the axons within the ventral root or primary rami of the peripheral nerve; (3), the neuro-effector junction (*i.e.,* the motor end-plates of the somatic efferent neurons); and (4) the effector (*i.e.,* the skeletal muscle).

Classifications of Spinal Reflex Arcs (Fig. 7-6)

As stated above, the classifications of the spinal reflex arcs given here are based on the course of the internuncial neuron with regard to the location and physiological properties of the afferent and efferent neurons connected.

A. Segment of Spinal Cord

(1) *Intrasegmental.* Those spinal reflex arcs which have the afferent and efferent neurons of that arc within the same segment (vertebral level) are intrasegmental.

(2) *Intersegmental.* Those reflex arcs which have the afferent and efferent neurons of the arc at different segmental levels of the cord are intersegmental. As an example, the afferent neuron may be at thoracic level 5 and the efferent at thoracic level 8.

B. Side of the Body

(1) *Ipsilateral.* Those reflex arcs in which the connector neuron serves as a link between afferent and efferent neurons on the same side of the body are identified as ipsilateral.

(2) *Contralateral* (*crossed*). Those reflex arcs in which the internuncial neuron connects an afferent neuron on one side of the body with efferent neuron(s) on the opposite side of the body are identified as contralateral (crossed).

C. Functional Components

The functional components of the neurons composing the spinal nerve fascicles have been discussed in the first portion of this Chapter. To repeat, it is the internuncial neuron that characterizes the functional aspects of the reflex, depending on the connections of the specific nucleus (column) of termination and the nucleus of origin for the reflex considered. Based on the functional components, each designation of a particular spinal reflex arc is made up of two elements, the first one indicating the type of sensory neuron and the second indicating the type of efferent neuron. On this basis, therefore, there are four types of spinal reflex arc:

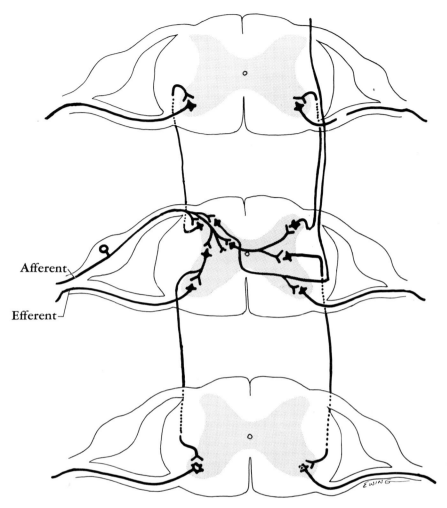

Afferent

Efferent

Fig. 7-6. Diagram of connections of the spinal nerve afferent terminations by synapses with internuncial neurons which pass to efferent neurons at various levels and on both sides of the spinal cord. The internuncial neuron plays the principal role in determining the particular category of classification described in the text that a specific reflex arc falls into.

(1) *Somatosomatic* (Sensory—GSA, Motor —GSE). Example of reflex—as a result of pinching a dog's paw, there is a withdrawal of the limb.

(2) *Somatovisceral* (Sensory—GSA, Motor —GVE). Example of reflex—as a result of pinching the dog's paw, there is an increase in heart rate.

(3) *Viscerovisceral* (Sensory—GVA, Motor —GVE). Example of reflex—as a re-

sult of noxious substance in the stomach there is an increase in gastric motility.

(4) *Viscerosomatic* (Sensory—GVA, Motor —GSE). Example of reflex—as a result of a noxious substance in the stomach there is contraction of the skeletal muscle of the abdomen (*e.g.,* during vomiting).

Obviously, the above examples of spinal reflexes are overly simplified. For the purpose of studying functional components and the classifications of spinal reflexes, it seems advisable to consider each component separately, although *in vivo* complex combinations of various components occur. As an example, a complex phenomenon such as vomiting actually involves all the following combinations: viscerovisceral, viscerosomatic; ipsilateral, contralateral, intrasegmental, and intersegmental spinal reflexes. To add to the complexity, it should be noted that brain levels influence the spinal levels of phenomena such as vomiting. There is an emetic center in the medulla oblongata of the brain stem.

Although the internuncial neurons are not necessarily totally confined to the gray matter, the synapses at each end of the neuron are always within gray matter. The functional integrity of a reflex arc or any multiple neuronal linkage is dependent on the synapses just as much as on the neurons composing the chain. A brief résumé of the importance of the synapse is included here.

The Synapse

Definition and Types of Synapses

A synapse is the junction between two or more neurons.

Based on the neuronal areas participating in the junction, there are three common types of synapses: (1) *axodendritic,* in which the junction is between the axon of the presynaptic neuron and the dendrites of the postsynaptic neuron; (2) *axosomatic,* in which the junction is between the axon of the first neuron and the cell body (soma) of the second neuron; and (3) *axosomatodendritic,* in which the axon of the first neuron joins both the cell body and the dendrites of the second neuron. The terms presynaptic and postsynaptic are applied to neurons in this context with reference to all synapses regardless of location and not in the restricted sense in which preganglionic and postganglionic are applied to autonomic neurons

in Chapter 9. In addition to the above three common types of synapses, the axon hillock and the initial nonmyelinated segment of certain axons may be sparsely covered with synaptic knobs (*axoaxonic* synapses), across which impulses may be received from other neurons.

Morphology of the Synapse

As a presynaptic fiber approaches the neuron with which it will contact, the terminal end of the fiber divides into fine filaments which radiate to meet the cell body and/or dendrites of the second neuron. At the contact points, presynaptic axonal filaments possess small swellings which are known by various names, for example, *end-feet, end-bulbs, synaptic buttons, synaptic knobs, boutons terminaux,* and *terminal buttons,* which are among those most commonly used. The proximal portions of the dendrites and the surface of the cell body are densely covered with synaptic knobs, as illustrated in Figure 7-7. Note that the axon hillock is the only area of the cell body that is relatively free of such knobs.

There is no protoplasmic continuity between the two (or more) neurons participating in the synaptic junction. Figure 7-7 illustrates the space (*synaptic cleft*) between the presynaptic terminal button and the postsynaptic cell body. The average width of the synaptic cleft for synapses on the cell bodies of mammalian somatic efferent neurons is approximately 200 Å (McLennan, 1970). According to McLennan, the cleft is a portion of the interstitial spaces and therefore is filled with extracellular fluid.

Figure 7-7 shows a *glial membrane* which extends from the surface of the presynaptic axon, over the terminal button to enclose the synaptic cleft, and continues onto the surface of the postsynaptic neuron. The cell membranes which form the neuronal surface walls of the cleft are the *presynaptic membrane* on the proximal side and the *subsynaptic membrane* on the distal side. Within

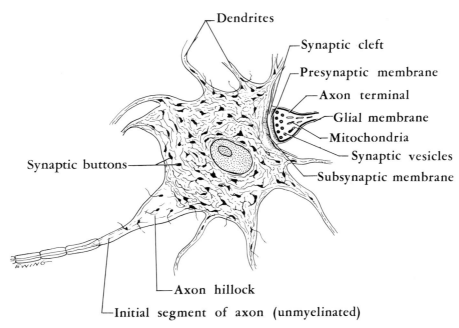

Dendrites

Synaptic cleft

Presynaptic membrane

Axon terminal

Glial membrane

Mitochondria

Synaptic vesicles

Subsynaptic membrane

Synaptic buttons

Axon hillock

Initial segment of axon (unmyelinated)

Fig. 7-7. Diagram showing proximal segments of dendrites and cell body of a postsynaptic neuron densely covered with synaptic buttons (knobs), defining axodendritic and axosomatic synapses. An enlarged view of an axosomatic synapse is shown, to illustrate the anatomic structures involved.

the terminal button are *mitochondria* and *synaptic vesicles.* It is thought that the vesicles (300–500 Å in diameter) are essential for transmission of nerve impulses across the synaptic cleft, in that they emit a chemical, presumably acetylcholine, which depolarizes the subsynaptic membrane in preparation for generation of the nerve impulse in the postsynaptic neuron (Everett, 1965). Further details of synapses are reported by McLennan (1970). He states that four-fifths of the motoneuronal (lower motor neuron) surface may be covered by synapses of the type described above, but these typical synapses are undoubtedly not the only type.

As discussed briefly earlier, the general phenomenon of convergence is well exemplified at the large somatic multipolar cell body of the lower motor neuron in the ventral horn of the spinal cord gray matter. The phenomenon which is opposite to convergence is logically called *divergence.* One example of this is within the autonomic ganglia, more specifically the sympathetic, where one presynaptic axon synapses with many cell bodies or dendrites (or both) of the postsynaptic neurons. Another example of divergence is at the entrance of the dorsal root into the spinal cord. As the sensory root reaches the spinal cord, it divides into many rootlets which join the spinal cord in a row over a considerable cephalocaudal distance (Fig. 5-4, p. 82). In turn, each nerve fiber (axon) subdivides in a radiating fashion to establish synapses with many postsynaptic neurons in the gray matter of the spinal cord.

Properties of Synapses

A basic understanding of the importance of the synapse may be gained from a brief summary of a few of its many properties.

1. There is no anatomical continuity of the protoplasm between the two or more neurons involved in the synapse.

2. There is a chemical medium between the axon and the cell body or dendrite in-

volved in the synapse. The chemical most commonly found at these locations is acetylcholine, but certain autonomic ganglia have norepinephrine at synapses (see Chapter 9).

3. It is the synapse that governs dynamic polarity. This means that *in vivo* the synapse regulates the correct direction of the impulse flow.

4. The synapse acts as a barrier to degeneration.

5. The synapse is a site greatly susceptible to hypoxia, drugs, and ischemia.

The synapse is sensitive to toxic substances that may be introduced artificially or formed internally as a result of metabolic disturbances, as well as to medicinal drugs that may be administered therapeutically. The effectiveness of general anesthetic agents is largely dependent on their ability to block transmission of impulses across the synapses.

The synapse is much more susceptible to oxygen deprivation (asphyxia) and reduction of blood supply (ischemia) than are the neuron processes.

6. The synapse is a site responsible for fatigue.

Relatively speaking, the nervous system is noted for its high resistance to fatigue; however, when fatigue does occur, the synapse is the site of failure.

7. The synapse is the site of 'synaptic delay.'

It takes the impulse a longer time to traverse the chemical medium at the synapse than it does to travel along the neuronal processes. Because this delay is measured in terms of milliseconds, it may appear unimportant to any one other than the neurophysiologist.

Location of Synapses

(1) *Central nervous system.* In the brain and spinal cord synapses occur within the gray matter.

(2) *Peripheral nervous system.* In the peripheral nervous system synapses occur within the autonomic (motor) ganglia, both sympathetic and parasympathetic.

The synapse has received a tremendous amount of attention among researchers. In recent years probably the most noteworthy studies have been presented by the electron microscopists, the neuropharmacologists, and the neurophysiologists.

Peripheral Nerve Endings

The peripheral terminations of the craniospinal nerves are within some organ by means of one or several types of endings.

The concept of sensory and motor nerve fibers within the same nerve, and even within the same fascicle, has been discussed earlier.

Motor Nerve Endings

Skeletal Muscle Innervation

The somatic efferent (lower motor) neurons which innervate skeletal muscle terminate by means of neuro-effector junctions labeled *motor end-plates*. This ending is frequently referred to as the *myoneural junction*.

The histologic and physiologic details of the myoneural junction are quite similar to those of the synapse. Electron microscopy of the myoneural junction has revealed mitochondria, synaptic vesicles, and presynaptic and post-(sub)synaptic membranes separated by the cleft (Rhodin, 1963), all similar to those in the synapse as discussed previously in this Chapter. In spite of these obvious similarities, the myoneural junction is not a synapse, which by definition is the junction of two or more neurons.

Three basic general characteristics of the motor end-plate are: (1) acetylcholine is emitted into the synaptic cleft between the nerve and muscle at the myoneural junction; (2) generally one skeletal muscle fiber has one motor end-plate; and (3) one motor neuron (motoneuron) usually innervates many muscle fibers. As a motor nerve approaches the muscle to be innervated, the individual nerve fibers (usually heavily myelinated A fibers) separate, radiate from the common nerve bundle, and repeatedly branch so that a single axon innervates many muscle fibers. A single motoneuron and the muscle fibers innervated by it are referred to as the *motor unit*. The quantitative gradation of normal skeletal muscular response is based on motor units. It is convenient to express the motor unit as the innervation ratio of one axon to the number of muscle fibers innervated. Muscles with a very precise voluntary controlled movement have fewer muscle fibers per motor unit than those which act in the general, gross manner. As an example, in the cat the extrinsic muscles of the eye may have a motor unit of 1:3, whereas in some muscles in the cat's hindleg the motor unit has been reported to be 1:150 (Ruch *et al.*, 1965). In a morphologic study of motor units in specific muscles in man, Feinstein *et al.* (1955) calculated 579 motor units with a mean of 1,934 muscle fibers per motor unit in the human gastrocnemius muscle.

Smooth Muscle Innervation

Smooth muscle cells (fibers) are innervated by autonomic (general visceral efferent) neurons which do not have motor end-plates at their terminations. The exact morphologic details of the autonomic axon terminals have not been determined satisfactorily. It is generally thought that naked axons ramify within the substance of the smooth muscle fibers, but not each fiber necessarily receives a contact point of innervation. According to Ruch *et al.* (1965) smooth muscle functions as an electrical syncytium. The propagation of excitation from cell to cell is favored by their anatomic contact. The degree of smooth muscle innervation is related to the muscular activity. Stretch depolarizes the smooth muscle membrane and thereby may initiate nerve impulses or may increase the rate of neuron firing.

Cardiac Muscle Innervation

The extrinsic innervation of the heart is by way of the autonomic nervous system as discussed in Chapter 9. The functional antagonism is such that sympathetic stimulation increases heart rate, whereas parasympathetic (vagus nerve) excitation results in cardiac deceleration.

Intrinsically within the heart there is an *impulse-conducting system* composed of specialized and modified bundles of cardiac muscle fibers which are unique in that they respond to stimulation by conduction rather than contraction. The overall function of the impulse-conduction system is to coordinate and regulate the timing and amplitude of the proper sequential contractions of the cardiac components necessary for a normal cardiac cycle. In many diseases or as a result of various acute noxious stimuli the orderly sequence of events in the cardiac cycle is disturbed. This results in a decreased efficiency of the heart which may range from insignificant to irreversible and fatal.

It should be realized that there are many controversies regarding terminology and opinions with reference to the impulse-conducting system. Species differences, difficulty in studying the system, and semantics account for much of the disagreement. According to Miller *et al.* (1964) it is impossible to identify any part of the conducting system in a preserved dog's heart by gross dissection. They consider only the first three of the four parts of the conducting system as discussed below, and refer to all parts as being composed of Purkinje fibers. This is not a common interpretation, and many authorities agree with Ham (1965) in considering it incorrect to refer to the entire specialized muscle conducting system as the Purkinje system when in fact only a portion of the system is composed of true Purkinje fibers. In agreement with the latter view, the various parts of the system will be described briefly here, without consideration of the many controversial opinions and terminology.

THE SINOATRIAL (S-A) NODE

This small mass of specialized cardiac muscle fibers is located near the junction of the precava and the right atrium. In the early part of this century it was found in the heart of the dog that the S-A node was the first area to be electrically active when a heart contraction developed. Because of its influence on regulating the rate of initiating heart contractions, the S-A node is referred to as the 'pace-maker' of the heart. It is the main recipient of extrinsic stimulation by means of both sympathetic and parasympathetic divisions of the autonomic nervous system. Parasympathetic (vagus) postganglionic cell bodies are present in close association with and directly inside the node.

THE ATRIOVENTRICULAR (A-V) NODE

Impulses that arise in the S-A node pass to the atrioventricular node, which in the dog is approximately 1.5 mm in diameter (Baird and Robb, 1950) and located in the lower part of the interatrial septum immediately craniodorsal to the septal cusp of the tricuspid (right atrioventricular) valve.

Nonidez (1943) reported that in the dog the A-V node is larger than the S-A node and has a much richer parasympathetic innervation. He observed the typical conducting system as discussed here and advocated the use of young puppies for anatomic study of the system.

THE ATRIOVENTRICULAR (A-V) BUNDLE

From the A-V node the attached bundle extends through the interventricular septum into the ventricular area and divides into right and left branches which pass deep to the endocardium of the two sides of the interventricular septum.

PURKINJE FIBERS

The A-V bundle fibers extend to the true Purkinje fibers which pass to the papillary muscles and then to the lateral walls of the ventricles to form a network in the subendocardium.

It should be noted that the conducting system and the myocardium are both heavily innervated. In the fifth of a series of papers on innervation of the human heart Hirsch (1963) reported a comparative study of the intrinsic innervation of the heart in many submammals and mammals. His study revealed that in the dog heart, which is similar to those of other mammals, there is abundant intrinsic innervation to the adventitia of the coronary arteries and to the myocardium. At the terminal levels the exceedingly rich, fine nerve plexuses extend over or wrap around individual bands of muscle.

Botár (1966) presented a detailed illustrated report with an extensive review of the literature on the modes of terminations of postganglionic fibers in cardiac and smooth muscle in dogs of different ages.

Sensory Nerve Endings

As a fundamental criterion with a functional significance, the relationship to the primary germ layers of the embryo of the different types of sensory receptors and the afferent nerve fibers associated with them are shown in Figure 7-8.

Details of the embryologic significance are not discussed, but Figure 7-2 illustrates the body locations of the afferent terminals in relation to the general organization of the body. It should be remembered that different types of nerve endings, both efferent and afferent, may be terminals of the same specific cranial or spinal nerve.

Exteroceptive Receptors

The specialized nerve endings that are stimulated by changes in the external envi-

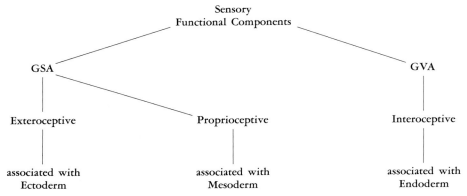

Fig. 7-8. Diagram showing the association of sensory functional components and primary germ layers of the embryo.

ronment are classified as exteroceptive. As one would expect, these are located near the surface of the body (*i.e.,* in the dermis and hypodermis). Morphologically and physiologically there are many types of exteroceptive receptors.

In general consideration of receptors as presented in many textbooks, the reader is erroneously led to believe that each morphologically distinct exteroceptive nerve ending responds to a single specific type of stimulation. Such a concept is easy to accept and neatly correlates the anatomic and physiologic characteristics of receptors which can even logically be correlated with body locations of the receptors. Researchers, however, tend to support the idea that nerve terminations are characteristically non-specific (Crosby *et al.,* 1962). Gardner (1968) states that perhaps sensation depends more on frequency and timing of impulses than on the morphologic type of receptor stimulated.

Frequently the errors are more of omission than of admission; for example, the Ruffini corpuscles in the subcutaneous connective tissue are considered to be probably receptors which generate impulses in response to warmth. Frequently one finds only this function stated unequivocally without even an implication that this same type of ending is found in joint capsules

where it may serve as a pressure receptor as reported by Gardner (1944) in his study of the knee joint of the cat. Most present-day investigators conclude that there is no one specific receptor which can be positively correlated with only response to temperature change.

Without belaboring the point with numerous specific examples, the main emphasis of the above discussion is that the evidence for correlation of morphologic and physiologic properties for a single type of receptor is inconclusive. Obviously, there is much to be learned about receptors, particularly those that respond to temperature change.

GENERAL CLASSIFICATION OF RECEPTORS

The following discussion of receptors is based on the properties of receptor types and is not intended to describe specific receptors as is done in most texts of human histology and neuroanatomy. The detailed knowledge of receptors in the dog is less well documented than that of those in man, but the general information given here applies to both.

Morphologically, receptors may be divided into two main groups: *encapsulated,* which have a great variety of shapes and sizes; and *nonencapsulated,* or *free,* sensory nerve endings. Both types occur among exteroceptors.

Encapsulated receptors. The encapsulated endings are subdivided morphologically into (1) those that are *thinly encapsulated,* with coiled and branched terminations, and generally are located near the body surface on the deep side of the epidermis, and (2) the *heavily encapsulated,* which are best exemplified by the Pacinian corpuscle. The capsule of this receptor consists of numerous concentric collagenous connective tissue laminae, which histologically in cross section give a general appearance analogous to that of the layers of an onion. It is characteristic for the heavily medullated sensory fibers to lose the myelin within the encapsulation (Fig. 7-9). These receptors are located in the subcutaneous connective tissue where they are thought to respond to pressure. The same receptor is also found in and around joint capsules, where they undoubtedly serve as proprioceptive endings. They are also found in places where their function is not clear, for example, in the pancreas of the cat.

The cutaneous, thinly encapsulated nerve endings which respond to light pressure or touch are referred to as tactile endings and are classified as *mechanoreceptors.* Those which are located deeper in the subcutaneous connective tissue also respond to *pressure.* These are heavily encapsulated. If the stimulus for mechanoreceptors causes a deformation of the nerve ending to a degree that its threshold is reached, an impulse is generated in the afferent nerve. Other names of nerve endings that convey their functional significance, and that seem sufficiently self-explanatory, are *thermoreceptors*

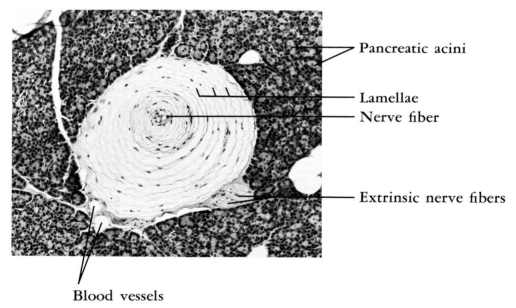

Pancreatic acini

Lamellae
Nerve fiber

Extrinsic nerve fibers

Blood vessels

Fig. 7-9. High-power photomicrograph of a Pacinian corpuscle in the pancreas of the cat. These receptors are located at various sites in the body, where apparently they are stimulated by different types of stimuli. Their specific function in such locations as the pancreas of the cat (and not other laboratory animals) is not known. Note the presence of numerous connective tissue lamellae with their nuclei, which identifies this receptor as the classic example of a heavily encapsulated nerve ending. The naked nerve fiber is identified in the central area of the corpuscle. These receptors are large enough to have their own vascular supply, as labeled above. They are large enough, particularly in the mesentery of the cat, to be seen with the naked eye.

and *chemoreceptors*. The receptors that respond to pain are often referred to as *nociceptors*. In contrast to the contact receptors described above, the receptors associated with the organs of special senses (eye, ear, and nose) respond to stimuli that originate away from the body rather than in physical contact with it. Such receptors therefore are named *teleceptors*.

Nonencapsulated receptors. The nonencapsulated receptors, or free nerve endings, do not have a capsule and are the terminals of thinly myelinated or nonmedullated fibers. These receptors are widely distributed throughout the body in connective tissue and epithelia. The free nerve endings are generally regarded as pain receptors and may be responsive to gross tactile stimuli.

Proprioceptive Receptors

In contradistinction to exteroceptive receptors which are stimulated by changes in the external environment, proprioceptors are stimulated by changes within the body wall itself, namely, within muscles, tendons, and joints (remember that from the embryologic standpoint the limbs are a part of the body wall). Proprioceptive impulses supply the central nervous system with 'information' from the peripheral areas that results in the senses of equilibrium, position, and movement of body parts, and pressure. These correlations should be realized when clinically testing these modalities during a neurologic examination. As previously stated, the Pacinian corpuscles and the corpuscles of Ruffini are located in and around joint capsules, as are the free nerve endings. In addition two types of complex encapsulated proprioceptors are associated with equilibrium, movement, and position. These are the *neuromuscular endings*, or *muscle spindles*, and the *neurotendinous endings*, or *tendon spindles*. The names are explanatory of their respective locations.

MUSCLE SPINDLES

The muscle spindle probably is better known than other types of proprioceptive receptors, and a general discussion may point up the basic importance of proprioception. The complexity of this structure is realized immediately when one notes that there are both afferent and efferent nerve fibers involved in its construction and function. A typical mammalian muscle spindle is depicted schematically in Figure 7-10, to serve as reference for the following discussion.

Muscle spindles, so named because of their shape, are found in most muscles embedded among the regular (extrafusal) skeletal muscle fibers. The spindle contains a few intrinsic modified skeletal muscle (intrafusal) fibers running parallel to the extrafusal fibers. Spindle size varies with species and the particular muscle, and the range in length of the intrafusal muscle fibers has been reported between 1.5 and 14 mm (Kidd, 1966).

The entire spindle is covered by a connective-tissue capsule which extends beyond each end for attachment to the connective-tissue components of the muscle. The intrafusal fibers are striated toward each pole (polar regions), but in the middle area of the spindle the fibers are nonstriated and their nuclei occupy a central position and become spherical in shape. This middle or equatorial region with the spherical nuclei is referred to as the *nuclear bag,* which is thought to be noncontractile (Barker, 1948). Barker also described a longitudinal chainlike extension of the spherical nuclei beyond the nuclear bag, termed the *myotube region.* The swollen equatorial region contains a lymph space which envelops the intrafusal fibers, and has connective tissue trabeculae between the fibers as well as between the nuclear bag and the inside wall of the capsule.

Innervation of the muscle spindle. Muscle spindles are very well supplied with

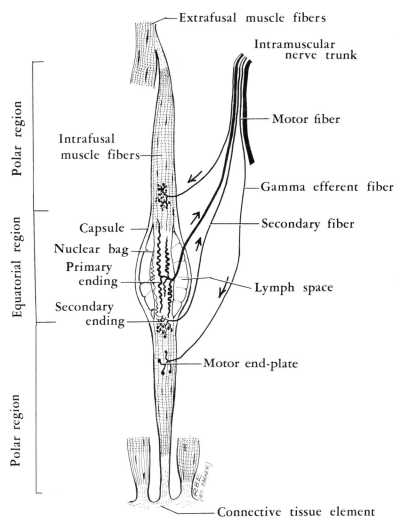

Polar region

Equatorial region

Polar region

Extrafusal muscle fibers

Intramuscular nerve trunk

Intrafusal muscle fibers

Motor fiber

Gamma efferent fiber

Secondary fiber

Capsule

Nuclear bag

Primary ending

Lymph space

Secondary ending

Motor end-plate

Connective tissue element

Fig. 7-10. Diagram of a muscle spindle. Note that this complicated nerve ending is associated with both sensory and motor neurons. The extrafusal muscle fibers are those of the muscle in which the spindle is located (see text for details of structure and function). (After Barker, 1948.)

nerve fibers. According to Boyd (1962), approximately two-thirds of the medullated afferent and efferent fibers within nerves to skeletal muscle innervate muscle spindles. He stated that as many as 25 nerve fibers may enter a single spindle within certain skeletal muscles in the cat.

Each polar (striated muscle) region is reported to have its own motor innervation by small *gamma efferent fibers* which terminate

by motor end-plates in the intrafusal muscle fibers (Fig. 7-10). Because of this innervation, the two polar segments are presumed to function as independent units of contraction.

The sensory nerve terminations are of two types: (a) *The primary annulospiral endings* located in the center of the nuclear bag are the terminations of the larger myelinated sensory fibers. As the nerve reaches the spindle capsule it loses its myelin, and the naked nerve fibers wrap around the individual muscle fibers and pass lengthwise in a spiral formation. (*b*) In most animals smaller myelinated fibers have their terminals outside the nuclear bag in a radiating formation which gives them the name of *secondary, flowerspray,* or *myotube endings.*

In addition to the major innervation of the spindle as discussed, there are thought to be vasomotor fibers to the intrinsic small blood vessels (Truex and Carpenter, 1969).

Functionally, the muscle spindle serves as the receptor of the stretch (myotatic) reflex discussed in the second portion of this Chapter, under Spinal Reflex Arcs. Because the spindles are parallel to the extrafusal muscle fibers, they are stimulated when the muscle fibers are stretched but not when they are contracted. The intrafusal fibers are contracted when the gamma efferents are stimulated, and this contraction also stimulates the afferent nerve fibers of the spindle by stretching the intervening noncontractile region containing the sensory nerve ending. Note that the muscle spindle is thus activated by both a stretch of the extrafusal and a contraction of the intrafusal muscle fibers. The result of this combined action is a proper regulation of the (extrafusal) muscle length for its normal participation in such complex muscular actions as reflexes for maintaining equilibrium and correct posture.

TENDON SPINDLES

The tendon spindles, or neurotendinous organs of Golgi, are located in tendons close to their junction with muscle. A tendon spindle is somewhat similar to the muscle spindle, except that this spindle consists of several tendon fascicles. The knowledge of the tendon spindle is much less detailed than that of the muscle spindle, and there seems to be less agreement about its characteristics. Apparently the tendon spindles are stimulated by increased tension in the tendons as a result of either muscular contraction or stretch.

Polácek (1966) studied species differences of joint receptors in correlation with evolutionary changes. He found that the types of joint receptors in the dog, a typical runner in that the joint movements are somewhat limited to a single plane of flexion-extension, are fewer and simpler than those in the cat, which has much more complex joint movements.

In this Chapter proprioception has been considered in relation to spinal nerves, with no mention of central nervous system pathways or brain involvement.

It should be realized that proprioception does involve higher levels. The special somatic afferent functional component of the vestibular nerve and the vestibular portion of the inner ear, as discussed in Chapter 19, combined with the vestibular nuclei of the brain stem, are the proprioceptive components most importantly related to proper orientation of the body in space (posture).

Proprioception is also considered with spinal cord pathways (Chapter 10) and by necessity is included in the discussion of cerebellar functions (Chapter 14). In addition, conscious proprioception must involve the cerebral cortex. It can be realized, therefore, that proprioception is indeed much more complicated than is indicated in the above discussion, and involves all levels of the nervous system for total integration and proper functioning.

Interoceptive Receptors (Visceroreceptors)

The general visceral afferent (GVA) receptors have received comparatively little attention by researchers. Within the walls of the viscera are free nerve endings that are stimulated by distention of the walls of the viscera, especially of the hollow tubular viscera. Within the walls of larger arteries (*e.g.,* at the carotid sinus near the bifurcation of the common carotid artery and the wall of the aortic arch) are relatively complex receptors (GVA) that are sensitive to increased blood pressure. Therefore these receptors are known as *pressoreceptors*. In the lungs there are GVA terminals that are stimulated by changes in size of the alveoli during respiration and function in respiratory reflexes. Pacinian corpuscles also function as GVA receptors when located in the connective tissue of the abdominal cavity and mesenteries.

Bibliography

Baird, J. A., and Robb, J. S.: Study, reconstruction and gross dissection of the atrioventricular conducting system of the dog heart. Anat. Rec. *108:* 747–763, 1950.

Barker, D.: The innervation of the muscle-spindle. Quart. J. Micr. Sci. *89:* 143–186, 1948.

Botár, J.: *The Autonomic Nervous System. An Introduction to Its Physiological and Pathological Histology.* Budapest, Akadémiai Kiadó, 1966.

Boyd, I. A.: The structure and innervation of the nuclear bag muscle fibre system and the nuclear chain muscle fibre system in mammalian muscle spindles. Phil. Trans. Roy. Soc. Lond., Ser. B *245:* 81–136, 1962.

Crosby, E. C., Humphrey, T., and Lauer, E. W.: *Correlative Anatomy of the Nervous System.* New York, Macmillan Co., 1962.

Everett, N. B.: *Functional Neuroanatomy,* 5th ed. Philadelphia, Lea & Febiger, 1965.

Feinstein, B., Lindegård, B., Nyman, E., and Wohlfart, G.: Morphologic studies of motor units in normal human muscles. Acta Anat. *23:* 127–142, 1955.

Gardner, E.: The distribution and termination of nerves in the knee joint of the cat. J. Comp. Neurol. *80:* 11–32, 1944.

Gardner, E.: *Fundamentals of Neurology,* 5th ed. Philadelphia, W. B. Saunders Co., 1968.

Ham, A. W.: *Histology,* 5th ed. Philadelphia, J. B. Lippincott Co., 1965.

Hirsch, E. F.: The innervation of the human heart. V. A comparative study of the intrinsic innervation of the heart in vertebrates. Exp. Molec. Path. *2:* 384–401, 1963.

Kidd, G. L.: The muscle spindle. Vet. Rec. *78:* 202–204, 1966.

McLennan, H.: *Synaptic Transmission,* 2nd ed. Philadelphia, W. B. Saunders Co., 1970.

Miller, M. E., Christensen, G. C., and Evans, H. E.: The Heart and Arteries. Chapter 4 in *Anatomy of the Dog (ibid.).* Philadelphia, W. B. Saunders Co., 1964. Pp. 267–388.

Nonidez, J. F.: The structure and innervation of the conductive system of the heart of the dog and rhesus monkey, as seen with a silver impregnation technique. Am. Heart J. *26:* 577–597, 1943.

Polácek, P.: Receptors of the joints, their structure, variability, and classification. Acta Fac. Med. Univ. Brun. *23:* 1–107, 1966.

Rhodin, J. A. G.: *An Atlas of Ultrastructure.* Philadelphia, W. B. Saunders Co., 1963.

Ruch, T. C., Patton, H. D., Woodbury, J. W., and Towe, A. L.: *Neurophysiology,* 2nd ed. Philadelphia, W. B. Saunders Co., 1965.

Truex, R. C., and Carpenter, M. B.: *Human Neuroanatomy,* 6th ed. Baltimore, Williams & Wilkins Co., 1969.

chapter 8

Degeneration and Regeneration of Nerve Tissue

By DUANE E. HAINES, B.A., M.S., PH.D.*

Degeneration of Nerve Tissue

When a neuron is so injured that portions of its fibers are morphologically disconnected from the cell body, the fibers and the cell body undergo characteristic changes in structure. These changes constitute the phenomenon of neuronal degeneration. In the peripheral nervous system, under certain favorable conditions, the injured neurons may regenerate to full functional capacity.

The degenerative changes resulting from injury to the nerve fiber affect the entire neural process distal to the site of injury and a short segment proximal to the site, and precipitate a certain degree of change in the cell body of the injured neuron. It should be noted that in higher mammals (*e.g.,* dog, cat, subhuman primates, and man), regeneration of nerve tissue to a functional state takes place only in the peripheral nervous system; it does not occur in the central nervous system.

*Present address: Department of Anatomy, Medical College of Virginia, Health Sciences Division, Virginia Commonwealth University, Richmond, Virginia 23219.

Degeneration Studies As a Research Technique

In a large dog pyramidal cells of the cortex can have axons two or three feet long, extending from cortical motor areas down to lower levels of the thoracic cord. This would be a single fiber extending through four divisions of the brain and much of the length of the spinal cord. The fact that the axon will degenerate after separation from the cell body affords the researcher an excellent method for study of the tracts of the central nervous system. As an example, if the fibers of the corticospinal tract are transected at the level of the internal capsule, they will degenerate below the level of the lesion. The tissue can then be stained for the degeneration products of the axon and axon terminals, by the Nauta-Gygax (1954) or Fink-Heimer (1967) method. This technique of experimental placement of lesions and the subsequent staining for degenerating fibers assists the researcher in anatomical and physiological investigations. Throughout the procedures mentioned above the investigator is utilizing specific staining techniques which use chemicals and stains that, in combination, have an affinity for the degeneration products of nerve tissue.

Injury to Nerve Bundle

In 1891 Waldeyer formulated the 'neuron doctrine' (see Chapter 3), part of which stated that the neuron is the trophic and functional unit of the nervous system. The neuron is the cell body and all of its processes. The cell body (perikaryon) is the nutritional center for the entire neuron. The myelin of the central and that of the peripheral nervous system are formed by the oligodendroglial and neurilemmal cells, respectively, and the material is not dependent on the neuron for nutrition. Nutrients pass throughout the processes of the neuron, as a proximodistal flow of axoplasm,

a process particularly evident during neuronal regeneration (Weiss and Hiscoe, 1948). Injury to a nerve bundle can be of one of three general types: (1) *neurotmesis* is a complete transection of the nerve; (2) *axonotmesis* is the condition in which the axons are severed but the supporting tissues (neurilemmal sheath cells and connective tissue) are not; and (3) *neurapraxia* refers to only a temporary blocking of function without transection of the nerve or its protective covering (Truex and Carpenter, 1969). The axon and myelin sheath distal to the site of injury are the structures most severely affected. The neurilemma (sheath of Schwann) has its individual cellular nuclei located throughout the length of the nerve fiber, generally one neurilemmal cell between each two nodes of Ranvier. The neurilemmal sheath does not depend on the cell body of the neuron as its trophic center, and degenerative alterations in the axon do not necessarily precipitate degenerative changes in the neurilemmal sheath.

Morphologic Changes of Degeneration

The general changes in the neuron are essentially the same, regardless of the type of degeneration under discussion. The initial reactions are: swelling of the neurofibrils and axon, and accumulations of mitochondria in the areas of the nodes of Ranvier. During the first 12 hours after injury to the nerve the gross morphology of the nerve fiber does not change significantly, although there is a small degree of variation among individual fibers (Ramon y Cajal, 1928). The myelin is practically unchanged, and the swelling of the axon is presumably due to the fragmentation of the neurofibrillae (Crosby *et al.*, 1962).

From 12 to 24 hours after injury, the axon becomes irregular in shape and begins to fragment. The myelin sheaths of small fibers begin to fragment into ellipsoidal-shaped droplets, and the myelin of larger fibers starts to retract from the nodes of Ranvier (Ramon y Cajal, 1928). Some larger fibers have irregular areas, and the myelin appears 'fractured' in some regions (Fig. 8-1).

Two to four days after injury, 'digestion chambers' containing folded axon remnants are formed (Ramon y Cajal, 1928), and the myelin continues to retract from the node (Fig. 8-2). The retraction of the myelin from the nodes begins close to the lesion and progresses distally as degeneration of the fiber continues (Causey and Palmer, 1953). The retraction of the myelin from the node and the development of 'digestion chambers' occur simultaneously. At this stage the neurofibrils no longer stain, and it is proba-ble that they have entirely disappeared. From three days after injury the myelin continues to fragment into smaller spheres of material which react positively to fat stains. Breakdown of the myelin sheath and axon is usually complete in two to three weeks. The axon fragments into finely granular material, and the myelin becomes reduced to small fatty masses.

As the degenerating debris accumulates, it is phagocytized by investing tissue macrophages and the neurilemmal cells of Schwann in the peripheral nervous system. The endoneurial fibroblast also has the ability to become phagocytic, and the neurilemmal cell of Schwann assumes a phagocytic role following injury. There is an increase in granular endoplasmic reticulum and mitochrondria so that the appearance of the cell changes to resemble that of a macrophage. The laminated appearance of the myelin is disrupted, and large clumps of osmiophilic material (interpreted as degenerating myelin) are ingested into the cytoplasm of the modified Schwann cell (Satinsky, Pepe, and Liu, 1964).

In the central nervous system the microglial cells presumably become phagocytic and constitute the 'gitter cells' (Smith and Jones, 1966). Maxwell and Kruger (1965) pointed out that vascular pericytes assume a phagocytic role, and that fatty particles seen in oligodendroglia may indicate a phagocytic ability of this type of cell. A type of cell that may be responsible for long-term phagocytic activity in the central nervous system is the astrocyte. The cell volume of the reaction astrocyte increases, vacuoles appear, and abnormal fibrils are noted (Eager and Eager, 1966). The vacuoles and intracellular lipid material are seen mainly in the enlarged perivascular processes.

The cell bodies of the neurons that are involved in the injury also undergo characteristic changes, even though they are at a great distance from the site of injury. The most characteristic change in the cell body

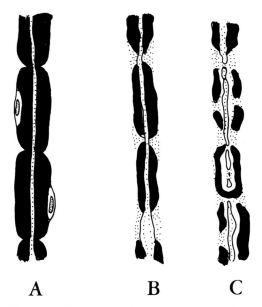

A   **B**   **C**

Fig. 8-1. Diagrammatic representation of the main morphologic changes that occur in the distal portion of the degenerating fiber. A. Normal fiber. B. The myelin and axonal changes at 2 to 4 days. C. The disruption of myelin and axon at 5 to 7 days. In osmium-stained tissue the myelin stains black, the axon and surrounding connective tissue, a light brown. As degeneration time increases the fat slowly disappears and the predominant color will change from black to a light brown.

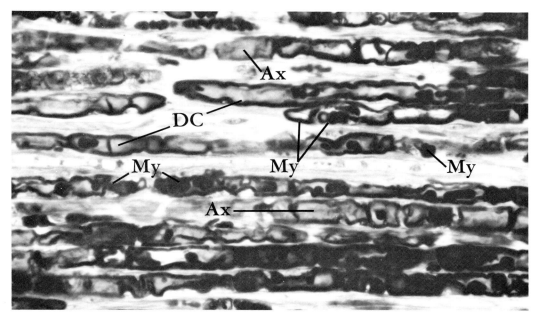

Fig. 8-2. High-power photomicrograph showing degeneration of the distal portion of the sciatic nerve of cat at 13 days after injury. Note the degeneration of the myelin (*My*) and the axon (*Ax*), and the development of 'digestion chambers' (*DC*). Fragments of degenerating axons can be seen in the myelin.

is a dissolution of the Nissl substance (*chromatolysis*) (Beresford, 1965). Accompanying this phenomenon (Fig. 8-3) are other changes, among the most obvious of which are a swelling of the cell body and a displacement of the nucleus toward the periphery of the cell body. Chromatolysis varies with: (1) the type of neuron, (2) the nature of the injury, and (3) the location of the injury. Among the three criteria, the location of the injury is probably the most important. The greater the percentage of neuron separated from its cell body, the greater the degree of chromatolysis.

Some authorities define *incomplete chromatolysis* as that which occurs in a cell body that is located a considerable distance from the point of injury. In such a cell body the Nissl substance only partially disappears and the remaining cytoplasm becomes vacuolated. Complete chromatolysis follows an injury close to the neuron cell body. The Nissl substance completely disappears, and

the nucleus becomes shrunken and irregular in shape and is displaced toward the periphery of the cell body. Complete chromatolysis is indicative of impending cell death.

In summarizing the morphologic changes that occur during degeneration, it should be noted that the initial microscopic changes occur rapidly and fairly constantly, but it may take several weeks for the debris to be entirely cleared away by phagocytic action. The Schwann cells undergo mitotic division along the entire length of the nerve from approximately 4 to 25 days. The hyperplasia and hypertrophy of the Schwann cells are accompanied by a mild proliferation of the endoneurium which results in a decrease in the diameter of the space remaining after degeneration of the axon. The myelin is greatly reduced, and completely disappears if regeneration does not take place.

If nerve tissue is not fixed soon after death, it will undergo postmortem degenera-

Fig. 8-3. Drawings illustrating chromatolysis. Note the eccentric nucleus and the dissolution of the Nissl bodies, beginning in the center of the cell body. A. Normal neuron. B. Incomplete chromatolysis. C. Complete chromatolysis. Following complete chromatolysis there can be a restoration of the Nissl bodies and subsequent regeneration of the neuron (return to A), or there can be complete dissolution of the cell body and all organelles, with subsequent removal by phagocytosis.

tion, or autolysis, resulting in distinct morphologic changes. The axon and myelin sheath change relatively little over a period of 48 hours after death; however, the cell body undergoes drastic change during the same period of time. The nucleolus becomes fragmented, the nucleus loses its integrity, the cell border is unclear, and the granular endoplasmic reticulum will no longer stain (Haines and Jenkins, 1968). In neurohistologic technique, care should be taken so that postmortem artifacts are not inadvertently created in specimens under study.

Direction of Degenerative Changes

From the point of injury degenerative changes can extend toward the cell body (retrograde) or distally from the point of injury (secondary, or Wallerian), and in unusual instances degenerative changes may cross the synaptic space (trans-synaptic degeneration).

Retrograde Degeneration

Retrograde degeneration can be defined as the degeneration process that progresses from the point of injury proximally toward the cell body (*i.e.,* against the flow of axo-

plasm). The extent of retrograde degeneration is relatively minor and depends on the type of injury to the nerve. If the transection is a clean cut, the degeneration will extend back the length of only two or three internodes. However, if the nerve is torn in an irregular cleavage plane, the degeneration may extend back for 20 or 30 mm.

Retrograde degenerative changes that extend to the cell body not only cause chromatolysis, but degeneration of the dendrites will also occur. Degeneration of the dendrites is especially pronounced in young animals (Grant, 1965).

Secondary, or Wallerian, Degeneration

Secondary, or Wallerian, degeneration is that which takes place from the point of transection distally toward the effector organ or the synapse. Regardless of the type of injury, this portion of the nerve will undergo complete degeneration. To repeat a point which is often misinterpreted, secondary, or Wallerian, degeneration occurs in both the central and the peripheral division of the nervous system, but, as stated previously, in higher mammals regeneration to the functional stage may occur only in the peripheral division.

Trans-synaptic Degeneration

Trans-synaptic degeneration occurs in certain isolated areas of the central nervous system but very rarely, if ever, in the peripheral division. Trans-synaptic degeneration occurs from the point of injury distally to the synapse and crosses the synapse to involve the next neuron. It is said to be common in the optic pathways, and in neurons of the second order within the corticopontocerebellar pathway (Crosby *et al.,* 1962). Trans-synaptic degeneration is also reported in the olfactory system (Matthews and Powell, 1962).

Loss of Function in Spinal Cord Degeneration

The spinal cord is composed of ascending and descending tracts, each functionally associated with certain general types of modalities. In all ascending tracts the cell body of origin is in either the spinal cord gray matter or the dorsal root ganglion, and in descending tracts the cell body of origin is located at some higher level.

Lesions of the lateral funiculus of the cord may result in degeneration of spinocerebellar and spinothalamic fibers above the level of the lesion, and degeneration of corticospinal and rubrospinal fibers below the level of the lesion. The cell bodies of origin of the dorsal and ventral spinocerebellar tracts are in Clarke's column and the lateral portion of the intermediate zone (Hubbard and Oscarsson, 1962). The cell bodies of origin for the rubrospinal and corticospinal tracts are in the contralateral red nucleus and in the contralateral frontal, precentral, and postcentral regions (Chambers *et al.*, 1966; Crevel and Verhaart, 1963).

Therefore, the portion of the fiber that degenerates depends on the location of its cell body of origin (Fig. 8-4). In the spinal cord, as an example, a lesion will cause a loss of function below the level of the lesion, even though the lesion may involve ascending fibers. In spinal cord lesions the direction of degeneration of fibers depends on the location of the cell body, whereas the loss of any particular function is manifest distal to the lesion.

Regeneration of Nerve Tissue

Regeneration of Peripheral Nerves

Regeneration of a peripheral nerve may take place after injury if the cell body has not undergone complete chromatolysis and dissolution, and if certain favorable conditions exist in the region of the injured

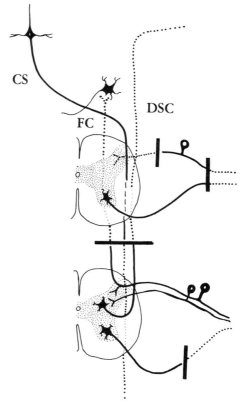

Fig. 8-4. Diagrammatic illustration of degeneration, in relation to cell bodies, of fibers in spinal cord, dorsal and ventral roots, and spinal nerves, caused by lesions at the levels represented by the transverse solid lines. In a descending tract (*e.g.*, corticospinal—CS) the fibers will degenerate below the level of the lesion. In ascending tracts (*e.g.*, dorsal spinocerebellar—DSC; fasciculus cuneatus—FC) the fibers will degenerate above the level of the lesion. In both of these examples the functional loss is below the level of the lesion. Degeneration following lesions of the dorsal roots is central to the lesion, whereas following lesions of the ventral roots the degeneration is peripheral to the lesion. Degeneration after lesions of a spinal nerve is peripheral to the lesion.

trunks. Among the conditions necessary to recovery are the approximation of severed endings, absence of connective tissue invasion, and immobilization of the limb involved. Two to six weeks after injury, the

Nissl substance begins to reappear within the cell body. As the cell body is recovering, the axoplasm begins to form small swellings at the end of the proximal segment of the severed nerve, and many minute branches are given off from each axonal stump (Fig. 8-5).

In the discussion on degeneration it was noted that the neurilemmal sheath of the degenerated distal portion has proliferated and consequently maintained some of its tubal integrity. The terminal swellings of the severed axons are a result of the proximo-distal flow of axoplasm (Weiss and Hiscoe, 1948). While the cell body is recovering from the injury the axoplasm enters minute newly formed branches of the axon. There is also proliferation and investment of connective tissue in the area of the severed endings.

Thus, the success of the regenerative attempts depends primarily on three major considerations: (1) the amount of connective tissue invasion, (2) the extent of the gap between the ends of the proximal and distal segments of the severed axons, and (3) the apposition of these ends. As the axoplasm flows peripherally toward the partially patent neurilemmal sheath cells, it can bridge a small gap and even pass through small amounts of connective tissue. However, not all of the advancing axonal sprouts will reach the distal neurilemmal sheaths. If the advancing axoplasm and neurofibrils within the axon are obstructed by invading connective tissue, they may be diverted away from the degenerating distal stump (Fig. 8-5). An abundance of invading connective tissue can prevent regeneration, resulting in the formation of a neuroma. A *neuroma* is an abnormal growth of nerve tissue; in a regenerating nerve the neuroma would consist of the dilated axonal tips thwarted by the invading connective tissue.

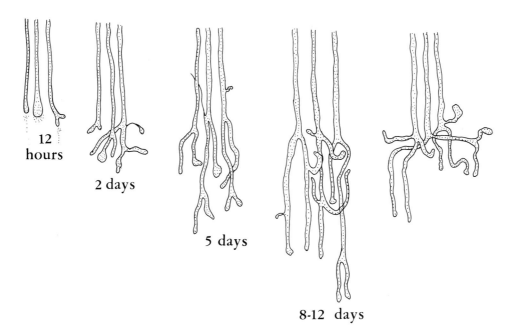

Fig. 8-5. Diagrammatic representation of regenerative courses followed by axons as they grow distally. Aberrant lateral and recurrent fibers will eventually degenerate, and only those fibers that reach the proper effector ending will remain. Axons that attempt to grow through dense connective tissue will follow aberrant courses as shown at the extreme right.

The second consideration cited above is extent of the gap between the ends of the segments of the severed nerve. If the gap is too great, the advancing axoplasm will not reach the distal portion of the nerve, not only because of connective tissue invasion as discussed above, but because of displacement of the distal portion of the neurilemmal tubes, making them 'out of line.' In both of the above-mentioned situations the axoplasm will advance for a time, then eventually stop and undergo degenerative changes.

The primary objective of neurosurgery in veterinary medicine is to approximate the segments of the severed nerve by suturing. Although it is beyond the scope of this work to discuss the details of peripheral neurosurgery, a few general concepts are mentioned. If the ends of the interrupted nerve trunk are jagged they should be trimmed and approximated by sutures, of nylon or silk, through the epineurium and perineurium covering the proximal and distal segments of the trunk (Hoerlein, 1965). Nerve tissue, once severed or torn, will retract from the site of injury. When suturing is attempted, care must be taken not to apply severe tension to the repaired portion of the nerve. After suturing, the repaired portion of the nerve trunk may be wrapped with dried muscle, fat, or a plasma clot, to help prevent adhesions and connective tissue invasion (Hoerlein, 1965).

If some of the neurofibrils reach the distal portion of the nerve and enter the neurilemmal tubes, functional regeneration may follow. Motor fibers may grow into distal sensory tubes, and it is generally accepted that, when such deviation happens, these fibers do not function and they eventually degenerate (Crosby et al., 1962). A high percentage of axons, however, grow into the neurilemmal tubes associated with the proper effectors. As the growing axon enters the neurilemmal tube, it continues along the tube as a thin filament intil it reaches either the terminal end-organ (e.g., motor end-plate or specialized sensory receptor) or the synapse. As it reaches this point, the efferent or afferent ending, or both, are also regenerated, with the motor endings reappearing more rapidly than the sensory endings. Myelination eventually is completed in a proximal-distal order, with the distance between nodes greatly decreased. Thus, in a regenerated nerve, the normal relationship between the diameter of the myelinated axon and the internodal distance does not exist. After the nerve fibers enter the distal stump, it is largely a matter of time before regeneration and clinical recovery are complete. Under favorable conditions a growth rate of 3 to 4 mm a day can be expected. In the clinical situation, however, one should be more conservative, and expect a regenerative growth rate of about 1 mm or less per day.

Regeneration of Central Nervous System

As mentioned previously, there is no central nervous system regeneration to the functional stage in higher mammals, including dogs and cats. Experimentally, partial regeneration has been achieved in severed cat spinal cords where the gap has been less than 1 mm. (Windle and Chambers, 1950). As encouraging as such reports seem to be, it is still necessary to state that in higher mammals there is no regeneration within the central nervous system to the clinically functional stage.

Regeneration of the Spinal Cord

Following accidental injury to the spinal cord of the dog or cat the caudal portion of the cord is partially or wholly separated from the brain from both the anatomical and the physiological standpoint. Ascending impulses including those causing sensations of pain and temperature and impulses from proprioceptive receptors cannot reach higher centers; similarly descending motor impulses cannot reach any portion of the body caudal to the site of transection.

Pertinent literature (of dog, cat, and other selected animals) is reviewed in the following discussion to show some of the factors affecting attempts at regeneration of a transected spinal cord. In animals such as fishes, amphibians, and reptiles the spinal cord will regenerate, following transection, to a state of relatively normal function (Windle, 1956). In lower vertebrates such as the rat and the chick the spinal cord will regenerate to a functional state only if the injury is induced in the prenatal or early postnatal animal (Clearwaters, 1954; Windle, 1956). These authors suggest that since a functional return is noted only in animals experimentally injured before or shortly after birth, the regeneration is dependent on the presence of primitive nerve cells (neuroblasts).

A wide variety of studies on regeneration of the spinal cord in dog and cat exist, but only a few will be considered here. It has been known from the early studies of Ramon y Cajal (1928) that the invasion of connective tissue and the formation of a glial barrier in spinal cord lesions would prohibit axons from regenerating across a gap in the spinal cord. One problem is to decrease or stop the invasion of connective tissue. Brown and McCouch (1947), studying the dog and cat, used pieces of gall bladder, aorta, omentum, and amnion to wrap the transected portion of the cord. They reported no return of function, noted a rapid invasion of connective tissue, and further postulated that the materials they used to wrap the transected region enhanced connective-tissue invasion instead of preventing it.

The formation of the connective-tissue barrier has been partially prevented by the intravenous or intraperitoneal injection of a bacterial pyrogen (Windle and Chambers, 1950; Arteta, 1956). The connective tissue barrier that forms is more reticulated (*vs* compact) and appears to be relatively vascular. The spaces in the reticular connective tissue allow the out-growing axon sprouts to extend distally. Windle and Chambers (1950) also noted that this technique apparently prevented the formation of a glial barrier. These authors reported that the connective-tissue scar, even if invaded by regenerating axons, would eventually contract and pinch any axons extending through the reticular network. Pinner-Poole *et al.* (1967), in a study on cats, induced lesions in the spinal cord without damaging the meninges. These authors suggested that connective-tissue scars and glial barriers will not form if the pia remains uninjured.

A porous membrane of cellulose esters* has been used to bridge the gap in spinal cord transections in the cat and monkey (Campbell and Windle, 1960; Bassett *et al.*, 1959). The severed ends of the spinal cord are encased in the Millipore membrane. The first reaction of the body to this foreign membrane is a proliferation of mesenchymal cells from the leptomeninges, which line the Millipore implant (Noback *et al.*, 1962). By 70 days (in the cat) numerous axons are seen to bridge the gap in the spinal cord protected by the Millipore. In the monkey the neurons did not bridge the gap, and there was no return of function. One difficulty in the use of Millipore *in vivo* is that it begins to undergo calcification 1 month after implantation, and by 12 to 18 months the implant is hardened and brittle (Bassett and Campbell, 1960). These authors suggest that this factor influences the poor long-term results following the use of Millipore.

The use of hypothermia to induce return of function to the injured spinal cord has been studied by Albin *et al.* (1968). These authors report excellent return of neurological function one month after injury. It is postulated that the cooling results in a reduction in edema in the injured area of

*Millipore—Millipore Filter Corporation, Watertown, Mass.

the cord. It is significant to note that this study dealt with impact injury to the cord, not a transection.

The problems associated with destructive injury to the nervous system are many, and they have been touched on only briefly in the preceding discussion. Animal experimentation is imperative as new efforts are made to better understand this relatively unexplored area of human neuropathology.

Bibliography

Albin, M. S., White, R. J., Acosta-Rua, G., and Yashon, D.: Study of functional recovery produced by delayed localized cooling after spinal cord injury in primates. J. Neurosurg. *29:* 113–120, 1968.

Arteta, J. L.: Research on the regeneration of the spinal cord in the cat submitted to the action of pyrogenous substances (5 OR 3895) of bacterial origin. J. Comp. Neurol. *105:* 171–184, 1956.

Bassett, C. A. L., and Campbell, J. B.: Calcification of Millipore in vivo. Transplant. Bull. *1:* 132–133, 1960.

Bassett, C. A. L., Campbell, J. B., and Husby, J.: Peripheral nerve and spinal cord regeneration: Factors leading to success of tubulation technique employing Millipore. Exp. Neurol. *1:* 386–406, 1959.

Beresford, W. A.: A Discussion on Retrograde Changes in Nerve Fibers. In *Degeneration Patterns in the Nervous System* (Singer, M., and Schadé, J. P., Eds.). Progress in Brain Research, Vol 14, pp. 33–56. New York, Elsevier, 1965.

Brown, J. O., and McCouch, G. P.: Abortive regeneration of the spinal cord. J. Comp. Neurol. *87:* 131–137, 1947.

Campbell, J. B., and Windle, W. F.: Relation of Millipore to healing and regeneration in transected spinal cords of monkeys. Neurology *10:* 306–311, 1960.

Causey, G., and Palmer, E.: The centrifugal spread of structural change at the nodes in degenerating mammalian nerves. J. Anat. *87:* 185–191, 1953.

Chambers, W. W., Liu, C. N., McCouch, G. P., and d'Aquili, E.: Descending tracts and

spinal shock in the cat. Brain *89:* 377–396, 1966.

Clearwaters, K. P.: Regeneration of the spinal cord of the chick. J. Comp. Neurol. *101:* 317–329, 1954.

Crevel, H. van, and Verhaart, W. J. C.: The 'exact' origin of the pyramidal tract. A quantitative study in the cat. J. Anat. *97:* 495–515, 1963.

Crosby, E. C., Humphrey, T., and Lauer, E. W.: *Correlative Anatomy of the Nervous System.* New York, Macmillan Co., 1962.

Eager, R. P., and Eager, P. R.: Glial responses to degenerating cerebellar cortico-nuclear pathways in the cat. Science *153:* 553–555, 1966.

Fink, R. P., and Heimer, L.: Two methods for selective silver impregnation of degenerating axons and their synaptic endings in the central nervous system. Brain Res. *4:* 369–374, 1967.

Grant, G.: Degenerative changes in dendrites following axonal transection. Experientia *21:* 722–723, 1965.

Haines, D. E., and Jenkins, T. W.: Studies on the epithalamus. I. Morphology of postmortem degeneration: the habenular nucleus in dog. J. Comp. Neurol. *132:* 405–418, 1968.

Hoerlein, B. F.: *Canine Neurology—Diagnosis and Treatment.* Philadelphia, W. B. Saunders Co., 1965.

Hubbard, J. I., and Oscarsson, O.: Localization of the cell bodies of the ventral spinocerebellar tract in lumbar segments of the cat. J. Comp. Neurol. *118:* 199–204, 1962.

Matthews, M. R., and Powell, T. P. S.: Some observations on transneuronal cell degeneration in the olfactory bulb of the rabbit. J. Anat. *96:* 89–102, 1962.

Maxwell, D. S., and Kruger, L.: Small blood vessels and the origin of phagocytes in the rat cerebral cortex following heavy particle irradiation. Exp. Neurol. *12:* 33–54, 1965.

Nauta, W. J. H., and Gygax, P. A.: Silver impregnation of degenerating axons in the central nervous system—A modified technic. Stain Techn. *29:* 91–93, 1954.

Noback, C. R., Thulin, C.-A., Bassett, C. A., and Campbell, J. B.: The role of mesenchymal cells in regeneration in the spinal cord of the adult cat. Acta Anat. *47:* 144–155, 1962.

Pinner-Poole, B., Tomasula, J. J., De Crescito, V., and Campbell, J. B.: Preliminary observations of the regenerative potential of the spinal cord after cryogenic surgery. Anat. Rec. *157:* 302, 1967 (abstract).

Ramon y Cajal, S.: *Degeneration and Regeneration of the Nervous System.* Vols. I & II (Facsimile of the 1928 edition, translated and edited by R. M. May). New York, Hafner Publishing Co., 1968.

Satinsky, D., Pepe, F. A., and Liu, C. N.: The neurilemma cell in peripheral nerve degeneration and regeneration. Exp. Neurol. *9:* 441–451, 1964.

Smith, H. A., and Jones, T. C.: *Veterinary Pathol-* *ogy,* 3rd ed. Philadelphia, Lea & Febiger, 1966.

Truex, R. C., and Carpenter, M. B.: *Human Neuroanatomy,* 6th ed. Baltimore, Williams & Wilkins Co., 1969.

Weiss, P., and Hiscoe, H. B.: Experiments on the mechanism of nerve growth. J. Exp. Zool. *107:* 315–395, 1948.

Windle, W. F.: Regeneration of axons in the vertebrate central nervous system. Physiol. Rev. *36:* 427–440, 1956.

Windle, W. F., and Chambers, W. W.: Regeneration in the spinal cord of the cat and dog. J. Comp. Neurol. *93:* 241–257, 1950.

chapter 9

Autonomic Nervous System

In addition to the total morphological makeup of the cranial and spinal nerves, the peripheral nervous system has a major subdivision, the autonomic nervous system. Note that it is such an important division that it is sometimes referred to as the 'visceral nervous system.'

It has been emphasized that the separation of the peripheral and central divisions of the nervous system is merely a convenience for purposes of anatomic study. Physiologically the two divisions are inseparable and certainly, for critical clinical evaluation of neurologic deficits and malfunction, proper knowledge of both divisions must be applied. Too frequently, in study of the autonomic nervous system, proper emphasis is given to its involuntary characteristics as a portion of the peripheral division of the nervous system, but the autonomic control by the brain is minimized or overlooked.

The hypothalamus is the most important 'head office' for autonomic regulation. The rostral hypothalamic area controls the parasympathetic division, whereas the caudal hypothalamus regulates sympathetic function. The hypothalamus is extremely complicated and has many connections with other brain areas which may influence autonomic activity via the hypothalamus. This is particularly true of the cerebral cortex.

One example seems sufficient to emphasize the importance of the hypothalamus in relation to autonomic innervation and influence on endocrine function. The pituitary gland (hypophysis) has long been considered the 'master gland' of the endocrine system, but, in turn, the pituitary is controlled by the hypothalamus (see Chapter 15). The study of the brain-pituitary interactions for controlling the endocrine system is the main objective in neuroendocrinology. For details of this subject, the reader is referred to the two-volume work edited by Martini and Ganong (1966).

As an introduction to the specific details of the autonomic (general visceral efferent) innervation, it seems appropriate to compare it with somatic motor innervation, which has been discussed in earlier chapters. Characteristics of autonomic innervation are summarized in Table 9-1 as compared with somatic efferent innervation. Figure 9-1 illustrates the general organization of the autonomic nervous system, which is elucidated in the following discussion.

TABLE 9-1. **Comparison of Somatic and Visceral Efferents**

General Somatic Efferent	General Visceral Efferent
1. Target tissue (effector): skeletal muscle.	1. Target tissues (effectors): (*a*) visceral muscle (smooth and cardiac); (*b*) glandular epithelium.
2. All levels of spinal cord have cell bodies of somatic efferent neurons.	2. GVE cell bodies absent in cervical, lower lumbar, and coccygeal levels of spinal cord.
3. Two types of regulation: (*a*) voluntary; (*b*) involuntary (reflex).	3. Only type of regulation is involuntary (reflex).
4. Effector receives only one type of efferent neuron.	4. Target tissues receive two types of efferent neurons (sympathetic and parasympathetic), which are physiologically antagonistic to each other.
5. One (lower motor) neuron between CNS and target tissue.	5. Two neurons between CNS and target tissues. Myelinated preganglionic neuron synapses with nonmyelinated postganglionic neuron in ganglion.
6. Target tissue reacts only by excitation (contraction).	6. Target tissues react by both excitation and inhibition.
7. Skeletal muscle is absolutely dependent on innervation (denervation—paralysis).	7. Visceral muscle exhibits myogenic principle.
8. Somatic effect—rapid adjustment to external environment.	8. Visceral effect—slow attempt to preserve constant internal environment at cellular level (homeostasis).

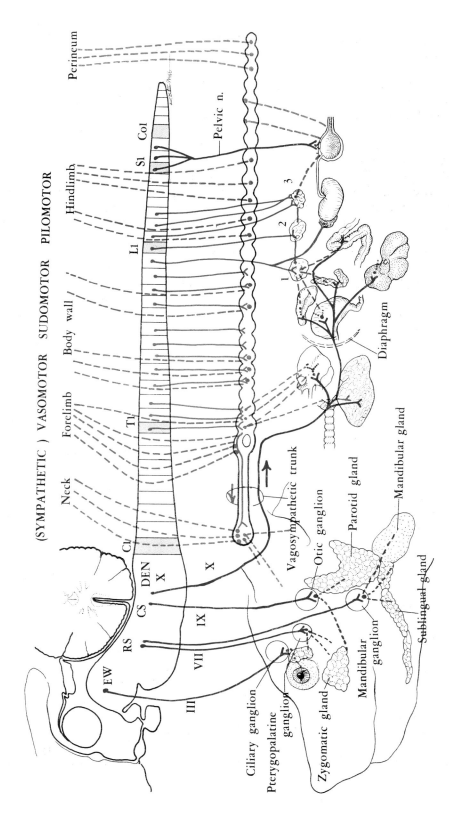

Fig. 9-1. Diagram illustrating the general organization of the autonomic nervous system. Sympathetic neurons are red, parasympathetic neurons blue. Solid lines represent presynaptic (preganglionic) neurons; broken lines, postsynaptic (postganglionic) neurons. *EW*, Edinger-Westphal nucleus; *RS*, rostral salivatory nucleus; *CS*, caudal salivatory nucleus; *DEN X*, dorsal efferent nucleus of X; *C1*, cervical 1 level; *T1*, thoracic 1 level; *L1*, lumbar 1 level; *S1*, sacral 1 level; *Co1*, coccygeal 1 level; *III*, oculomotor nerve; *VII*, facial nerve; *IX*, glossopharyngeal nerve; *X*, vagus nerve; *1*, celiac ganglion; *2*, cranial mesenteric ganglion; *3*, caudal mesenteric ganglion.

Characteristics of the Autonomic Nervous System

1. The target tissues of the autonomic nervous system are: (*a*) visceral muscle (cardiac and smooth), and (*b*) glandular epithelium.

Anywhere in the body only these two types of cells respond to autonomic innervation. Therefore, the effect of autonomic stimulation on viscera is a direct result of response in visceral muscle and/or glandular cells constituting the anatomic components of the organs.

2. The autonomic nervous system is entirely motor (efferent).

By definition, the autonomic nervous system, with the sympathetic and parasympathetic divisions, contains only one functional component, general visceral efferent (GVE). The autonomic nervous system is frequently called the 'visceral nervous system.' This can be confusing, because the general visceral afferent (GVA) neurons are not included as a part of the autonomic nervous system. Since the visceral afferent neurons generally are found within the autonomic plexuses and accompany the visceral efferent neurons (often within the same nerve sheaths) throughout their peripheral distribution, as illustrated in Figure 9-2, it may seem illogical to exclude the visceral afferent neurons from the autonomic nervous system.

The GVA neurons are excluded from the autonomic nervous system primarily because the main criterion for this classification is functional component of the nerves, which is defined as general visceral efferent for this system. In addition, the visceral afferent fibers do not necessarily synapse with general visceral efferent neurons; that is, there are viscerosomatic reflexes as well as viscerovisceral reflexes (see Chapter 7).

3. Two neurons exist between the central nervous system and the innervated target.

The two neurons meet peripherally at a synapse within an autonomic ganglion. The first neuron arises within the gray matter of the central nervous system. Its axon passes peripherally within the ventral root of a spinal nerve or within certain cranial nerves as the *presynaptic*, or *preganglionic*, neuron. The second one is labeled the *postsynaptic*, or *postganglionic*, neuron. Although the terms presynaptic and postsynaptic neurons are used as synonyms for preganglionic and postganglionic neurons, the latter names are more commonly used.

The main purpose for introducing presynaptic and postsynaptic is for clarity in designation of the relation to the site of synapse. As an example, in Figures 9-2 and 9-3, the splanchnic nerves contain 'preganglionic' axons even though the axons leave the paravertebral ganglion and extend to the prevertebral ganglion for synapse. Application of the term 'presynaptic' to the splanchnic nerve axons specifically designates that the neuron does not synapse as it passes through the paravertebral ganglion. Similarly, in Figure 9-4 'presynaptic' seems more proper for the neurons that arise from spinal levels T2 and T4 and pass through numerous paravertebral ganglia and the vagosympathetic trunk before synapsing within the cranial cervical ganglion. It is not a question of one term being correct and the other wrong, but it is advisable for one to understand the basic reasoning for differences in terminology.

One exception in the body to the two-neuron characteristic is the innervation of the suprarenal (adrenal) gland, which is innervated only by the preganglionic sympathetic neuron. Embryonically, the medulla of the suprarenal gland is formed from the same anlage, the neural crest, as the autonomic ganglia. The preganglionic neuron innervating the suprarenal gland actually terminates within the suprarenal medulla as the activator for the release of epinephrine when it is stimulated.

4. Autonomic effects are involuntary.

In general the autonomic effects are not

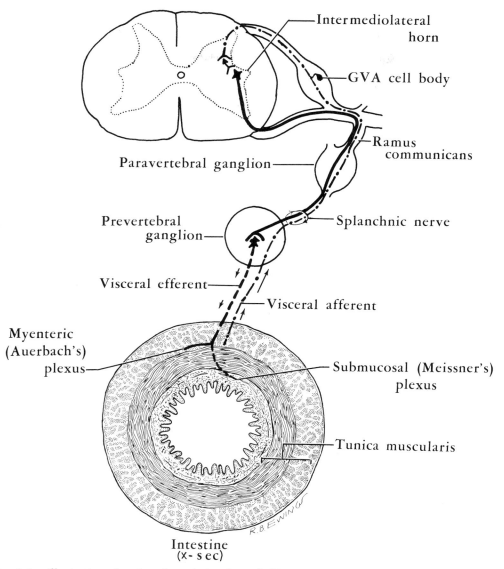

Fig. 9-2. Illustration showing the relationship of GVE sympathetic and GVA (sensory) fibers concerned with spinal reflexes of the intestine. Note that the splanchnic nerves contain both motor (autonomic) and sensory fibers. The pathway shown illustrates an ipsilateral intrasegmental viscero-visceral spinal reflex (see Chapter 7).

under voluntary control, but fortunately they occur automatically depending on the physiological changes that occur within the body. As an example, due to the automaticity of the autonomic nervous system, the individual (man or dog) does not have to voluntarily increase heart rate and respiration, and make other physiologic changes when they are needed during physical exercise.

5. Each target organ is usually supplied by two sets of general visceral efferent

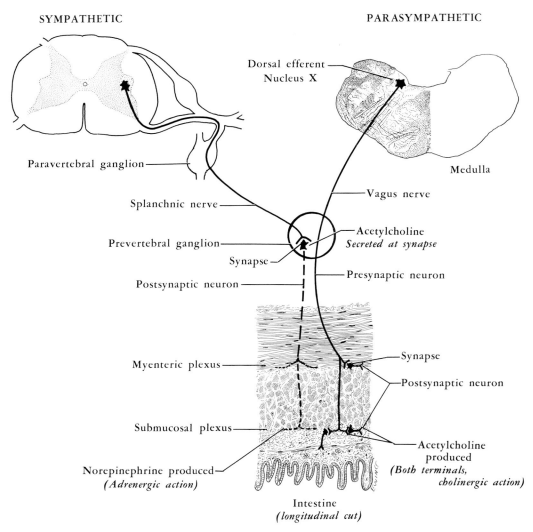

SYMPATHETIC   PARASYMPATHETIC

Dorsal efferent
Nucleus X

Paravertebral ganglion

Medulla

Splanchnic nerve

Vagus nerve

Prevertebral ganglion

Acetylcholine
Secreted at synapse

Synapse

Postsynaptic neuron

Presynaptic neuron

Myenteric plexus

Synapse

Postsynaptic neuron

Submucosal plexus

Norepinephrine produced
(Adrenergic action)

Acetylcholine
produced
*(Both terminals,
cholinergic action)*

Intestine
(longitudinal cut)

Fig. 9-3. Sympathetic and parasympathetic innervation of the upper intestine. Note the differences in the chemical productions at the postsynaptic terminals of sympathetic and parasympathetic neurons, which are correlated with adrenergic and cholinergic activity.

nerves, sympathetic and parasympathetic, which act physiologically in a manner antagonistic toward each other.

This phenomenon of double innervation of GVE neurons is referred to as *duplicity of autonomic innervation.* Exceptions to the rule of duplicity of innervation are suprarenal glands, sweat glands, arrectores pilorum muscles, and blood vessels, all of which are innervated by only sympathetic efferent neurons.

The physiological effect of one type (either sympathetic or parasympathetic) is not the same on all organs. For example, the sympathetic division stimulates heart rate and increases cardiac output, but retards gastric motility; the parasympathetic division, on the other hand, decreases heart rate

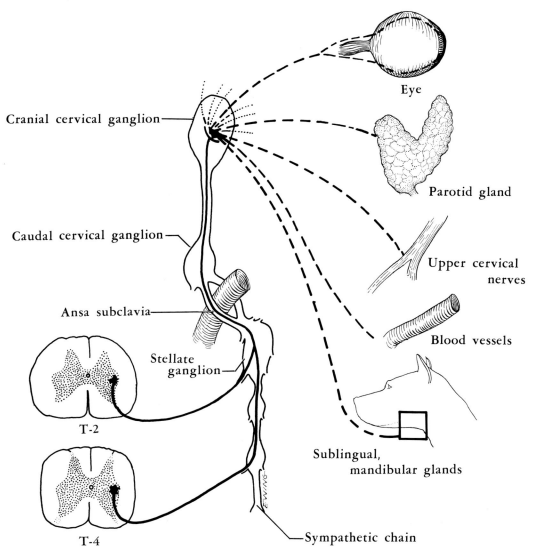

Fig. 9-4. Schematic diagram showing that presynaptic (preganglionic) sympathetic neurons from the upper thoracic levels enter the sympathetic chain and pass cephalically to the cranial cervical ganglion. Here they synapse with postsynaptic neurons which innervate the target organs within the head and in the neck region.

but increases gastric motility and intestinal peristalsis (Table 9-2). This antagonism of physiological effect on target organs is correlated with the pharmacological aspects of the chemical elaboration at the respective nerve endings.

As illustrated in Figure 9-3, within both sympathetic and parasympathetic ganglia at the synapse of the presynaptic and postsynaptic neurons, acetylcholine is secreted. At the neuro-effector junctions, however, norepinephrine is produced at the termination of the postganglionic sympathetic neurons, whereas acetylcholine is secreted at the termination of the postganglionic parasympathetic neurons. The physiological

TABLE 9-2.   Visceral Effects of Sympathetic and Parasympathetic Stimulation

Target Organ	Sympathetic	Parasympathetic
Heart	accelerates rate	decreases rate
Bronchi	dilates	constricts
GI motility and secretion	inhibits	stimulates
GI sphincters	stimulates (constricts)	inhibits (relaxes)
Suprarenal medulla	stimulates (cholinergic)	little or no effect
Sex organs	vasoconstriction contraction of muscular portions (ejaculation)	vasodilation erection
Urinary bladder	little or no effect	contracts bladder wall, promotes emptying

Arteries
  Coronary
  To skeletal muscles
  Peripheral (dermal)
	
	dilate
	dilate
	constrict
		
		little effect
		little or no effect
		little or no effect

Iris	dilates pupil (mydriasis)	constricts pupil (miosis)
Salivary glands	secretion reduced and more viscid	secretion increased and watery

effect of acetylcholine is generally opposite to that of norepinephrine.

Because of the epinephrine (adrenalin)-like effect at the sympathetic neuro-effector junction, the physiological action at this site is referred to as *adrenergic*. Based on the elaboration of acetylcholine at the parasympathetic neuro-effector, the physiological action is considered *cholinergic.*

Drugs that are administered specifically to affect the autonomic nervous system are classified as *sympathomimetic* if they 'mimic' the effects of the sympathetic division and therefore stimulate and enhance activity of that division. Likewise, *parasympathomimetic* drugs produce a response similar to that

produced by parasympathetic stimulation. There are drugs which specifically inhibit either the sympathetic or the parasympathetic action on a specific organ. Note that a sympathetic response may be brought about physiologically by administration of either a drug that stimulates sympathetic neurons or one that inhibits parasympathetic effect.

In brief summary of the general functions of each of the two divisions of the autonomic nervous system, the sympathetic division predominates in the physiological preparation of the body for a state of emergency. This phenomenon is the basis for the so-called 'fight or flight' theory which so concisely but adequately describes the general dominating influence of the sympathetic innervation throughout the body.

Table 9-2 summarizes the specific effects of the sympathetic and parasympathetic innervation on target organs. Note that in contradistinction to the sympathetic preparation of the body for the excitatory state of 'fight or flight,' parasympathetic domination results in restoration of the animal to a normal physiological state of quiescence or *status quo*.

Central Nervous System Origins of General Visceral Efferent Neurons

Anatomically the sympathetic division of the autonomic nervous system is referred to as the *thoracolumbar division,* and the parasympathetic is called the *craniosacral*

division. These designations are based on the specific levels of the central nervous system in which the cell bodies of the preganglionic neurons of the respective divisions are located (see Fig. 9-1).

Sympathetic preganglionic neurons have their cell bodies in the intermediolateral horn of the spinal cord gray matter at levels T1 through L4 or L5 (Stromberg, 1964).

Parasympathetic preganglionic neurons of cranial origin are from specific brain stem nuclei with axons contributing the GVE component to cranial nerves III, VII, IX, and X only (see Table 9-3). The sacral segment of the craniosacral division of the autonomic nervous system has cell bodies of its preganglionic neurons in the gray matter of the sacral segments of the spinal cord, with the corresponding axons contributing to all three sacral nerves.

TABLE 9-3. **Parasympathetic Distribution of Cranial Nerves**

Cranial Nerve	Nucleus of Origin in Brain Stem	Ganglion	Target Organ and Response
III	Edinger-Westphal (parasympathetic nucleus of III)	Ciliary	ciliary muscles—regulate lens curvature; muscles of iris—pupillary constriction (miosis)
VII	Rostral salivatory (parasympathetic nucleus of VII)	Pterygopalatine (sphenopalatine)	lacrimal, nasal, and palatine glands— secretion, vasodilation
		Sublingual and Mandibular	sublingual and mandibular glands— secretion, vasodilation
IX	Caudal salivatory (parasympathetic nucleus of IX)	Otic	parotid and orbital salivary glands— secretion and vasodilation
X	Dorsal efferent nucleus of X (parasympathetic nucleus of X)	Terminal (intramural)	cervical, thoracic, abdominal viscera (see Table 9-2; Fig. 9-2)

The following points should be noted with reference to the central nervous system nuclei of origin.

1. No GVE neurons arise within the central nervous system in the cervical, lower lumbar, and coccygeal levels.

2. Not all cranial nerves receive parasympathetic axons directly from a nucleus.

3. Peripheral areas of the body (viscera and skin areas) are not restricted to innervation by preganglionic neurons with cell bodies in specific levels of the cord; for example, the esophagus within the cervical region is innervated by neurons of both the thoracolumbar and craniosacral divisions (sympathetic and parasympathetic neurons), and the blood vessels, sweat glands, and arrectores pilorum muscles within the skin area of the cervical region are innervated by neurons of the thoracolumbar division (sympathetic neurons).

Autonomic Ganglia

As stated above, all GVE innervation (both sympathetic and parasympathetic), except that of the suprarenal gland, is composed of two neurons which synapse in a peripheral ganglion. These autonomic synapses are the only ones to be found outside of the central nervous system, since the sensory ganglia (dorsal root and cranial nerve ganglia) are composed of unipolar or bipolar cell bodies with no synapses.

According to location and distribution of the postganglionic neurons, there are two types of sympathetic ganglia and two types of parasympathetic ganglia.

Sympathetic Ganglia

Based on the above-stated criteria, the two types of sympathetic ganglia are: (1) paravertebral and (2) prevertebral.

Paravertebral Ganglia

Paravertebral ganglia are best demonstrated in the body regions caudal to the first rib. In these regions the ganglia are located at segmental levels near the intervertebral foramina, therefore along (*para*—Gk. next to) the vertebral column. They are connected cephalo-caudally by thin nerve fascicles in a chainlike fashion and therefore compose the *sympathetic chain.* Although the sympathetic presynaptic neurons arise from the intermediolateral horn of the spinal cord within the thoracic and lumbar regions, the sympathetic chain is not restricted to these areas. Caudally the sympathetic chain extends to the sacrum, where the two chains approach each other near the midline.

Opinions differ as to whether or not the dog has a *ganglion impar,* which would be a total fusion of the right and left L7 or S1 ganglia in the midline. The basis for this disagreement is that in man the ganglion impar designates the caudal termination of the two sympathetic trunks; in the dog both sympathetic trunks continue beyond this level and there may be one or more additional fused sacral ganglia (Mizeres, 1955).

The rostral termination of the paravertebral chain is the *cranial (superior) cervical ganglion.* This is located high in the cervical region near the petrobasilar fissure, in close association with the nodose sensory ganglion of the vagus nerve. As shown in Figure 9-4, the cranial cervical ganglion is the site of the cell bodies for all postganglionic neurons innervating target organs within the head and in the proximal cervical area.

The paravertebral chain within the cervical region differs from that of the thoracic and lumbar regions in that there is not a sympathetic ganglion for each cervical segment. The eight sympathetic ganglia within the cervical region in the embryo have coalesced to form the resulting three ganglia within this same region in the adult.

The nomenclature of these three ganglia differs between man and dog. In man, from cranial to caudal these ganglia are named: superior cervical ganglion, middle cervical ganglion, and inferior cervical ganglion

(Crosby *et al.*, 1962). The inferior ganglion, which lies at the level of the thoracic inlet, commonly fuses with the first thoracic sympathetic ganglion to form the cervico-thoracic, or stellate, ganglion.

In the dog there are also three cervical paravertebral ganglia. The *cranial cervical ganglion* is comparable in location to the superior cervical ganglion found in man. This likewise determines the rostral termination of the paravertebral chain, and is the site of cell bodies of postganglionic sympathetic neurons innervating target organs of the head and high cervical levels. In the dog there is no middle cervical ganglion. The *caudal cervical ganglion* is in the middle position and lies at the level of the seventh cervical vertebra. The *stellate ganglion* is considered the third cervical ganglion and is actually a fusion of the first two or three thoracic paravertebral ganglia (Stromberg, 1964). The caudal cervical and stellate ganglia are connected by two prominent nerve fascicles which envelop the subclavian artery to form the *ansa subclavia* (Fig. 9-4).

Immediately caudal to the nodose (sensory) ganglion of the vagus the sympathetic trunk and vagus nerve fuse and are found in a common sheath throughout the entire cervical region. This common trunk (grossly) of the vagus and sympathetic trunk is referred to as the *vagosympathetic trunk.*

RAMI COMMUNICANTES

In the thoracic and lumbar regions, the paravertebral ganglia are typically connected to the spinal nerve proper by the rami communicantes. The *white ramus communicans* is composed of the preganglionic myelinated fibers that traverse the ventral root and spinal nerve to reach the paravertebral ganglion. The *gray ramus communicans* is composed of postsynaptic nonmyelinated sympathetic fibers that arise at the synapse in the paravertebral ganglion and

re-enter either the ventral or the dorsal division of the spinal nerve for innervation of the peripheral muscular and dermal blood vessels, sweat glands, and arrectores pilorum muscles in the skin areas.

In the human there are two anatomically separate rami communicantes in the thoracolumbar levels. In the dog, however, the two, the gray and the white, are usually fused into one stalk, resulting in only one ramus communicans morphologically, although both gray and white rami communicantes are present (Stromberg, 1964; Fig. 9-5).

Rostral and caudal to the thoracolumbar levels of origin for the sympathetic preganglionic neurons there are no white rami communicantes between the sympathetic chain and the spinal nerves. The peripheral nerve divisions of the cranial, cervical, lower lumbar, sacral, and coccygeal levels do contain gray rami which have the cell bodies of their neurons located in the cephalic and caudal extensions of the paravertebral chain.

Figure 9-4 illustrates white rami communicantes containing preganglionic axons from T2 and T4 which pass rostrally within the vagosympathetic trunk to the cranial cervical ganglion. The nonmyelinated postganglionic axons which arise within the ganglion and enter the upper cervical nerves are considered gray rami communicantes. These rami may be considerably longer than their counterparts in the T1 to the L4 or L5 levels, which have both white and gray rami communicantes connecting the ganglia directly to the spinal nerves as illustrated in Figure 9-5.

The caudal end of the sympathetic chain beyond L4 or L5 consists of descending preganglionic axons from cell bodies located mostly in the intermediolateral horn of the upper lumbar segments of the spinal cord. The descending axons synapse with postsynaptic cell bodies in the chain which give rise to nonmyelinated postsynaptic fibers (gray rami communicantes) that enter the

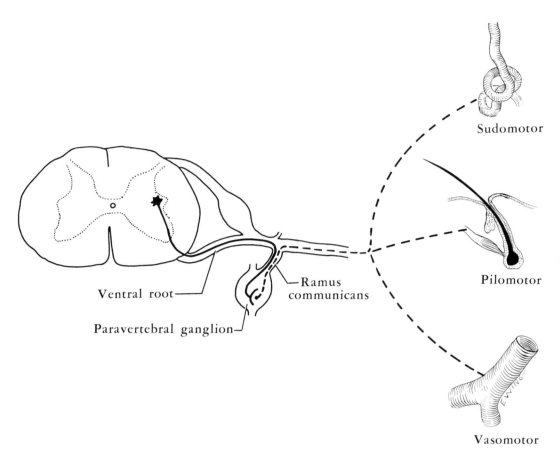

Sudomotor

Pilomotor

Vasomotor

Ventral root

Ramus communicans

Paravertebral ganglion

Fig. 9-5. Diagram of a presynaptic (preganglionic) sympathetic neuron which synapses in a paravertebral ganglion of the sympathetic chain. The nonmyelinated postsynaptic neuron then enters the ramus communicans and continues peripherally in either the ventral or the dorsal division of the spinal nerve to the dermal areas. Here it innervates three target organs: (1) sweat glands (the neuron is referred to as sudomotor); (2) arrectores pilorum muscles (pilomotor); and (3) blood vessels (vasomotor). Recall that in the thoracic and cervical regions, where there are no prevertebral ganglia, the paravertebral ganglia serve a dual role: in addition to distributing postsynaptic axons to the periphery, as shown above, they distribute postsynaptic axons to thoracic and cervical viscera.

lower lumbar, sacral, and coccygeal nerves. These sympathetic fibers pass distally within the nerves as their GVE component.

Summarizing the above discussion: all spinal nerves have gray rami communicantes, but only spinal nerves T1 to L4 or L5 have white rami communicantes.

FUNCTIONAL SIGNIFICANCE OF THE PARAVERTEBRAL GANGLIA

As a presynaptic sympathetic axon leaves its cell body in the intermediolateral horn of the spinal cord gray matter, it passes through the ventral root of the thoracic or lumbar spinal nerve and enters the paravertebral ganglion via the ramus communicans (Fig. 9-5). The presynaptic axon may then do one of four things: (1) synapse immediately within the ganglion; pass (2) cranially or (3) caudally within the chain prior to synapse; or (4) continue through the paravertebral ganglion to synapse in a prevertebral ganglion in the abdominal region.

1. As illustrated in Figure 9-5, the presynaptic axon may terminate within the ganglion by synapsing with a postsynaptic neuron. In this event the postsynaptic fiber is destined to pass back into the (gray) ramus communicans and proceed within either the dorsal or the ventral division of the spinal nerve peripherally to the arteries supplying skeletal muscle, or to blood vessels, sweat glands, and arrectores pilorum muscles in the skin. Those sympathetic axons that innervate the dermal blood vessels are called *vasomotor fibers*. The fibers innervating sweat glands are referred to as *sudomotor fibers,* and those innervating the arrectores pilorum muscles are *pilomotor fibers*.

The sympathetic effect on the dermal target organs in the dog is best observed in the pilomotor action resulting in the dog's hair 'standing up' at a time of 'fight or flight,' which is a physiological state of sympathetic domination. The vasomotor effect produced by sympathetic stimulation is not the same throughout the body (see Table 9-2).

During physiological states of emergency of 'fight or flight,' vasodilation occurs in the arteries supplying the organs which must become more active, whereas the same anatomic type of innervation (sympathetic) causes constriction of the vessels supplying organs not activated during the period of emergency. For example, the coronary arteries and arteries to skeletal muscles in the limbs are dilated, whereas the afferent vessels to the skin are constricted (Table 9-2).

Experimental studies show that sympathetic stimulation causes an *active* vasodilation in the hindlimb skeletal muscles of the dog (Petkovic *et al.,* 1958). The sympathetic fibers to vessels in skeletal muscle are cholinergic (Uvnäs, 1960), rather than adrenergic, as typical for sympathetic innervation. The anatomical and physiological details for sweat-gland innervation in the dog are inconclusive. In man, innervation of the sweat glands is also unlike the general autonomic innervation, in that it is solely sympathetic and cholinergic in action (Truex and Carpenter, 1969).

2. A second alternative for the presynaptic fiber as it enters the paravertebral ganglion is to pass through the ganglion without synapsing and pass up the sympathetic chain (cranially) to a more cephalic ganglion for synapse (Fig. 9-4). The nonmyelinated postsynaptic fiber may then follow the course discussed under the first alternative above.

3. A third alternative is for the presynaptic axon to pass down the sympathetic chain (caudally) to synapse in a chain ganglion at a more caudal level.

4. The last alternative for the presynaptic fiber as it enters the paravertebral ganglion is to pass directly through the ganglion and continue to one of the abdominal *prevertebral* ganglia for synapse. Botár (1966) made an extensive study of the termination of the preganglionic fibers in the prevertebral ganglia in dogs of different ages and in different states of health. He was unable to morphologically distinguish the exact mode of termination of preganglionic fibers in the celiac ganglia. The identification was particularly difficult in senile dogs and in those with pathologic conditions.

As illustrated in Figures 9-2 and 9-3, the myelinated presynaptic sympathetic axons that pass through the paravertebral ganglia and continue to the prevertebral ganglia for synapse form a major portion of the *splanchnic nerves*. Grossly, these nerves connect the paravertebral and prevertebral ganglia. Functionally, the splanchnic nerves contain the GVE sympathetic neurons as discussed, but also contain GVA fibers passing to the spinal cord for the completion of viscerovisceral and viscerosomatic spinal reflex arcs as defined in Chapter 7.

It should be remembered that the visceral afferent neurons are sensory neurons (and, by definition, not autonomic), they have no synapses peripherally, and their unipolar cell bodies are located in the dorsal root ganglia (Fig. 9-2).

The number and levels of origin of the

thoracic splanchnic nerves in the dog are not comparable to those in man. From the thoracic sympathetic trunk in the dog there emerges only one large abdominal sympathetic nerve, usually from the thirteenth thoracic ganglion (Mizeres, 1955). This single nerve is the *thoracic splanchnic nerve*, which is larger than the sympathetic trunk.

The canine lumbar splanchnic nerves vary somewhat in number and anatomic details. There are usually five, but there may be a full quota of seven lumbar splanchnic nerves in the dog (Mizeres, 1955).

As stated previously, there are no prevertebral ganglia in the thoracic and cervical regions. Therefore, the pattern of sympathetic innervation as described above for abdominal viscera does not apply. The paravertebral ganglia in the upper thoracic and cervical regions of the sympathetic chain must serve a dual function in that they (1) are the sites of postganglionic cell bodies for vasomotor, sudomotor, pilomotor (gray ramus communicans) neurons, as in the abdominal region, and in addition (2) contain cell bodies of postganglionic neurons that innervate the viscera of the thorax and neck.

Prevertebral Ganglia

Prevertebral ganglia are located in the abdominal cavity, ventral to the vertebral column and close to the dorsal aorta at the origins of the major blood vessels, the three largest ganglia bearing the same names: *celiac, cranial mesenteric,* and *caudal mesenteric.* These three ganglia are the largest, but not the only, prevertebral ganglia; for example, the *gonadal* and the *aorticorenal* are other prevertebral sympathetic ganglia (Fig. 9-6).

The ganglia are connected by a myriad of nerve fibers forming plexuses (*e.g.,* the connectors between the cranial and caudal mesenteric ganglia form the *intermesenteric plexus;* the fibers attached to the adrenal ganglion form the *adrenal plexus*). The celiac and cranial mesenteric ganglia are also strongly connected by a myriad of fibers, forming a combined plexus. Individual variations exist among dogs with regard to the number and size of the prevertebral ganglia, especially the celiac and cranial mesenteric. Some dogs may have only one of each, and others have two of each; still others may have two celiac and only one cranial mesenteric.

The hypogastric nerves (Fig. 9-6) extend caudad from the caudal mesenteric ganglion over the common iliac arteries to enter the pelvic cavity. On the lateral surface of the rectum they anastomose with rami of the pelvic nerves to form a dense *pelvic plexus.*

Parasympathetic Ganglia

The parasympathetic (craniosacral) ganglia are also of two general types. Just as the sympathetic ganglia and fibers are not confined to the thoracic and lumbar regions of the body, the parasympathetic ganglia and fibers are not confined to the cranial and sacral regions. The two general anatomic types referred to above are: (1) the grossly dissectible parasympathetic ganglia of cranial nerves III (oculomotor), VII (facial), and IX (glossopharyngeal); (2) the microscopic intramural (*intra*—L. within, and *mural*—wall), or terminal, ganglia (X, vagus) that are located within the walls of the viscera.

In the dog, all three sacral levels of the cord contribute to the pelvic nerve which ramifies as presynaptic fibers to the pelvic plexus, where some fibers synapse in the plexus ganglia and others continue to the viscera for terminal synapses. In a detailed study of the nerve endings in the urinary bladder of the cat, Fletcher *et al.* (1969) reported a majority of the terminal cell bodies were located in the vesical adventitia, although some cell bodies were located within the muscular wall, always distributed among the fibers of relatively large nerves.

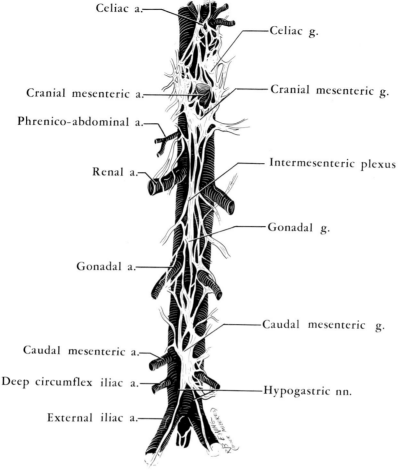

Celiac a.

Celiac g.

Cranial mesenteric a.

Cranial mesenteric g.

Phrenico-abdominal a.

Renal a.

Intermesenteric plexus

Gonadal g.

Gonadal a.

Caudal mesenteric g.

Caudal mesenteric a.

Deep circumflex iliac a.

External iliac a.

Hypogastric nn.

Fig. 9-6. The prevertebral sympathetic ganglia and plexuses in relation to the dorsal aorta and its major abdominal branches in the dog. (After Mizeres, 1955.)

Cranial Portion of Parasympathetic Division

Table 9-3 summarizes the parasympathetic distribution of cranial nerves.

It has been emphasized that only 4 of the 12 pairs of cranial nerves contain GVE axons at their origin. It may be confusing if one refers to a basic source on gross anatomy and reads that the parasympathetic ganglia in the head region are associated with terminal branches of the fifth cranial (trigeminal) nerve. We know that the trigeminal nerve does not give rise to GVE fibers directly from the brain stem. The grossly dissectible relationships of the parasympathetic ganglia should not be confused with the functional aspects of the specific GVE nuclei of cranial nerves as given in Table 9-3.

Truly, in gross dissection one uses the branches of the trigeminal nerve as landmarks for locating the parasympathetic ganglia of the head. The ciliary ganglion is associated grossly (not functionally) with the ophthalmic division of the trigeminal nerve within the caudal portion of the orbit.

The maxillary division of cranial nerve V is used as a guide for locating the pterygopalatine ganglion, and the lingual nerve (a branch of the mandibular division of the trigeminal) has the submandibular GVE ganglion attached to it.

Deep within the parotid gland, one can locate the otic ganglion (GVE parasympathetic for nerve IX), which is associated grossly with the mandibular division of the trigeminal nerve near its emergence from the skull via the oval foramen.

The *vagus nerve* (X) is a special nerve with reference to the autonomic ganglia. As stated above, the vagus is the only one of the four cranial nerves with a parasympathetic nucleus that does not have a grossly dissectible parasympathetic ganglion for synapses of presynaptic and postsynaptic neurons. This same nerve, however, has two dissectible sensory ganglia (nodose and superior), located near the base of the skull, which should not be confused with autonomic ganglia. As illustrated in Figure 9-3, the parasympathetic ganglia of the vagus nerve are microscopic in size and are labeled *terminal,* or *intramural,* as discussed above.

A stereotyped histologic section of the gastrointestinal tract, as illustrated in Figures 9-2 and 9-3, reveals the positions of two intramural plexuses. The *myenteric (Auerbach's) plexus* is between the outer longitudinal and the inner circular smooth muscle layer of the tunica muscularis. The *submucosal (Meissner's) plexus* is within the submucosa. These are considered parasympathetic plexuses, but as illustrated in Figures 9-2 and 9-3 postsynaptic sympathetic fibers are within these plexuses.

The main point of emphasis is that the cell bodies located within these plexuses are only those of parasympathetic (vagal) postsynaptic neurons. The cell bodies of the sympathetic postsynaptic neurons are within the prevertebral ganglia. Figure 9-3 illustrates the difference in the chemical productions at the terminations of the sympathetic and parasympathetic neurons, but the anatomical differences in the length of the postsynaptic neurons of the two divisions of the autonomic nervous system should also be noted.

The parasympathetic (vagus) presynaptic neuron is extremely long; the cell body in the dorsal efferent nucleus of cranial nerve X within the medulla oblongata projects its axon caudally through the cervical vagosympathetic trunk and the thoracic cavity to terminate within the abdominal viscera. This is one cell (neuron) that extends from the brain stem to the abdomen, where it contributes to the synapse with the postganglionic intramural neurons. Because of their origin at the synapses within the prevertebral ganglia as illustrated in Figure 9-3, the postganglionic sympathetic neurons that innervate the intestine are very long in comparison with those of the (microscopic) parasympathetics.

Figure 9-3 also illustrates a prevertebral (sympathetic) ganglion, with the preganglionic parasympathetic (vagus) neuron passing through the ganglion on its way to its termination within the intramural plexuses and ganglia. The cardinal point of emphasis is that there are no parasympathetic cell bodies or synapses within the prevertebral (sympathetic) ganglia, but there are many preganglionic parasympathetic axons passing through the sympathetic ganglia and plexuses.

Figure 9-3 illustrates only one neuron labeled 'vagus nerve.' Obviously, this is only one of many presynaptic neurons, and not all axons pass through the sympathetic prevertebral ganglion as illustrated. Many parasympathetic presynaptic axons will contribute to the prevertebral plexuses without necessarily passing through the ganglia.

In concentrating on the parasympathetic distribution by cranial nerves it should be remembered that none of the cranial nerves is solely autonomic and that the parasympathetic GVE component is only one among possibly five others, as in the vagus nerve.

All of the functional components for the cranial nerves are discussed fully in Chapter 17.

SUMMARY OF PREVERTEBRAL AND INTRAMURAL GANGLIA AND PLEXUSES (FIG. 9-3)

1. The prevertebral ganglia and plexuses are classified as sympathetic because only sympathetic cell bodies and synapses occur in the ganglia, although parasympathetic presynaptic axons may pass through each of them.

2. The prevertebral plexuses contain both presynaptic and postsynaptic sympathetic axons, but only presynaptic parasympathetic axons.

3. The intramural (myenteric and submucosal) ganglia and plexuses are classified as parasympathetic because only parasympathetic cell bodies and synapses occur in the diffuse microscopic ganglia; sympathetic postsynaptic axons, however, contribute to the plexuses.

4. The intramural plexuses contain both presynaptic and postsynaptic parasympathetic axons, but only postsynaptic sympathetic axons.

Sacral Portion of Parasympathetic Division and Pelvic Visceral Innervation

All three sacral nerves in the dog contain presynaptic parasympathetic fibers. The neurons arise within the gray matter of the sacral segments of the spinal cord and the axons pass distally through the ventral roots of the sacral nerves. Within the pelvic cavity the parasympathetic axons unite and form the main component of the *pelvic nerve* (*nervus erigens*), which is the parasympathetic contribution to the pelvic plexus.

The name 'nervus erigens' is no longer accepted, because the pelvic nerve contains efferent and afferent fibers for all pelvic viscera and is not confined solely to erectile tissue, to which the term applies (Nomina Anatomica Veterinaria, 1968). In cat, the parasympathetic cell bodies that contribute preganglionic axons to the pelvic nerve are located in the intermediolateral and intermediomedial cell columns of spinal cord levels S2 and S3 (Oliver *et al.*, 1969).

The pelvic plexus is composed of a myriad of both sympathetic and parasympathetic fibers which are essential for such functions as defecation, urination, erection, and ejaculation. This nerve network is located on the sides of the distal rectum and of the pelvic urogenital organs. Subdivisions of the plexus are named with reference to the specific organ innervated. Obviously such plexuses as vaginal and prostatic plexuses differ between sexes, whereas urinary bladder, urethral, rectal, and hemorrhoidal plexuses are the same in both sexes.

Similar to the autonomic innervation in the abdominal region, there are intramural ganglia and plexuses in the pelvic viscera. In addition, however, there are small ganglia within the pelvic plexus exterior to the walls of the viscera. A majority of these are thought to be parasympathetic in function. Therefore, the parasympathetic innervation of pelvic viscera seems to be more complex than that of abdominal viscera which do not have the extramural parasympathetic ganglia. Postganglionic neurons from these ganglia ramify to innervate the pelvic viscera and extrapelvic genital organs. Mizeres (1955) reported that sacral parasympathetic fibers ascend within the wall of the rectum. Presumably these would extend to the distal colon for its parasympathetic innervation.

SYMPATHETIC CONTRIBUTIONS TO THE PELVIC PLEXUS

The pelvic plexus contains sympathetic fibers from two sources: (1) the caudal mesenteric ganglion via the hypogastric nerves, and (2) the paravertebral ganglia in the sacral portion of the sympathetic chain. When compared with the innervation of the abdominal viscera, the sympathetic contribution to the pelvic viscera appears more complicated, in that there are two sources, rather than merely one.

Clinical Considerations of Autonomic Lesions

Since autonomic fibers are so widely distributed throughout the body, the effects of autonomic disturbances should be considered in all lesions. As an example, the changes that occur in vasomotor functions may play a significant role in the healing of peripheral lesions which may not primarily involve the nervous system.

A basic principle that applies directly to lesions of the autonomic nervous system is the *law of denervation,* as proposed by Cannon (1939). This law refers to the hypersensitivity of denervated structures to chemical agents. More specifically, section of preganglionic fibers results in an increased sensitivity of the isolated neurons and innervated structures to epinephrine. It is known that destruction of postganglionic fibers innervating an organ result in a greater sensitivity to epinephrine than does preganglionic section. In experimental studies on the nictitating membrane of the cat, Hampel (1935) showed that response to postganglionic denervation was maximal in about eight days and was approximately double the response to preganglionic denervation.

Horner's syndrome is a lesion of the sympathetic division of the autonomic nervous system which occurs occasionally in dog (McGrath, 1960). He cites some causes that have been observed as: tumors of the anterior mediastinum, cervical cord lesions, especially tumors, and tumors of the hypothalamus. Neck injuries that destroy the cervical sympathetic ganglia may also produce the syndrome. Note that this may result from a lesion of either the central or the peripheral nervous system. The three cardinal symptoms of this syndrome are the effects demonstrated in the ipsilateral eye, as: miosis, ptosis, and enopthalmos.

PELVIC VISCERAL DYSFUNCTION DUE TO LESIONS

Spinal cord lesions or peripheral lesions that involve the autonomic innervation to pelvic viscera result in serious dysfunctions of those viscera. The effects of such lesions are manifested in malfunction of urinary and reproductive organs, and intestinal (rectal and anal) disorders. All of these are important, but the urinary bladder malfunction probably requires the most urgent attention. The symptoms vary with location of the lesion, its extent, and its duration. A paralyzed bladder at first causes urinary retention and bladder distention, which commonly leads to uncontrolled overflow or dribbling incontinence. This does not effectively empty the bladder. For detailed information and better understanding of these very important clinical problems, the reader should consult a text of neurology. Hoerlein (1965) discusses the clinical management of urinary problems in dogs.

In a functional anatomic study of the micturition reflex in the cat, the preganglionic sympathetic neurons were found to originate from spinal cord segments L2 to L4 or L5. The parasympathetic preganglionic cell bodies that contributed to the innervation of the urinary bladder were distributed in the spinal cord over approximately one and one-half segments, centered near the junction of segments S2 and S3 (Oliver *et al.,* 1969).

Bibliography

Botár, J.: *The Autonomic Nervous System. An Introduction to Its Physiological and Pathological Histology.* Budapest, Akadémiai Kiado, 1966.
Cannon, W. B.: A law of denervation. Am. J. Med. Sci. *198:* 737–750, 1939.
Crosby, E. C., Humphrey, T., and Lauer, E. W.:

Correlative Anatomy of the Nervous System. New York, Macmillan Co., 1962.

Fletcher, T. F., Hammer, R. F., and Bradley, W. E.: Nerve endings in the urinary bladder of the cat. J. Comp. Neurol. *136:* 1–19, 1969.

Hampel, C. W.: The effect of denervation on the sensitivity to adrenine of the smooth muscle in the nictitating membrane of the cat. Am. J. Physiol. *111:* 611–621, 1935.

Hoerlein, B. F.: *Canine Neurology—Diagnosis and Treatment.* Philadelphia, W. B. Saunders Co., 1965.

McGrath, J. T.: *Neurologic Examination of the Dog,* 2nd ed. Philadelphia, Lea & Febiger, 1960.

Martini, L., and Ganong, W. F., Eds.: *Neuroendocrinology.* New York, Academic Press, 1966.

Mizeres, N. J.: The anatomy of the autonomic nervous system in the dog. Am. J. Anat. *96:* 285–318, 1955.

Nomina Anatomica Veterinaria, Adopted by the General Assembly of the World Association of Veterinary Anatomists, Paris, 1967. Vienna, Adolf Holzhausen's Successors, 1968.

Oliver, J. E., Jr., Bradley, W. E., and Fletcher, T. F.: Identification of preganglionic parasympathetic neurons in the sacral spinal cord of the cat. J. Comp. Neurol. *137:* 321–328, 1969: Spinal cord representation of the micturition reflex. *Idem. 137:* 329–346, 1969.

Petkovic, D. S., Husni, E. A., and Simeone, F. A.: Sympathetic control of the circulation in the hind leg of the dog. Am. J. Physiol. *192:* 106–110, 1958.

Stromberg, M. W.: The Autonomic Nervous System. Chapter 12 in *Anatomy of the Dog* (Miller, M. E., Christensen, G. C., and Evans, H. E.). Philadelphia, W. B. Saunders Co., 1964. Pp. 626–644.

Truex, R. C., and Carpenter, M. B.: *Human Neuroanatomy,* 6th ed. Baltimore, Williams & Wilkins Co., 1969.

Uvnäs, B.: Central Cardiovascular Control. In *Handbook of Physiology,* Sec. 1, Neurophysiology (Field, J., Magoun, H. W., and Hall, V. E., Eds.). Washington, D.C., American Physiological Society, 1960. Vol. II, pp. 1131–1162.

chapter 10

Spinal Cord

Development

As discussed in Chapter 2, the entire central nervous system (brain and spinal cord) is formed from ectoderm and is the first body system to begin differentiation and the last to complete functional anatomic development. The progressive developmental stages of ectodermal proliferation and differentiation for both brain and spinal cord are illustrated in Figure 10-1 in sequence as: neural plate, neural groove, and neural tube. The changes that occur between development of the hollow neural tube and the spinal cord as the end-product are less than between that of the neural tube and the brain. The most obvious evidence of this difference is that the spinal cord retains the original cylindrical shape, whereas the brain region of the neural tube expands and bends in the complex manner described in Chapter 2.

Histogenesis of Spinal Cord and Nerves

Figure 10-1 D illustrates the primitive neural tube in cross section as it is located in the dorsal midline above the notochord. Remember that the entire nervous system, with the exception of one type of glial cell, is derived from ectoderm; the notochord is generally believed to be of mesodermal origin and contributes nothing *per se* to the nervous system. The notochord probably

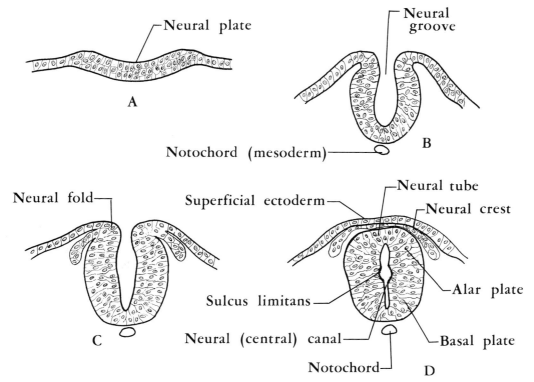

Fig. 10-1. The basic stages in the formation of the neural tube from ectoderm. A. The thickening of ectoderm as the *neural plate.* B. The indentation of the plate in the longitudinal axis of the future body, the *neural groove.* C. Deepening of the groove, with approximation of the neural folds (lips of the groove) in the dorsal midline. D. Closure of the groove dorsally to complete the *neural tube.* Between the dorsolateral surface of the tube and the superficial ectoderm the *neural crests* are formed.

gives support to the developing fetus, but atrophies and loses its identity, to be represented in the adult as the *nucleus pulposus* of each fibrocartilaginous intervertebral disc.

Figure 10-2 schematically shows that histologically the wall of the neural tube is composed of three layers. (1) The *ependymal layer* is a single row of cells which lines the lumen. In the adult the ependymal cells retain their embryonic elongated shape and ciliated surface, and also their position lining the central canal (Fig. 10-2). (2) The thick middle layer of the neural tube, the *mantle layer,* is very cellular; it gives origin to the gray matter of the spinal cord and brain. (3) The *marginal layer* of the neural tube is the outermost layer which contains

relatively few cells; it gives rise to the white matter of the spinal cord and brain.

Figures 10-1 D and 10-2 show the *sulcus limitans,* the lateral groove at the midpoint of each lateral wall of the large lumen of the neural tube. The sulcus limitans of the neural tube indicates the horizontal division of the mantle layer into a dorsal *alar plate* and a ventral *basal plate.* The alar plate forms the dorsal sensory horn of the spinal cord gray matter, and the basal plate forms its ventral motor horn.

It should be recalled that for spinal nerves the two functional columns of termination (general somatic afferent and general visceral afferent) are located in the dorsal horn, and the two columns of origin (general

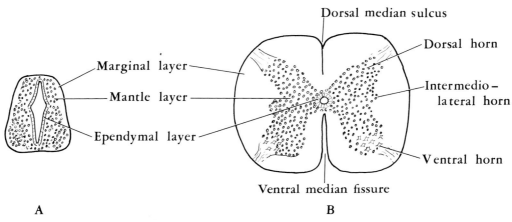

Fig. 10-2. Diagram of spinal cord development from the three primitive histologic layers of the neural tube. A. Neural tube containing the three basic layers. B. Cross section of spinal cord indicating the fate of each primitive layer: the ependymal layer remains fundamentally in the same position (*i.e.,* it lines the central canal), the mantle layer forms the gray matter of the spinal cord, and the marginal layer forms the white matter of the spinal cord. The three 'horns' of gray matter of the lumbosacral spinal cord are labeled on the right.

visceral efferent and general somatic efferent) are located in the ventral horn. In the brain stem region of the neural tube the alar plate contains the four sensory nuclei and the basal plate contains the three motor nuclei (see Fig. 11-3). Further details of functional components of spinal nerves are elucidated in Chapter 7 and of cranial nerves in Chapter 17.

Neuroblasts and Spongioblasts

The developing mantle layer contains cells which proliferate profusely and specialize into two major types to establish the classification of cells of the central nervous system. Those embryonic stem-cells that give rise to neurons are the *neuroblasts;* the other undifferentiated stem-cells that give rise to neuroglia are the *spongioblasts.* Recall that the peripheral sensory and autonomic ganglia are derived from the neuroblasts of the neural crests. With the formation of the alar and basal plates in the mantle layer and the primordial dorsal root ganglionic cells, the typical spinal nerves are formed as processes from these sources, as illustrated

schematically in Figure 10-3. The undifferentiated dorsal root ganglionic cells are bipolar and later form the unipolar type neurons that are characteristic of sensory ganglia.

The central processes enter the spinal cord to terminate chiefly within the alar plate, whereas the terminals of the peripheral processes are within the body wall and viscera. The multipolar cell bodies in the basal plate give origin to axons which emerge from the spinal cord via the ventral root as somatic motor and autonomic fibers. As indicated in Figure 10-3, the autonomic ganglia are formed and innervation patterns are established to complete the construction of a spinal nerve as discussed for the adult in Chapters 6 and 7.

Gross Anatomy

In the adult dog, the spinal cord extends from its junction cephalically with the medulla oblongata at the level of the foramen magnum to its caudal termination (*conus medullaris*) at the level of the junction of the sixth and seventh lumbar vertebrae.

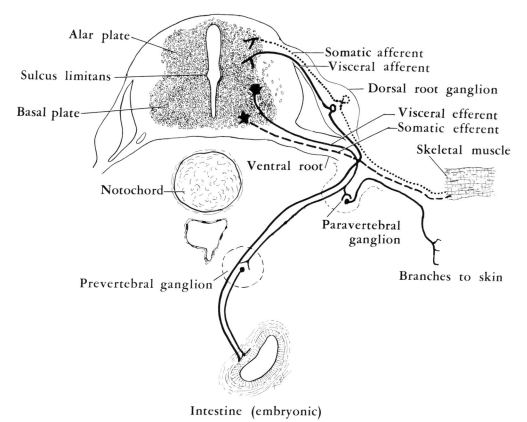

Alar plate

Sulcus limitans

Basal plate

Somatic afferent

Visceral afferent

Dorsal root ganglion

Visceral efferent

Somatic efferent

Skeletal muscle

Ventral root

Notochord

Paravertebral
ganglion

Branches to skin

Prevertebral ganglion

Intestine (embryonic)

Fig. 10-3. Schematic illustration of the anatomic pattern of the embryonic spinal nerves and the establishment of the two columns of termination (sensory) and the two columns of origin (motor) in the mantle layer of the undifferentiated spinal cord. Note that the cell bodies of the neurons of the dorsal root are at the intermediate stage between the bipolar and the final unipolar form.

The spinal cord is not uniformly cylindrical throughout its length, but has two slight enlargements at the attachments of the nerves forming the brachial and lumbosacral plexuses. The enlargements are not dramatically obvious, in that the increase in spinal cord width is merely 2 mm in each location.

According to Fletcher and Kitchell (1966) the width of the spinal cord is 9 mm directly cephalic to the brachial plexus and increases to 11 mm at the enlargement; the width is 8 mm immediately cephalic to the lumbosacral plexus and increases to 10 mm at the enlargement. The same authors reported that the average extent of the brachial enlargement is 5 cm and includes the rootlets of C6 through T1; the lumbosacral enlargement averages 4 cm in length and includes segments L5 through S1.

Relationships of Spinal Cord Segments, Nerves, and Vertebrae

As stated previously, the total number of spinal nerves in the dog is 36 pairs, whereas in man there are 31 pairs. The term 'spinal cord segment,' as used in the discussion that follows, refers to the area of spinal cord to which the rootlets of a single specific spinal nerve are attached. Obviously the spinal cord segments are identical in number to the spinal nerves, but not all segments are identical in length. Vertebral segments referred to below are synonymous with the vertebral numbers. The vertebral formula for dog is: 7 cervical, 13 thoracic, 7 lumbar, 3 sacral, and approximately 20 coccygeal.

Chapter 6 contains a general discussion of the attachments of the spinal nerve rootlets to the spinal cord in relation to the vertebral levels. The same general growth pattern based on the inability of the spinal cord to increase in length at a pace comparable to that of the vertebral column explains the level of the conus medullaris at the junction of vertebrae L1 and L2 for man and of vertebrae L6 and L7 for the dog. The general phenomenon of the vertebral column outgrowing the spinal cord in length is not a passive relative cranial migration of the spinal cord in dog, and other domestic animals, but is a non-uniform differential growth of specific spinal cord areas (Fletcher and Kitchell, 1966). They report that the only cord segments found entirely within the corresponding vertebral segments are the first one or two cervical, the last two thoracic, and the first two or three lumbar.

The relatively long cervical vertebrae enclose all of the cervical and the first thoracic cord segments. Therefore, most cervical cord segments are beneath the vertebra one segment above (cephalad to) the corresponding number, that is, the dorsal and ventral roots of each cervical nerve descend within the vertebral canal one vertebral level before they unite and exit through the intervertebral foramen.

Caudal to the midthoracic levels the cord segments gradually shift caudad as far as L4, which is partially within vertebral level L3. Caudal to this level, however, there is a progressively more cranial shift of cord segments with respect to the corresponding vertebrae. The three sacral cord segments are within vertebral segment L5, and the five coccygeal nerves attach to the cord beneath the cranial three-fourths of vertebra L6.

Obviously, the caudal spinal cord segments are not as long as the more cephalic ones.

There is inherent variability among individuals, but the above-cited authors report that within medium-sized dogs the variation is independent of size, breed, age, and sex. In estimating the cord segments in small dogs (less than 7 kg), however, it is recommended that approximately one-half vertebral caudal displacement of the lumbosacral plexus levels be applied to the above specific spinal cord–vertebral segment relationships.

Surface Anatomy of the Spinal Cord

Throughout its length the spinal cord is partially divided into two hemisections. The *dorsal median sulcus* (Fig. 10-2) is a relatively shallow narrow groove in the dorsal midline and the *ventral median fissure,* which is much wider and deeper than its dorsal counterpart, extends throughout the length of the cord to separate the ventral left and right halves. The ventral spinal artery is located at the surface of the open portion of the ventral fissure.

The dorsal rootlets of the spinal nerves attach to the spinal cord in a longitudinal shallow groove, the *dorsolateral sulcus,* which is lateral to the dorsal median sulcus. In the upper thoracic and cervical region there is a shallow sulcus on each side of the cord, between the dorsal median and dorsolateral sulci. This sulcus, the *intermediate sulcus,* is adjacent to the dorsolateral sulcus at its origin in the thoracic region and then in ascending the cord gradually migrates mediad due to the intervening fasciculus cuneatus in the lateral portion of the dorsal funiculus (area) of the cord.

The ventral rootlets of the spinal nerves are attached to the spinal cord in a longitudinal manner at the ventrolateral surface, but there is no ventrolateral sulcus comparable to the groove in which the dorsal rootlets are attached.

Meninges of the Spinal Cord

The meninges are discussed in Chapter 5 and therefore the discussion will not be repeated here in detail. It should be recalled that the dura mater covering the brain is in direct contact with the bones of the skull and serves as their periosteum (endosteum), whereas that covering the spinal cord is separated from the walls of the vertebral canal by a large extradural space which contains adipose tissue and blood vessels (see Fig. 6-1, p. 90). The dura mater of the spinal cord does not have prominent internal projections into sulci as in the cranial area, nor does the spinal dura contain venous sinuses. The leptomeninges of brain and cord are structurally similar, but the choroid plexuses and the arachnoid villi are not present in the spinal cord region.

The Spinal Cord in Cross Section

The basic changes that have occurred in the development from the three layers of the neural tube to the internal structure of the adult spinal cord are illustrated in Figure 10-2.

Gross Aspects

Gross cross sections of the spinal cord reveal the general organization of the gray and white matter. The internally placed gray matter is characteristically in the form of a butterfly, or H-shape, as illustrated in Figures 10-2 and 10-4. Referring to these figures, one can observe grossly the dorsal (sensory) and ventral (motor) horns (columns) and the narrow connecting structure of gray matter which immediately surrounds the inconspicuous central canal. This connecting gray matter dorsal and ventral to the central canal constitutes the *dorsal* and *ventral gray commissures.* Within the white matter between the ventral gray commissure and the dorsal end of the ventral median fissure is the *ventral white commissure.* This contains commissural fibers to the contralateral fasciculi. There is no dorsal white commissure.

The white matter, primarily white because of the dominance of myelinated fibers, completely surrounds the gray matter. Cross sections through different levels of the cord reveal variations in the gray–white matter relationships. This is because of the variation in the size and number of fibers composing the spinal nerves which attach to the cord at different levels (*e.g.,* at the brachial and lumbosacral enlargements of the cord, correlated with the attachment of the large nerves of the respective plexuses). The bilateral symmetry of the internal organization of the spinal cord is apparent at all levels.

The white matter is conveniently divided into areas referred to as *funiculi.* To emphasize the bilateral symmetry, one may refer to three funiculi on each side of the cord, namely, the dorsal, lateral, and ventral (Fig. 10-6). The *dorsal funiculus* extends from the dorsal median sulcus to the entrance of the dorsal rootlets of the spinal nerves in the dorsolateral sulcus. The *lateral funiculus* is the area between the entrance of the dorsal rootlets and the exit of the ventral rootlets of the spinal nerves. The *ventral funiculus* is bordered laterally by the emerging ventral rootlets of the spinal nerves and medially by the ventral median fissure. As discussed in detail below, the funiculi contain the spinal cord *pathways,* or *tracts,* which may be referred to as *fasciculi.*

A brief statement of summary for organization of the white matter may be: "Each spinal cord funiculus contains many fasciculi." An individual fasciculus cannot be observed by gross dissection. Therefore, special techniques such as degeneration experiments followed by neurohistologic staining are used for the localization of pathways and nuclei.

Microscopic Structure

Gray Matter

Recall that the gray matter is composed primarily of neuron cell bodies, nonmyelinated fibers, neuroglia, and capillaries. In addition, the neuronal dendrites and axons plus the fine glial processes compose the background meshwork, the *neuropile*. It is of fundamental importance that one understands the true significance of the visceral and somatic afferent and efferent components of spinal nerves and their representation within the spinal cord as discussed in Chapter 7 and illustrated in Figure 10-3.

Rexed (1952, 1954) subdivided the gray matter of the spinal cord of the cat into 10 laminae, based on differential cytoarchitectonics. Figure 10-4 illustrates the general plan of the architectonic map on the basis of his work. It is emphasized here that Figure 10-4 is merely an example based on

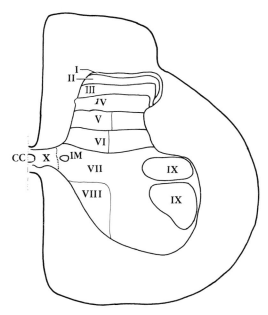

Fig. 10-4. Schematic representation of the structural lamination of the gray matter of segment L7 of the spinal cord of the cat. *CC,* central canal; *IM,* intermediomedial nucleus. (After Rexed, 1952.)

the subdivisions of spinal cord segment L7. The organization and relative positions of the laminae are not uniform throughout the cord. It should be apparent to the reader that the identification of exact subdivisions of the dorsal and ventral horns (columns) of the gray matter of the spinal cord provides a precise criterion for localization of specific axon terminations and origins. Romanes (1951) was able to correlate specific cell bodies in the ventral horn of gray matter of the spinal cord of the cat with innervation of individual muscles in its hindlimb.

GENERAL LANDMARKS OF GRAY MATTER CORRELATED WITH FUNCTION

The following general features of spinal cord gray matter are of fundamental importance in understanding the mammalian spinal cord.

1. The ventral horn (column) contains the large (80 to 120 micra) multipolar cell bodies that compose the nuclei of origin for the somatic efferent neurons which send axons peripherally through the ventral root and spinal nerve divisions for innervation of skeletal muscle of the limbs and body wall. These large multipolar cell bodies and their axons are referred to as the *lower motor neurons,* in contrast to the *upper motor neurons* which have their axons typically within the descending fasciculi and are totally within the central nervous system.

As discussed in Chapter 7, the lower motor neuron is labeled the *final common path* because it is the only normal exit from the central nervous system for the influence of converging fibers from the spinal cord tracts and reflex arcs (Fig. 10-5). These large somatic motor cell bodies are easily found throughout the entire length of the spinal cord.

2. The dorsal horn (column) is capped by a band, the *substantia gelatinosa (Rolandi),* which appears gelatinous in that it does not stain (see Plates 1 through 6 in the Atlas).

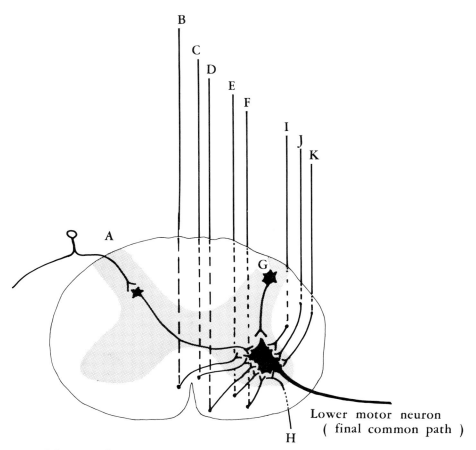

Fig. 10-5. Schematic illustration to show the multiple converging fibers that influence the lower motor neuron. Since this neuron virtually is the only normal exit from the central nervous system for stimuli of skeletal muscle activity, it is referred to as the final common path. A majority of the converging fibers synapse with a small internuncial neuron to the lower motor neuron, rather than directly, as implied in the schema. *A*, contralateral sensory neuron input (*e.g.*, in spinal reflexes); *B*, ventral corticospinal tract; *C*, medial vestibulospinal tract; *D*, tectospinal tract; *E*, pontine reticulo-spinal tract; *F*, lateral vestibulospinal tract; *G*, ipsilateral internuncial neuron (*e.g.*, in spinal reflexes); *H*, fasciculus proprius (ascending fibers of intersegmental reflexes); *I*, fasciculus proprius (descending fibers of intersegmental reflexes); *J*, lateral corticospinal fibers; *K*, rubrospinal fibers.

It is generally thought that the non-staining properties are due to the fact that the substantia gelatinosa is composed of extremely small cell bodies and contains a great quantity of fine nonmyelinated fibers (Ariëns Kappers *et al.*, 1967). The functional significance of the substantia gelatinosa as a portion of the pathway for pain and temperature is discussed later. As illustrated in the Atlas, the substantia gelatinosa extends the entire length of the cord.

3. In cross sections of the cord from levels T1 descending through the upper lumbar segments there is a small lateral evagination of gray matter slightly below the horizontal level of the central canal. This is the *intermediolateral horn*, which indicates the horizontal stratum of cell bodies as a

nucleus of origin for the general visceral efferent sympathetic neurons as discussed in Chapter 9.

4. The *dorsal nucleus,* or *Clarke's column* (Plate 4), is dorsolateral to the central canal at the junction of the dorsal horn and the gray commissure and extends from T3 and sometimes T2 or T1 caudally to segment L3 in the cat (Grant and Rexed, 1958). Although the above names are more commonly used, McClure (1964) prefers the term *thoracic nucleus.* Thoracic nucleus (nucleus thoracicus) is also the name given the structure in *Nomina Anatomica Veterinaria* (1968). This nucleus is the origin of the dorsal spinocerebellar tract as discussed

below in the coverage of the white matter of the spinal cord.

5. In the dog and cat the *lateral cervical nucleus* (Fig. 10-6) is well developed (Kitai *et al.,* 1965; Truex *et al.,* 1970). This nucleus extends caudally from approximately the obex to the third cervical segment of the spinal cord as a lateral projection of the dorsal column at the ventrolateral border of the substantia gelatinosa (Fig. 10-6). The afferents are reported to be mainly collaterals of the dorsal spinocerebellar tract and partially of the ventral spinocerebellar tract. Some have considered the lateral cervical nucleus a relay nucleus activated primarily by impulses from tactile, pressure, joint and

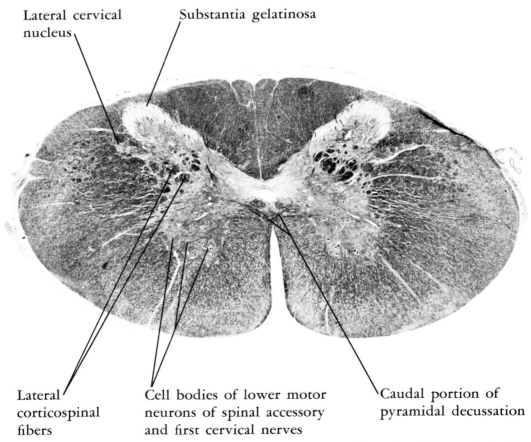

Lateral cervical nucleus

Substantia gelatinosa

Lateral corticospinal fibers

Cell bodies of lower motor neurons of spinal accessory and first cervical nerves

Caudal portion of pyramidal decussation

Fig. 10-6. Low-power photomicrograph of a transverse section at the cord–medulla junction in the dog, to show particularly the lateral cervical nucleus. A few additional structures are labeled for orientation. (Luxol fast blue–cresyl violet stain.)

mechanoreceptors. Rather than being a relay nucleus to the cerebellum (Rexed and Brodal, 1951; Brodal and Rexed, 1953), however, recent evidence suggests that efferents of the lateral cervical nucleus cross in the upper cervical region and ascend with those of the medial lemniscus (Busch, 1961, cited in Brodal, 1969).

White Matter (Fig. 10-7)

The organization of the white matter of the spinal cord has been discussed in general terms, and it has been established that the three funiculi on each side are subdivided into separate bundles of fibers, each of which traverses the spinal cord as an *ascending* (transmitting impulses toward the brain) or *descending* (conveying impulses away from the brain) fasciculus, or tract.

The axons within each tract tend to have the same origin, general course, and termination.

In a general sense the long ascending tracts referred to above are sensory. The long descending tracts contain the axons of the so-called 'upper motor neurons.' Brodal (1969) maintains that it is no longer permissible to refer to the descending pathways collectively as a 'motor' system. Recent research reveals that, among other functions, the descending tracts regulate the transmission of sensory impulses in the ascending tracts. This indicates the close interaction of the two systems.

In a study of the schematic cross section of the spinal cord presented in Figure 10-7, one may be misled into believing that the spinal cord tracts are very well circumscribed and functionally well documented.

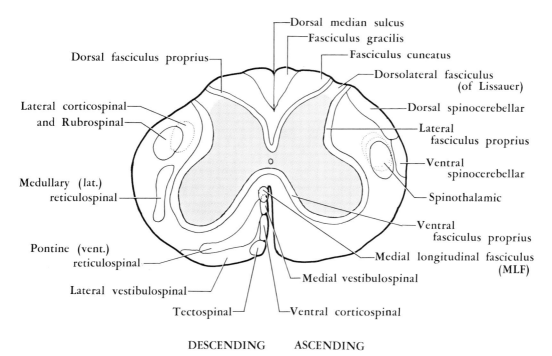

DESCENDING   ASCENDING

Fig. 10-7. Diagram of cross section of the spinal cord showing the relative positions of descending (left) and ascending (right) pathways. The fasciculus proprius contains both ascending and descending fibers. Dotted line of the lateral corticospinal tract indicates the intermixing of its fibers with those of the rubrospinal tract. The dotted lines of the spinothalamic tract indicate the diffuse area of this pathway. The medial vestibulospinal tract and the medial longitudinal fasciculus overlap.

Even in man the tracts of the spinal cord are not clearly delineated, but the relative positions and the specific functions are known sufficiently well to permit successful performance of specific spinal cord surgery in the clinical management of certain selected patients. Probably the best example of this spinal cord surgery is the bilateral anterolateral chordotomy (cutting of the ventrolateral funiculi of the spinal cord), performed to relieve intractable pain that has not responded to any other treatment. Actually the lateral spinothalamic tract, which transmits impulses for pain and temperature sense, is the specific tract that is severed in the above-cited surgical procedure, in the hope of providing relief, but surrounding pathways must also be sacrificed. In man the somatotopic organization of the lateral spinothalamic tract is documented to the extent that selective portions of the tract are sometimes cut to cause a loss of pain from a small body area, as from an arm, leg, or even a foot (Crosby *et al.*, 1962).

In man the lateral spinothalamic tract is more discrete than in animals, so that a unilateral anterolateral chordotomy will result in loss of pain and temperature sense in regions of the body wall and extremities supplied by fibers entering the cord approximately one segment below the cut on the opposite side of the body. Very briefly this is because the thinly myelinated peripheral fibers, with cell bodies in the dorsal root ganglion, enter the spinal cord and traverse within the dorsolateral fasciculus of Lissauer mainly upward for one segment to terminate in the dorsal horn gray matter and the substantia gelatinosa. From synapses in these areas the neurons of the second order cross in the ventral white commissure and ascend in the opposite lateral funiculus within the lateral spinothalamic tract.

The main purpose for presenting such seemingly unnecessary details is to emphasize that, although sophisticated knowledge of some of the human pathways may have been acquired, the reader is strongly cautioned to avoid the great temptation of blindly transferring detailed knowledge of the human central nervous system directly to animals, as well as the reverse (*i.e.,* of assuming that data obtained from animals are always applicable to man).

Before consideration of other individual spinal cord pathways, a few brief remarks pertaining to their phylogenetic evolvement as correlated with position and function may be helpful in understanding salient characteristics of tracts, particularly with regard to their position in the spinal cord (Fig. 10-7).

As one studies the comparative neuroanatomic characteristics of the spinal cord among animals he finds that there are differential growth patterns that occur. According to Nyberg-Hansen (1966) the reticulospinal fibers provide the only pathway by which the spinal cord receives impulses from higher centers in cyclostomes, which are thought to possess the first phylogenetic signs of cephalization.

Added to this, in passing to higher animals, are the vestibulospinal, the tectospinal, and finally the corticospinal system. Therefore, in lower animals, the ventral funiculus of the spinal cord is relatively much larger than the lateral funiculus which contains the phylogenetically younger rubrospinal and corticospinal fibers.

It is interesting to note that in man the sequence of myelination of the above-named tracts parallels the order in which they became differentiated phylogenetically, that is, the vestibulospinal fibers are among the first to become myelinated, whereas the corticospinal are among the last.

Functionally, the older supraspinal fibers, for example, the pontine reticulospinal and vestibulospinal tracts, which descend in the ventral funiculus, excite the extensor mechanisms and thereby contribute to the maintenance of postural tone. In contradistinction, the phylogenetically younger rubrospinal and corticospinal fiber systems, which descend in the lateral funiculus, chiefly excite the flexor mechanisms and tend to develop beyond that of the extensor systems, especially in man, where they permit the very specialized skilled voluntary movements of the fingers and arms.

Cortical dominance of the corticospinal system is evidenced anatomically in passing from lower to higher animals; in dogs the corticospinal tracts form about 10 per cent of the total white matter of the cord, in monkeys, about 20 per cent, and in man, about 30 per cent (Ariëns Kappers *et al.*, 1967).

Spinal Cord Pathways

The ascending fasciculi within the spinal cord are the first portions of the tracts that continue cephalad into the brain stem. Some terminate in brain-stem nuclei, others pass through the brain stem without synapses and continue to the cerebellum (spinocerebellar tracts), and a third type synapse within the brain stem for relay to the cerebral cortex. The descending spinal cord fasciculi are the terminal portions of various tracts from the brain which converge on and influence the lower motor neurons (Fig. 10-5).

A few of the many fundamental factors to consider in correlating the anatomical and physiological properties of tracts are: (1) There are many more collaterals given off from the main axons within a tract to various loci than most investigative methods reveal. (2) One must be particularly careful in attempting to properly evaluate spinal cord lesions because of the close proximity of so many pathways which may be poorly delineated one from another. Therefore lesions which appear to involve discrete fibers of a specific pathway may in fact include fibers from adjacent tracts. (3) Physiological data are sometimes erroneously interpreted because of insufficient knowledge of the interconnections of a specific nucleus or

tract which may profoundly influence its function and dysfunction. (4) Extreme caution should always be exercised when knowledge acquired on the basis of one species is sought to be applied to another species.

In the following discussion not all of the known tracts are included, but only the more important ones based on the current knowledge of neuroanatomy and neurophysiology, and especially those tracts that are well documented for the cat. As alluded to previously, the cat is used as the carnivore of choice because there have been so many reports for that animal in comparison with the extremely small number in recent years for the dog.

Dorsal Funiculus (Fig. 10-7)

For our purposes, the dorsal funiculus may be considered as purely ascending (sensory), in that two large ascending tracts (fasciculi) compose the whole funiculus.

Fasciculus Gracilis

The fasciculus gracilis is the medial tract which borders the dorsal median sulcus and extends the entire length of the cord. This tract is composed chiefly of axons from the cell bodies in the dorsal root ganglia of the nerves innervating the caudal trunk and hindlimbs. The somatotopic organization is such that the more caudal fibers that enter the cord are the most medial, and as fibers are added cephalically up to the midthoracic level they are added laterally, as shown in Figure 10-8.

Fasciculus Cuneatus

The fasciculus cuneatus is located lateral to the fasciculus gracilis, with which it shares many characteristics. Like the fasciculus gracilis, this tract is composed of axons from cell bodies in the dorsal root ganglia which enter the dorsal funiculus directly

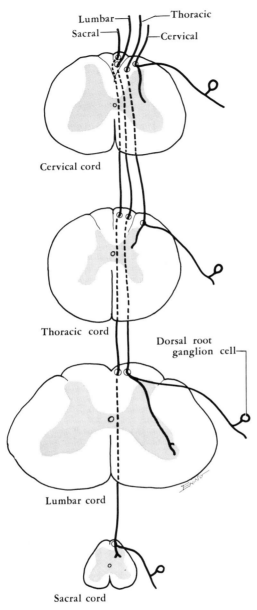

Lumbar — Thoracic
Sacral — Cervical

Cervical cord

Thoracic cord

Dorsal root
ganglion cell —

Lumbar cord

Sacral cord

Fig. 10-8. Diagram showing the organization of fibers in the dorsal funiculus in various regions of the spinal cord. Note that axons which enter from caudal levels are located medially (fasciculus gracilis), and as fibers enter cephalically they are added laterally (fasciculus cuneatus).

without synapses. Cuneate fibers, however, transmit impulses from the forelimb, upper trunk, and neck. The arrangement of fibers is similar to that in the fasciculus gracilis, in that the longer fibers from the mid-thoracic levels are medial to the shorter ones from the forelimb and cervical regions (Fig. 10-8).

GENERAL REMARKS ABOUT THE FASCICULI GRACILIS AND CUNEATUS

The fasciculi of the dorsal funiculus are among the newer acquisitions of the nervous system (Truex and Carpenter, 1969). As indicated previously, they receive impulses chiefly from the forelimbs (the fasciculus cuneatus) and hindlimbs (the fasciculus gracilis).

Both of the tracts convey impulses to the brain, and the first neuron terminates in its respective nucleus (nucleus gracilis or cuneatus) in the medulla oblongata (Plate 9). A brief account of this complete pathway (Fig. 10-9) should aid in understanding the two fasciculi as conveying impulses of conscious proprioception. From the nuclei in the medulla axons of the second order arch ventrolateromedially through the reticular formation as the *internal arcuate fibers* to cross the midline and enter the medial lemniscus on the opposite side (Plate 10). The fibers ascend within the medial lemniscus to the ventral posterolateral nucleus of the thalamus, to synapse with neurons of the third order. These neurons originating in the thalamus enter the internal capsule to ultimately reach the sensory cortex.

Based on the phylogenetic significance cited previously, it is not surprising that these pathways are functionally superior in man. In both dog and man it is understandable that they would convey impulses of conscious proprioception for the sensations of position and movement. Admittedly, however, it is doubtful if one can include the dog when considering the specialized functions known for man, for ex-

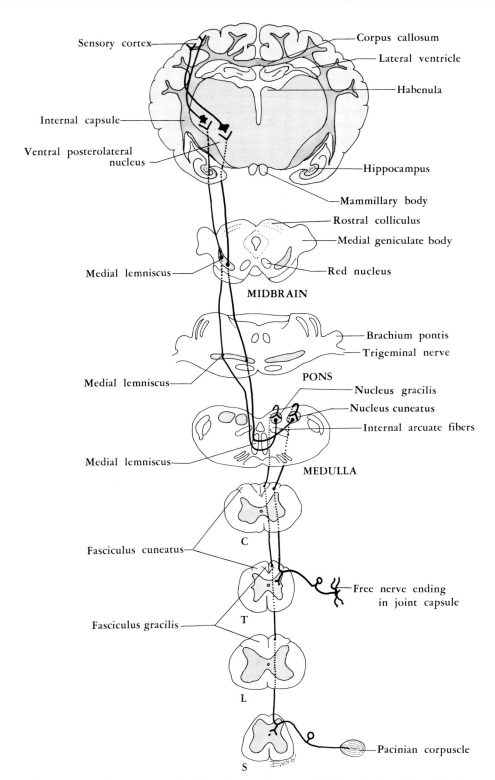

Fig. 10-9. Diagram of fibers in fasciculi gracilis and cuneatus in the spinal cord, medial lemniscus in medulla, pons, and midbrain, which synapse in the ventral posterolateral thalamic nucleus with neurons of the third order which project to the sensory cortex. Note the ipsilateral neurons I to the medulla. Neurons II from nuclei gracilis and cuneatus in the medulla pass as the internal arcuate fibers and cross to form the opposite medial lemniscus. A principal modality with which this pathway is concerned is conscious kinesthetic sense (*i.e.*, sense of position and movement). Letters indicate segmental levels of spinal cord (cervical, thoracic, lumbar, and sacral).

ample, discriminative (epicritic) touch, spatial discrimination (two-point tactile), exact tactile localization, and vibratory sensation. These modalities are very well documented as functions of the dorsal funiculus in man and are included in human neurologic examinations, but the obvious difficulties in attempting to test them in the dog, or other animals, need not be elaborated.

Dorsal Fasciculus Proprius

The dorsal fasciculus proprius is a portion of the fasciculi proprii, which form a continuous band of fibers bordering the gray matter in all six funiculi. The fasciculus proprius of each funiculus acquires the name designating its location, that is, dorsal, lateral, and ventral fasciculus proprius. The axons within each fasciculus proprius originate and terminate within the spinal cord to connect the spinal neurons of intrasegmental and intersegmental levels. This pathway is composed of ascending and descending fibers, and thereby differs from the typical spinal cord tract.

Lateral Funiculus (Fig. 10-7)

The lateral funiculus is more complicated than the dorsal in that within the lateral funiculus there are both ascending and descending fasciculi.

Ascending Tracts

Dorsolateral Fasciculus of Lissauer

The dorsolateral fasciculus of Lissauer (Plate 5) is peripheral to the substantia gelatinosa at the point of entrance of the dorsal root fibers at the dorsolateral sulcus. This tract is composed of thinly myelinated and nonmyelinated fibers of the lateral division of the dorsal root which principally convey impulses of pain and temperature sense. Most of the fibers within the dorso-lateral fasciculus terminate by synapses with the small cell bodies of the substantia gelatinosa (Pearson, 1952).

It is postulated that these small cells of the substantia gelatinosa serve chiefly as interneurons connecting the dorsolateral fasciculus with the dorsal horn gray matter as the origin of the lateral spinothalamic tract. As mentioned earlier in the example of the specificities of the lateral spinothalamic tract, the important fibers of the dorsolateral fasciculus ascend only one segment on the same side before either terminating within the substantia gelatinosa or passing through it to terminate directly in the dorsal horn gray matter.

Dorsal Spinocerebellar Tract

The dorsal spinocerebellar tract (Flechsig's fasciculus) is composed of heavily myelinated fibers which have their cell bodies within the dorsal nucleus of Clarke (Plate 4). From this nucleus the axons pass to the periphery of the same side of the cord ventral to the dorsal root entrance and lateral to the lateral corticospinal (pyramidal) tract. Caudally the tract first appears at segment L3 or L4, and it increases in size cephalically throughout the levels of Clarke's column, which extends most commonly in cat from L3 to T3 (Grant and Rexed, 1958). Since Clarke's column is not present caudal to L3 or L4, impulses destined for the cerebellum which enter the cord via the dorsal roots of the lower lumbar and sacral nerves traverse ascending branches (probably via fasciculus gracilis) several segments to terminate in Clarke's column when the latter is reached (Austin, 1961). The dorsal spinocerebellar tract ascends to enter the cerebellum via the restiform body.

According to Verhaart and van Beusekom (1958), the dorsal spinocerebellar tract contains an admixture of fibers from the pyramidal and other tracts. The same authors believe that not all of the dorsal spinocerebellar fibers reach the cerebellum, as some terminate either in the lateral cervical nucleus of the cord or in medullary nuclei.

The dorsal spinocerebellar tract also conveys proprioceptive impulses, but rather than the conscious type as carried by the dorsal funiculus which ultimately terminates in the cerebral cortex, the spinocerebellar tracts transmit impulses from muscle and tendon receptors to the cerebellum, which exerts its tonic and synergizing influence on the motor pathways for regulation of voluntary motor activity. It should be recalled that proprioceptive neurons are sensory (general somatic afferent), but clinically their dysfunction is tested by testing motor performance—there is no paralysis, but dyssynergia (ataxia) is the key clinical sign when they have been injured.

Ventral Spinocerebellar Tract

The ventral spinocerebellar tract is located at the periphery of the lateral funiculus ventral to the dorsal spinocerebellar tract and bordered on its medial side by the lateral spinothalamic tract. The spinal cord origin is diffuse, but many cell bodies are found in the gray matter in the base and neck of the dorsal horn in the lumbar segments in the cat (Hubbard and Oscarsson, 1962). There is physiological evidence that this tract transmits impulses which arise in the hindlimbs and caudal trunk only (Oscarsson, 1965). From the above-stated dorsal horn origins some axons cross to the opposite lateral funiculus, whereas others remain on the same side.

In its ascending course in the cat spinal cord this tract tends to shift dorsally from the lumbosacral through the thoracic and lower cervical segments, and then shifts ventrally again in the upper cervical segments (Grant, 1962). This tract ascends through the brain stem to the periphery of the rostral cerebellar peduncle (Plate 15), which it follows into the cerebellum.

Recall that the dorsal spinocerebellar tract is related only to the hindlimb and lower trunk, whereas the cuneocerebellar tract is related in a similar way to the forelimb and upper body regions. In comparison, the ventral spinocerebellar tract is related to only the hindlimb and caudal trunk regions, and a forelimb equivalent, the *rostral spinocerebellar tract,* has been identified by Oscarsson and Uddenberg (1964). Knowledge of the anatomic details of this tract is incomplete, but it is known that it terminates in the forelimb regions of the cerebellar cortex and appears to correspond closely to the ventral spinocerebellar tract.

Clinically, cerebellar influences are ipsilateral, and proprioceptive projections from cord levels into the cerebellum are chiefly ipsilateral. Those fibers which decussate in the cord cross again within the cerebellum so that they too conform to the ipsilateral dominance (see Chapter 14).

Lateral Spinothalamic Tract

The lateral spinothalamic tract lies mostly medial to the ventral spinocerebellar tract within the lateral funiculus. Because of the great clinical importance of this pathway by reason of the transmission of pain to higher centers, this tract was discussed above.

Kennard (1954) performed degeneration experiments with Marchi staining of sections from the spinal cord of the cat to determine the location of pathways for pain. She found a great difference between the cat and man in that the transmission of painful stimuli is much more diffuse in the spinal cord of the cat. She found that the response to painful stimuli cannot be altered in the cat by hemisection, bilateral ventral section, or unilateral dorsal section of the spinal cord. She concluded that pain is conveyed in the cat bilaterally and chiefly by way of the dorsolateral pathways (*i.e.,* the lateral spinothalamic tract is more dorsal in the cat).

Breazile and Kitchell (1968) proposed two ascending fiber systems in the pig which convey impulses originating from painful stimuli: one system which remains on one side but sends collaterals to the opposite side every three or four segments, and one system which crosses and recrosses the spinal cord every three or four segments.

The diffuse and bilateral representation of somatic pain fibers in the cord of domestic animals apparently is the cardinal difference when compared with man. It is well known that, even in man, the visceral pain fibers are represented bilaterally. The axons of the lateral spinothalamic tract synapse in the ventral posterolateral nucleus of the thalamus, from which final axons pass to the cerebral cortex via the internal capsule.

Lateral Fasciculus Proprius

The lateral fasciculus proprius is similar to the dorsal fasciculus proprius discussed previously. It borders the gray matter and contains ascending and descending short fibers.

Descending Tracts (Fig. 10-7)

The first two tracts discussed below are the most important motor pathways within the lateral funiculus. Their functional significance in domestic animals is not of the same magnitude as that in man.

Lateral Corticospinal Tract

The lateral corticospinal (crossed pyramidal) tract occupies the central area of the upper quadrant of the lateral funiculus, medial to the dorsal spinocerebellar tract, slightly dorsal to but largely intermingled with the rubrospinal tract, lateral to the lateral fasciculus proprius, and ventral to the dorsolateral fasciculus of Lissauer. As the name indicates, the lateral corticospinal tract originates in the cerebral cortex and terminates by synapse with the cell bodies of lower motor neurons (somatic efferent) in the ventral horn of the spinal cord gray matter. Direct synapses of corticospinal

fibers within the ventral horn region of the spinal cord of the cat have not been found either anatomically or physiologically (Buxton and Goodman, 1967).

The general anatomic plan of the corticospinal tract, as identified in the Atlas, is as follows: from the cerebral cortex the axons converge in the corona radiata and descend through the internal capsule (Plate 23), the basis pedunculi of the midbrain (Plate 20), and the basal portion of the pons (Plate 17), and into the medulla where the corticospinal tract becomes visible on the ventral surface as the pyramids (Plate 13).

In the medulla most of the fibers cross at the pyramidal decussation (Plate 9) and descend into the cord as the lateral corticospinal tract in the lateral funiculus on the opposite side of the cord. The minority of fibers, which do not cross at the medullary level, descend in the ventral funiculus of the cord as the ventral corticospinal tract which borders the ventral median fissure.

Verhaart and van Beusekom (1958) reported that in the cat the lateral pyramidal (corticospinal) tract is mixed with a great number of foreign fibers which increase in number in the caudal segments of the spinal cord. They found that the pyramidal fibers terminated in large numbers in the reticular formation at the cord–medulla junction. Only rarely could they identify a ventral pyramidal tract in the cat spinal cord.

There is much controversy in the literature about the size, extent, or even the existence of the ventral corticospinal tract in animals, and there is disagreement regarding the details of the tract in man. Since our main interest here is the spinal cord of the dog and cat, the following comments are based on the report of Nyberg-Hansen and Brodal (1963).

After producing extensive lesions of the sensorimotor cortex in cats, these authors found descending degenerating fibers in both lateral and both ventral funiculi of the cord—even to the lowest lumbar segments. The crossed lateral tracts contained the majority of degenerating fibers and the ventral tracts very few. The authors emphasized that the only safe procedure for determining the presence of degenerating fibers in the cord is the use of longitudinal sections. They admitted that only by this method could they recognize degenerating fibers in the lateral tract on the same side as the lesion and in both ventral tracts.

From both the functional and anatomic standpoints it seems unrealistic to isolate the lateral corticospinal tract and ignore the ventral corticospinal tract. Therefore, the two tracts generally are considered as belonging to the *corticospinal system*. The logic of this is further supported by the fact that, above the medullary decussation, the fibers which will ultimately become the two tracts are inseparable.

FUNCTIONAL CONSIDERATIONS OF THE CORTICOSPINAL SYSTEM

In contradistinction to the basic anatomic plan given above as being applicable to both carnivores and man, the physiologic properties and the clinical applications are significantly different for animals and man. As alluded to previously, the corticospinal system is phylogenetically a newer acquisition in the nervous system and therefore it is not surprising that its greatest physiological expression is found in the higher primates. As expected, the clinical deficits which result from lesions in the system produce a much greater incapacitation in man than in subprimates (*e.g.*, dog and cat).

Without considering the entire clinical syndrome following a unilateral brain lesion of the pyramidal tract (and extrapyramidal fibers) in man, it may be said that the cardinal symptoms are: a contralateral spastic paralysis with increased deep reflexes, loss of superficial reflexes, and presence of the positive sign of Babinski (a pathologic reflex). This is the typical upper motor neuron paralysis as seen in many hemiplegics who may have suffered a stroke.

In contrast to the paralysis and incapacitation described as resulting from cortico-

spinal lesions in man, Buxton and Goodman (1967) reported that following unilateral removal of motor cortex in dogs (aged 1 month, 3 months, and adult) there was no overtly abnormal deportment, that is, the animals appeared to walk, sit, and recline normally. Qualitative testing of these dogs revealed, however, that unilateral motor cortex destruction caused: (1) hypotonia, hypokinesia, and dysmetria on the opposite side; (2) greater functional deficits in the hindlimb than forelimb; and (3) a progressively increased severity of symptoms with increasing age of the experimental animal. To repeat, it should be noted that there was no paralysis in the dogs.

Rubrospinal Tract

The rubrospinal tract (von Monakow's bundle) (Fig. 10-10) arises from cells throughout the red nucleus located in the midbrain tegmentum. The fibers immediately cross the median raphe as the *ventral tegmental decussation of the midbrain* and descend in the lateral brain stem tegmentum and continue in the lateral funiculus of the spinal cord.

In this location the rubrospinal fibers are ventrolateral to and partially intermingled with the descending fibers of the lateral corticospinal tract, medial to the dorsal spinocerebellar tract. In the cat the red nucleus and the rubrospinal tract possess a somatotopic organization (Pompeiano and Brodal, 1957). This is correlated with the corticorubral somatotopic organization as illustrated in Figure 10-10.

In man the rubrospinal tract is so inconspicuous that its existence has been questioned by some (Austin, 1961), although, based primarily on the work of Stern (1938), it frequently is described as descending as far as the mid-thoracic segments of the cord. The relatively conspicuous tract in the cat has been traced throughout the entire cord as far caudally as the sacral segments (Nyberg-Hansen, 1966; Hinman and Carpenter, 1959).

FUNCTIONS OF THE RUBROSPINAL TRACT

Clark (1965) considers the rubrospinal tract the most important motor pathway in animals. It is probably true that this pathway is more important in the dog than in man, but a review of the pertinent literature does not reveal full agreement with Clark's view that voluntary motor paralysis in animals is a direct result of a lesion that severs only the rubrospinal tract, as he states.

It is generally agreed that the cardinal function of the rubrospinal tract is the con-

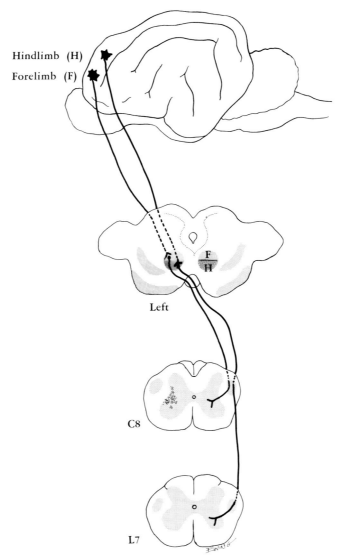

Fig. 10-10. Diagrammatic illustration of the hindlimb and forelimb somatotopic patterns of the corticorubral and rubrospinal fibers in the cat. Note that the corticorubral fibers remain on the same side, whereas those of the rubrospinal tracts cross (as the ventral tegmental decussation) in the midbrain. (Modified from Brodal, 1969.)

trol of muscle tone in the flexor muscles (Massion, 1967). In cats decerebrated precollicularly, stimulation of the dorsomedial parts of the red nucleus produced flexion of the contralateral forelimb with inhibition of extensor rigidity. Stimulation of the ventrolateral area of the red nucleus resulted in corresponding effects in the hindlimb. In the cat the terminations of the rubrospinal fibers in the gray matter of the spinal cord largely correspond to the sites of termination of the corticospinal fibers from the 'motor cortex' (Brodal, 1969).

There are conflicting reports concerning the relation of the rubrospinal tract to tonic and righting movements (*i.e.*, complex postural reflexes involved in changing body positions). Bilateral discrete lesions in the red nucleus result in hypokinesis. The voluntary motor system is not impaired, but animals show disinclination to move.

Medullary Reticulospinal Tract

The medullary (lateral) reticulospinal tract arises from the medullary reticular formation and descends in the ventromedial portion of the lateral funiculus as described below. The pontine reticulospinal fibers are within the ventral funiculus. Because of their positions, the medullary reticulospinal tract may be referred to as the lateral reticulospinal tract, and the pontine as the ventral reticulospinal tract.

Lateral Fasciculus Proprius

The lateral fasciculus proprius contains descending as well as ascending fibers, as discussed previously for the dorsal fasciculus proprius.

Ventral Funiculus (Fig. 10-7)

The ventral funiculus of the cord, which extends from the edge of the ventral median fissure laterally to the emergence of the ventral rootlets of the spinal nerves, primarily contains descending phylogenetically older tracts. With the exception of the ventral corticospinal tract, which is relatively insignificant in carnivores, the motor tracts in this funiculus descend from subcortical levels (*i.e.*, vestibular, tectal, and reticular nuclei) to exert a regulatory control over reflex activity, muscle tone, and posture, and automatic movement patterns characteristic of the species.

Ventral Corticospinal Tract

The ventral corticospinal tract has been considered previously in the discussion of the lateral corticospinal tract.

Lateral Vestibulospinal Tract

The lateral vestibulospinal tract within the ventral funiculus of the spinal cord in the cat contains descending fibers from the lateral vestibular nucleus (of Deiter) in the same side of the medulla (Nyberg-Hansen and Mascitti, 1964). After placing lesions in the lateral vestibular nucleus in the cat these authors were able to determine that the pathway extends the whole length of the cord, with the cervical and lumbar enlargements receiving more fibers than the gray matter of the thoracic segments.

Medial Vestibulospinal Tract

The medial vestibulospinal tract arises from the medial vestibular nucleus and descends on both sides, but mostly on the same side, within the medial area of the ventral funiculus in the area of the medial longitudinal fasciculus. Compared with the lateral vestibulospinal tract, the medial tract is minor and extends only to the midthoracic segments of the spinal cord. The nomenclature is inconsistent, but the above descriptions are based on the report of Nyberg-Hansen (1966) for the cat.

Functionally the vestibulospinal tracts,

possibly acting synergistically with the reticulospinal tracts, increase the extensor tonus of muscles on the same side and decrease the flexor and extensor tonus of muscles of the opposite side, according to the work cited by Nyberg-Hansen and Mascitti.

Tectospinal Tract

The tectospinal tract arises chiefly from the superior colliculus of the midbrain tectum. The fibers pass to the opposite side and descend in the periphery of the ventral funiculus bordering the ventral median fissure (Nyberg-Hansen, 1966). A majority of the fibers terminate in the upper four cervical segments by synapses with internuncial neurons to the lower motor neurons.

Reticulospinal Fiber System

The reticulospinal fiber system is complex and is described differently by different authors. There are even differences of opinion regarding how many reticulospinal tracts there are in the spinal cord. The following description is based on the work of Nyberg-Hansen (1966) and other recent reports.

The two reticulospinal tracts are thought to originate independently: one in the caudal pontine reticular area, and the other from the medulla in the nucleus reticularis gigantocellularis at the dorsomedial border of the inferior olive. The *pontine (ventral) reticulospinal fibers* descend almost exclusively ipsilaterally within the entire ventral funiculus caudally as far as the lumbosacral segments. The *medullary (lateral) reticulospinal fibers* (described previously) descend within the ventromedial portion of the lateral funiculus immediately lateral to the lateral fasciculus proprius. This tract is chiefly ipsilateral and extends the whole length of the spinal cord, but is very sparse as it approaches the sacral segments of the spinal cord.

Ventral Spinothalamic Tract

The ventral spinothalamic tract in animals is of no clinical significance. Its existence as a separate entity is questionable and therefore will not be considered. Verhaart (1953) reported that the ventrolateral funiculus of the cat spinal cord showed no circumscribed tracts in the low cervical segments and he was unable to locate any spinothalamic tract. Brodal (1969) believes it is of little importance to distinguish between a ventral and a lateral spinothalamic tract.

Medial Longitudinal Fasciculus

The medial longitudinal fasciculus (MLF) phylogenetically is a very old tract that extends from the level of the rostral midbrain caudally throughout the brain stem and continues into the ventral funiculus of the spinal cord. In the dog the MLF has been traced as far caudally as lumbar spinal levels; in the cat the majority of fibers end at the level of the second thoracic segment and only the vestibulospinal fibers extend to lumbar levels (Verhaart and van Beusekom, 1958).

The cells of origin have been determined in the dog, cat, and monkey as follows: nucleus of Darkschewitsch, interstitial nucleus of Cajal, and periaqueductal cells (all within the area of the diencephalon-midbrain junction); nuclei of cranial nerves III, IV, VI, VII, and XI; the descending nucleus of the trigeminal nerve (V); all four vestibular nuclei; the dorsal efferent nucleus of the vagus nerve (X); nucleus ambiguus; and reticular cells of the mesencephalon and pons (Palmer, 1965). The large vestibular input to the MLF is discussed in Chapters 12 and 19.

The MLF serves as a connector between vestibular and motor nuclei of cranial nerves III, IV, and VI to account for vestibular reflexes involving eye movements (*e.g.,* nystagmus resulting from abnormal vestibular activation). Nausea, vomiting, and vasomotor reactions, which commonly follow excessive labyrinthine stimulation, result from vestibular connections chiefly with the parasympathetic nucleus of the vagus nerve. In addition to MLF descending fibers such as tectospinal and reticulospinal, there are afferents to the (caudal, or inferior) olivary complex. Recall that the olive contributes a major input to the cerebellum via the olivocerebellar fibers to the contralateral restiform body. By means of this circuit the cerebellum receives a feedback from many areas of the brain stem.

Olivospinal Tract

Existence of the olivospinal tract mentioned in many texts is not substantiated by experimental studies (Brodal, 1969).

Ventral Fasciculus Proprius

The anatomic and physiologic properties of the ventral fasciculus proprius are similar to those of the dorsal and lateral fasciculi proprii described above.

Bibliography

Ariëns Kappers, C. U., Huber, G. C., and Crosby, E. C.: *The Comparative Anatomy of the Nervous System of Vertebrates, Including Man.* New York, Hafner Publishing Co., 1967.

Austin, G.: *The Spinal Cord. Basic Aspects and Surgical Considerations.* Springfield, Ill., Charles C Thomas, 1961.

Breazile, J. E., and Kitchell, R. L.: A study of fiber systems within the spinal cord of the domestic pig that subserve pain. J. Comp. Neurol. *133:* 373–382, 1968.

Brodal, A.: *Neurological Anatomy in Relation to Clinical Medicine,* 2nd ed. New York, Oxford University Press, 1969.

Brodal, A., and Rexed, B.: Spinal afferents to the lateral cervical nucleus in the cat. An experimental study. J. Comp. Neurol. *98:* 179–211, 1953.

Busch, H. F. M.: *An Anatomical Analysis of the White Matter in the Brain Stem of the Cat.* Assen, Van Gorcum & Co., 1961. Cited in Brodal, 1969.

Buxton, D. F., and Goodman, D. C.: Motor function and the corticospinal tracts in the dog and raccoon. J. Comp. Neurol. *129:* 341–360, 1967.

Clark, C. H.: Basic Concepts in Neurologic Diagnostics. Chapter 2 in *Canine Neurology—Diagnosis and Treatment* (Hoerlein, B. F.). Philadelphia, W. B. Saunders Co., 1965. Pp. 7–24.

Crosby, E. C., Humphrey, T., and Lauer, E. W.: *Correlative Anatomy of the Nervous System.* New York, Macmillan Co., 1962.

Fletcher, T. F., and Kitchell, R. L.: Anatomical studies on the spinal cord segments of the dog. Am. J. Vet. Res. 27: 1759–1767, 1966.

Grant, G.: Spinal course and somatotopically localized termination of the spinocerebellar tracts. An experimental study in the cat. Acta Physiol. Scand. 56: Suppl. 193, 1962.

Grant, G., and Rexed, B.: Dorsal spinal root afferents to Clarke's column. Brain *81:* 567–576, 1958.

Hinman, A., and Carpenter, M. B.: Efferent fiber projections of the red nucleus in the cat. J. Comp. Neurol. *113:* 61–82, 1959.

Hubbard, J. I., and Oscarsson, O.: Localization of the cell bodies of the ventral spinocerebellar tract in lumbar segments of the cat. J. Comp. Neurol. *118:* 199–204, 1962.

Kennard, M. A.: The course of ascending fibers in the spinal cord of the cat essential to the recognition of painful stimuli. J. Comp. Neurol. *100:* 511–524, 1954.

Kitai, S. T., Ha, H., and Morin, F.: Lateral cervical nucleus of the dog. Anatomical and microelectrode studies. Am. J. Physiol. *209:* 307–311, 1965.

McClure, R. C.: The Spinal Cord and Meninges. Chapter 9 in *Anatomy of the Dog* (Miller, M. E., Christensen, G. C., and Evans, H. E.). Philadelphia, W. B. Saunders Co., 1964. Pp. 533–543.

Massion, J.: The mammalian red nucleus. Physiol. Rev. 47: 383–436, 1967.

Nomina Anatomica Veterinaria, Adopted by the General Assembly of the World Association of Veterinary Anatomists, Paris, 1967. Vienna, Adolf Holzhausen's Successors, 1968.

Nyberg-Hansen, R.: *Functional Organization of Descending Supraspinal Fibre Systems to the Spinal Cord.* New York, Springer-Verlag, 1966.

Nyberg-Hansen, R., and Brodal, A.: Sites of termination of corticospinal fibers in the cat. An experimental study with silver impregnation methods. J. Comp. Neurol. *120:* 369–387, 1963.

Nyberg-Hansen, R., and Mascitti, T. A.: Sites and mode of termination of fibers of the vestibulospinal tract in the cat. An experimental study with silver impregnation methods. J. Comp. Neurol. *122:* 369–387, 1964.

Oscarsson, O.: Functional organization of the spino- and cuneocerebellar tracts. Physiol. Rev. *45:* 495–522, 1965.

Oscarsson, O., and Uddenberg, N.: Identification of a spinocerebellar tract activated from forelimb afferents in the cat. Acta Physiol. Scand. *62:* 125–136, 1964.

Palmer, A. C.: *Introduction to Animal Neurology.* Philadelphia, F. A. Davis Co., 1965.

Pearson, A. A.: Role of gelatinous substance of spinal cord in conduction of pain. Arch. Neurol. Psychiat. *68:* 515–529, 1952.

Pompeiano, O., and Brodal, A.: Experimental demonstration of a somatotopical origin of rubrospinal fibers in the cat. J. Comp. Neurol. *108:* 225–251, 1957.

Rexed, B.: The cytoarchitectonic organization of the spinal cord in the cat. J. Comp. Neurol. *96:* 415–495, 1952.

Rexed, B.: A cytoarchitectonic atlas of the spinal cord in the cat. J. Comp. Neurol. *100:* 297–379, 1954.

Rexed, B., and Brodal, A.: The nucleus cervicalis lateralis. A spino-cerebellar relay nucleus. J. Neurophysiol. *14:* 399–407, 1951.

Romanes, G. J.: The motor cell columns of the lumbo-sacral spinal cord of the cat. J. Comp. Neurol. *94:* 313–363, 1951.

Stern, K.: Note on the nucleus ruber magnocellularis and its efferent pathway in man. Brain *61:* 284–289, 1938.

Truex, R. C., and Carpenter, M. B.: *Human Neuroanatomy,* 6th ed. Baltimore, Williams & Wilkins Co., 1969.

Truex, R. C., Taylor, M. J., Smythe, M. Q., and Gildenberg, P. L.: The lateral cervical nucleus of cat, dog and man. J. Comp. Neurol. *139:* 93–104, 1970.

Verhaart, W. J. C.: The fiber structure of the cord in the cat. Acta Anat. *18:* 88–100, 1953.

Verhaart, W. J. C., and Beusekom, G. T. van: Fibre tracts in the cord in cat. Acta Psychiat. Neurol. Scand. *33:* 359–376, 1958.

chapter 11

Lower Medulla Oblongata

Introduction to the Brain Stem

The brain stem is the cephalic continuation of the spinal cord into the cranial cavity as a median core of the brain. An exact line of demarcation between spinal cord and brain stem is not evident, but the level of the foramen magnum is a convenient landmark for the separation. This level is between the attachments to the neuraxis of the rootlets of the hypoglossal and the first cervical nerve.

The caudal four divisions of the embryonic brain contribute to the *brain stem,* which is composed of the medulla oblongata, pons, midbrain, and diencephalon. Therefore, the brain stem (Fig. 11-1) can be exposed by removal of the two superstructures, the cerebral hemispheres and the cerebellum.

This Chapter and the remainder of the book focus on the brain stem as a great suprasegmental conveyor and coordinator for pathways and nuclei involved with vital regulatory and protective processes which affect the whole body. Control of such complex phenomena obviously is not solely the domain of the brain stem, but requires connections between the brain stem and the rest of the nervous system (both the central and the peripheral division).

The discussion of the spinal cord in Chapter 10 offers a basis for understanding

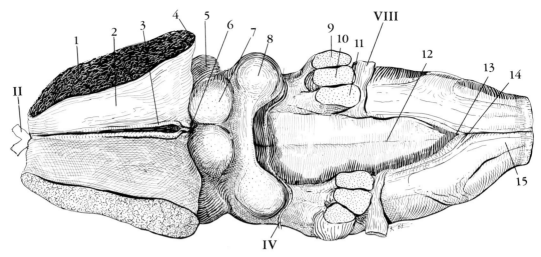

Fig. 11-1. Dorsal view of the brain stem of dog. 1, cut surface between cerebrum and brain stem; 2, thalamus; 3, stria medullaris of thalamus; 4, lateral geniculate body; 5, medial geniculate body; 6, pineal body; 7, rostral colliculus; 8, caudal colliculus; 9, middle cerebellar peduncle; 10, caudal cerebellar peduncle; 11, rostral cerebellar peduncle; 12, dorsal median sulcus in rhomboid fossa; 13, obex; 14, fasciculus gracilis; 15, fasciculus cuneatus; II, optic nerve; IV, trochlear nerve; VIII, vestibulocochlear nerve.

the increased complexities of the internal functional anatomic organization of the brain stem. Some of the basic comparative characteristics of spinal cord and brain stem are listed.

1. The spinal cord retains the tubular shape of the original neural tube more than does the brain stem. This results in the relatively uniform external and internal configuration of the spinal cord throughout most of its length. In the brain stem, added complexities are evident in structural changes which occur within short distances throughout its length.

2. All of the cranial nerves except the olfactory nerve are attached grossly to the brain stem in a general rostro-caudal succession from nerves II through XII (see Fig. 2-5, p. 20). The brain stem is an integration center for cranial nerves as is the spinal cord for spinal nerves, but the brain stem is much more complex.

3. The four functional component columns of the typical spinal cord gray matter (two sensory in the alar and two motor in the basal plate) are not continuous as solid columns within the brain stem. Instead, the four columns are spatially interrupted to produce smaller, isolated nuclei. In addition, three *special functional components* (special somatic afferent, special visceral efferent, and special visceral afferent) are added to some cranial nerves, as schematically illustrated in Figures 11-2 and 11-3.

The dorsal opening of the spinal cord central canal and the lateral migration of the dorsal sensory columns result in the general organization of the medullary gray matter in such a way that the sensory nuclei are located lateral to the sulcus limitans and the motor nuclei are medial (Fig. 11-3).

Anatomic Changes in Transition from Spinal Cord to Medulla

The caudal end of the medulla, internally, retains the spinal cord configuration of gray and white matter (Fig. 11-4). Rostrally, at the level of the caudal (inferior) olive, this resemblance is lost. Rather than the well-

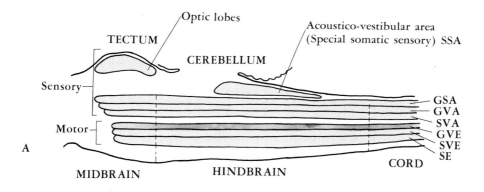

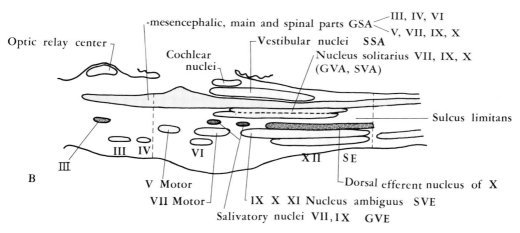

Fig. 11-2. Schematic diagrams of longitudinal views illustrating the pattern of sensory and motor nuclei. A. A hypothetical primitive (submammalian) stage in which spinal cord columns are continuous into the brain stem. B. The mammalian pattern in which the original columns are broken into individual nuclei. Note that the relative positions of nuclei having the same functional component remain in alignment according to the primitive pattern. As discussed in detail in the text, some nuclei represent two functional components, and certain nuclei serve more than one cranial nerve. (Modified from Romer, 1970.)

delineated central butterfly-shaped disposition of the gray matter, as in the spinal cord, the medulla has a diffuse area of intermixed cells and fibers, the reticular formation, which appears as the medium in which the specific nuclei and tracts are embedded.

The central canal of the spinal cord extends rostrally into the area of the medulla, where it rises to the dorsum and expands to form the fourth ventricle. In the floor of the fourth ventricle (rhomboid fossa) the sulcus limitans is present as a longitudinal

separation of the lateral sensory nuclei from the medial motor nuclei, as illustrated in Figure 11-5. In general terms, the tracts of the dorsal funiculi of the spinal cord continue into the dorsal area of the caudal medulla. Rostral to the obex these tracts (gracilis and cuneatus) and their nuclei are pushed laterally as the fourth ventricle widens (see Atlas).

It is advisable for the student to learn the changes in relative positions of specific structures at various levels of the neuraxis.

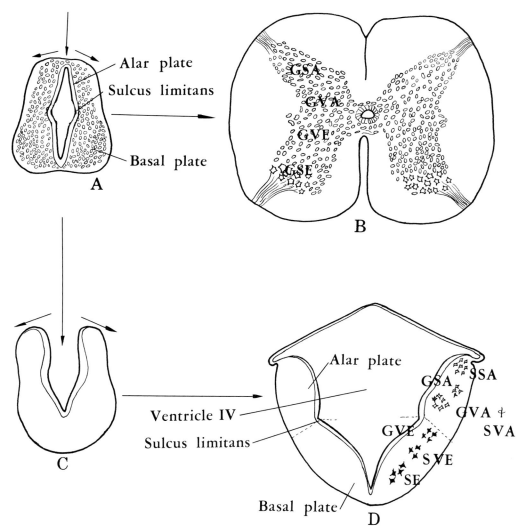

Fig. 11-3. Diagrammatic illustration of the differentiation of the alar and basal plates of the mantle layer of the neural tube (A). The four functional components of a typical spinal nerve are specifically represented in the spinal cord gray matter (B). Development from the neural tube to medulla (D) is more complicated and passes through an intermediate stage (C). The mid-dorsal area of the neural tube splits and the dorsal walls migrate laterally so that the somatic afferent components (which were located dorsally as in B) become located dorsolaterally in the medulla. The original neural canal (in A) becomes the fourth ventricle in the rhombencephalon and the sulcus limitans remains visible in the adult and continues to separate sensory (alar plate) from motor (basal plate) nuclei (D).

In study of serial sections of the brain stem, such as those presented in the Atlas, one should be able to correlate the gross and microscopic structure of the brain stem with function. One should be aware of specific levels where structures appear for the first time, or change position, and be able to recognize their functional significance. The remainder of the book is devoted to helping the reader accomplish these goals.

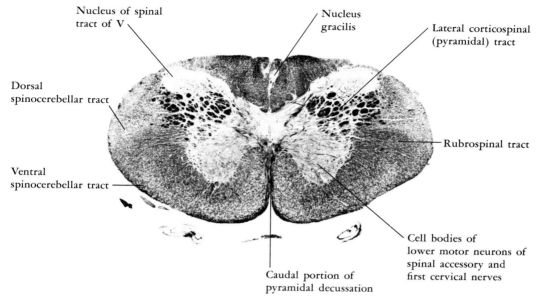

Nucleus of spinal tract of V

Nucleus gracilis

Lateral corticospinal (pyramidal) tract

Dorsal spinocerebellar tract

Rubrospinal tract

Ventral spinocerebellar tract

Cell bodies of lower motor neurons of spinal accessory and first cervical nerves

Caudal portion of pyramidal decussation

Fig. 11-4. A photomicrograph of a transverse section at the level of the spinal cord–medulla junction in the dog. Note characteristics of spinal cord which are continued into medulla: (1) "butter-fly-shaped" gray matter is present with dorsal and ventral horns indicated; (2) the dorsal funiculus of the spinal cord is retained as the fasciculi gracilis and cuneatus (a small portion of the nucleus gracilis is visible in the figure); (3) the substantia gelatinosa of the spinal cord is directly continuous with the nucleus of the spinal tract of the trigeminal nerve; (4) the fasciculi (tracts) of the spinal cord appear in approximately the same relative positions in the medulla.

Internal Organization of Lower Medulla

The separation of the lower medulla (as discussed in this Chapter) from the upper medulla and pons (discussed in Chapter 12) is somewhat arbitrary. Considering the medullary attachments of cranial nerves, the lower medulla is defined as the area of the medulla caudal to the attachment of cranial nerve VIII. This somewhat indefinite separation is merely for convenience, and is not meant to be a significant feature. In addition to cranial nerves IX through XII, other structures of importance will be considered in this Chapter as they contribute to the internal anatomy of this segment of the medulla.

The Medullary Nuclei of the Cranial Nerves

A more comprehensive consideration of all cranial nerve nuclei and their inner con-

nections, plus their peripheral distribution, is presented in Chapter 17. For the reader who has forgotten the names and numbers of the cranial nerves, they are listed with a mnemonic on page 266. The nuclei and pathways concerned with cranial nerves are emphasized here because they involve a goodly portion of the internal anatomy of the brain stem.

Although this Chapter is concerned with the details of cranial nerves IX through XII, in the following discussion the cranial nerves which attach to the upper medulla (VI through VIII) and cranial nerve V, of the pons, will be mentioned because some of the nuclei are shared by many nerves.

Figure 11-2 illustrates the relationships of the functional components of the cranial and spinal nerves. In addition to the maximum four general functional components of the spinal nerves, three special functional components are distributed among the cranial nerves, namely, special somatic afferent (SSA), special visceral afferent (SVA), and special visceral efferent (SVE). The somatic efferent column does not divide into general and special components; therefore, in this text the somatic efferent nuclei are designated as SE (without stating general or special).

Table 11-1 summarizes the relationships of the functional components and names of the nuclei functionally involved with the caudal eight cranial nerves. Of the seven functional components for cranial nerves, the general components (and SE) are defined the same as for spinal nerves, that is, the neuron processes of the designated component innervate the same type of target for both cranial and spinal nerves. The medullary nuclei of origin that give rise to the somatic efferent (SE) component are associated with cranial nerves XII (hypoglossal) and VI (abducens).

The autonomic parasympathetic nuclei (GVE) in the medulla give rise to axons for cranial nerves X (vagus), IX (glossopharyngeal), and VII (facial). The GVA (visceral sensory) neurons, like their counterparts within spinal nerves, tend to accompany the GVE fibers. The GSA nucleus of termination in the medulla and pons receives axons from cranial nerves V, VII, IX, and X.

Within the medulla the special somatic afferent (SSA) neurons compose the entire vestibulocochlear nerve (VIII) (*i.e.,* both the vestibular and the cochlear part). The special visceral afferent (SVA) neurons convey taste within nerves VII, IX, and X. Special visceral efferent (SVE) neurons serve motor innervation to the facial muscles via nerve VII, and muscles of the pharynx and larynx via nerves IX, X, and XI. All of the muscles innervated by SVE are derived embryonically from the mesoderm of the third and fourth branchial arches. Therefore these

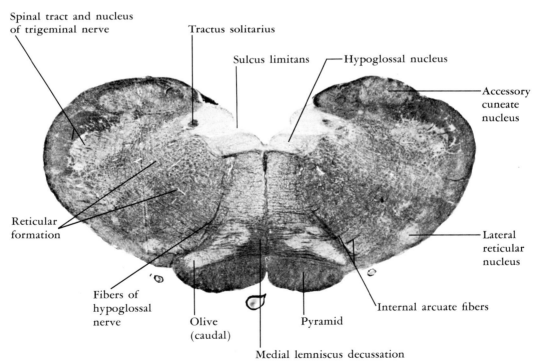

Spinal tract and nucleus of trigeminal nerve

Tractus solitarius

Sulcus limitans

Hypoglossal nucleus

Accessory cuneate nucleus

Reticular formation

Lateral reticular nucleus

Fibers of hypoglossal nerve

Olive (caudal)

Pyramid

Internal arcuate fibers

Medial lemniscus decussation

Fig. 11-5. A photomicrograph of a transverse section through the medulla of dog, showing some of the internal structures at the level of the hypoglossal nucleus.

muscles are referred to as branchiomeric muscles.

Figure 11-2 illustrates the isolated nuclei and their functional components which have been referred to previously. In addition, specific nuclei of cranial nerves that possess axons originating or terminating in the caudal three segments of the brain stem are labeled.

Special attention is directed to the facts that: (*a*) some nuclei are shared by portions of more than one cranial nerve, for example, *solitarius* (GVA and SVA of VII, IX, X), *ambiguus* (SVE of IX, X, XI), and *nucleus of the spinal tract of the trigeminal nerve* (GSA exteroceptive of V, VII, IX, X); (*b*) one nucleus shares more than one of the seven functional components, that is, *solitarius* (GVA and SVA); and (*c*) some nuclei have longitudinal fibers extending within or next to the nucleus as commonly indicated in the

name, for example, *fasciculus solitarius and nucleus, spinal trigeminal tract and nucleus.*

Figure 11-5 is a photograph of a transverse section through the caudal medulla at the level of the hypoglossal nucleus. This shows some of the nuclei discussed above. Since not all structures may be seen in any one section, the reader is referred to plates in the Atlas for serial study. Chapter 17 includes many more specific details of each individual nerve, and Part II summarizes many of the tracts discussed previously.

The medulla contains structures that are not directly associated with cranial nerves. Ascending (sensory) pathways from the spinal cord to higher centers pass through the medulla, and descending (motor) tracts from higher centers to the spinal cord can be identified. Discussions of some of the more important structures in the lower medulla follow.

TABLE 11-1. Functional Components and Nuclei of Cranial Nerves* Attached to the Pons and Medulla.

SE	SVE	GVE	GVA SVA	GSA	SSA
Abducens nucleus	Motor nucl. trigeminal	Parasym. nucl. of VII (Rost. salivatory)	Solitarius nucleus	Nucl. spinal tract of V	Vestibular nuclei (4)
VI	V	VII	VII, IX, X	V, VII, IX, X	VIII (vest.)
Hypoglossal nucleus	Facial nucl.	Parasym. nucl. of IX (Caudal salivatory)			Cochlear nuclei (2)
XII	VII	IX			VIII (coch.)
	Ambiguus nucleus	Parasym. nucl. of X (Dorsal eff. nucl. X)			
	IX, X, XI	X			

*V, trigeminal nerve; VI, abducens nerve; VII, facial nerve; VIII, vestibulocochlear nerve; IX, glossopharyngeal nerve; X, vagus nerve; XI, spinal accessory nerve; XII, hypoglossal nerve.

Corticospinal Decussation

The corticospinal (cerebrospinal, pyramidal) decussation is the most conspicuous structure in a transverse section of the brain stem in passing from spinal cord to medulla (Plates 7, 8, and 9). In sections of the medulla rostral to the decussation, the pyramidal tract borders the most ventral portion of each side of the ventral median fissure (Plates 11 to 15). At the level of the decussation most of the descending pyramidal fibers cross the midline ventral to the central gray matter and extend dorsolaterally to assume the position of the lateral or crossed pyramidal tract as it reaches the lateral funiculus of the spinal cord.

Chapter 10 contains a more comprehensive discussion of this tract, including some of its phylogenetic significance and particularly a comparison of the anatomic and physiologic features of this tract in animals and man.

Fasciculi and Nuclei Gracilis and Cuneatus

The main fasciculi of the dorsal funiculus of the spinal cord retain the same relative positions in the caudal medulla (Plates 8 and 9). These nuclei are the terminations of the long ascending axons from the dorsal root ganglia of the spinal nerves of the same side. The transverse somatotopic arrangement of the nuclei and fasciculi is identical to that of the fasciculi of the spinal cord as discussed in Chapter 10, that is, the nucleus and fasciculus gracilis contain the axons of the ipsilateral hindlimb and caudal body regions and are medial to the respective cuneate components which contain the axons of the forelimb and upper body regions.

In addition, there is an *accessory* (*lateral*) *cuneate nucleus* which lies dorsolateral to the nucleus cuneatus (Plate 10). The accessory cuneate nucleus is thought to receive axons primarily from the cervical spinal nerves and gives rise to the *dorsal external arcuate* fibers which enter the restiform body to convey proprioceptive impulses to the cerebellum. For this reason and others, including the histologic similarities of the two nuclei, the accessory cuneate nucleus is considered to be the medullary equivalent of the spinal cord dorsal nucleus of Clarke, which gives origin to the ipsilateral dorsal spinocerebellar tract.

Recall that this tract also ascends to convey proprioceptive impulses to the cerebellum via the inferior cerebellar peduncle. The external arcuate fibers and the accessory cuneate nucleus may be identified grossly as they are illustrated and discussed in Chapter 2.

Axons of secondary neurons that arise in the nuclei gracilis and cuneatus curve ventromedially through the reticular formation as the *internal arcuate* fibers (Plate 10) to cross the median raphe as the *decussation of the medial lemniscus* (Plates 10 and 11) dorsal to the pyramids and bend rostrally to form an ascending fasciculus known as the *medial lemniscus* (Plates 10 and 18).

This fasciculus may be identified throughout the medulla as an ascending bundle which borders the median raphe dorsal to the pyramids and ventral to the tectospinal tract. The medial lemniscus is associated with other ascending fibers as it passes through the brain stem to terminate in the ventral posterolateral nucleus of the thalamus.

The main function of the nuclei gracilis and cuneatus and their efferents into the medial lemniscus is conscious proprioception or kinesthetic sense. As an example, when this pathway is damaged in a dog, either at spinal cord or brain levels, the animal may stand and hold one leg in an abnormal position (*e.g.,* flexed or out to the side) because he is unaware that the limb is in this abnormal position. In addition, the dog may be expected to show incoordination (Clark, 1965). As mentioned earlier, discriminatory tactile sense and its localization, which are modalities conveyed by this pathway in man, are difficult to evaluate with much validity in a routine neurologic examination in animals.

Caudal Cerebellar Peduncle (Fig. 11-1; Plates 12, 13)

Grossly the caudal (inferior) cerebellar peduncle (restiform body) is between the rostral peduncle (brachium conjunctivum) on its medial side and the middle peduncle (brachium pontis) on its lateral side (Fig. 11-1). Functionally the caudal cerebellar peduncle contains mostly afferent axons to the cerebellum from the medulla and spinal cord.

The major afferents in this peduncle are: (1) *olivocerebellar fibers* from the contralateral inferior olivary complex in the medulla, (2) *dorsal spinocerebellar fibers* from the lateral funiculus of the spinal cord, (3) *reticulocerebellar axons* from the lateral reticular nucleus in the medulla which form the ventral arcuate fibers, (4) *cuneocerebellar axons* from the lateral (accessory) cuneate nucleus of the same side, which form the dorsal external arcuate fibers, and (5) *vestibulocerebellar fibers* from the vestibular nuclei.

In addition, there are indefinite axons from smaller nuclei of the reticular formation, some of which enter the ipsilateral peduncle, whereas others pass to the opposite side. The other two peduncles are discussed in Chapter 12.

Spinal Trigeminal Tract and Its Nucleus (Descending Root of Trigeminal Nerve and Its Nucleus)

The spinal cord general somatic afferent (GSA) exteroceptive representation as the substantia gelatinosa (Rolandi) and its closely associated dorsolateral fasciculus of Lissauer are continuous in the medulla with the nucleus and spinal (descending) tract of the trigeminal nerve, respectively (Fig. 11-5). In the brain stem the functional com-

ponent remains the same, but the spinal trigeminal tract contains descending fibers in contradistinction to the ascending fibers which constitute most of the dorsolateral fasciculus of the spinal cord. As illustrated in Figure 11-2, the spinal trigeminal tract and its nucleus receive primarily the GSA exteroceptive components of the fifth cranial nerve, but they also receive contributions from cranial nerves VII, IX, and X.

The Caudal Olivary Nuclear Complex

Rostral to the decussation of the pyramidal tracts the caudal (inferior) olivary nuclear complex appears on the dorsolateral border of the pyramids and mostly medial to the descending root fibers of the hypoglossal nerve (Fig. 11-5).

The complex is composed of three nuclei. In serial transverse sections of the medulla from caudal to rostral the *medial accessory olivary nucleus* is the first to appear as a gray band of cells at the dorsolateral border of the pyramids. Almost immediately the *dorsal accessory olivary nucleus* appears dorsolateral to the medial accessory nucleus, and lastly the *principal inferior olivary nucleus* appears between the other two nuclei. The relationships of the nuclei within the complex are similar to those described for the cat by Taber (1961). The arrangement of the three nuclei within the inferior olivary complex as described above for the dog and cat appears similar to that in the pig, sheep, and goat, but somewhat different from that in the cow and horse as shown in the atlas of brains of domestic animals by Yoshikawa (1968).

Connections of the Caudal Olivary Complex

The descending *afferents* to the caudal olivary complex from supra-olivary structures are abundant. In the cat there are descending connections from the cerebral cortex, caudate nucleus, globus pallidus, red nucleus, and periaqueductal gray matter

(Walberg, 1956). The same author subsequently reported the reticular formation as an origin of descending fibers to the inferior olive (Walberg, 1960). The majority of the efferent fibers that originate in the inferior olive are the olivocerebellar fibers (Plate 13) which pass from the olive through the ventrolateral reticular formation to the opposite side of the cerebellum via the caudal cerebellar peduncle.

In man the inferior olivary complex is much larger, with more predominant fiber connections. In the human brain the inferior olive is grossly obvious as a nodular swelling on the ventromedial side of the medulla, bordered ventrally and dorsally by the well-defined pre-olivary and post-olivary sulci on the ventral and dorsal olivary borders, respectively. These gross structures are not outstanding landmarks in the brain stems of animals.

The Reticular Formation (Substance)

The reticular formation is defined differently by various specialists. It is common to consider the reticular formation as anatomically diffuse, ill defined with no sharp borders, consisting of a variety of cell types and intrinsic nuclei engulfed in a diffuse fiber network. According to such anatomic investigations as those of Brodal (1958), however, the reticular formation is not diffusely organized, anatomically or physiologically. It can be subdivided into specific areas which differ according to the cytoarchitecture, fiber connections, and intrinsic organization. Brodal (1958) presents a very comprehensive coverage of the reticular formation, based primarily on the author's investigations of the brain stem of a cat. The anatomic structure of the reticular formation is so consistent that, according to Olszewski (1954), if one knows its subdivisions in any one of the four mammals, rabbit, cat, monkey, or man, he will have very little diffi-

culty in recognizing the corresponding cell groups in any of the other three. Brodal (1958) points out that this may be true, but there are species differences with regard to the degree of development of individual cell groups, and some cell groups may be present in one species but absent in others.

In the discussion below of the structure and functions of the reticular formation, the entire reticular substance is considered, in the belief that this is more meaningful than relating the information to the separate divisions of the brain. Therefore, the following discussion is not restricted to merely the medullary reticular formation.

Structure and Connections of Reticular Formation

The general meshwork appearance of the reticular formation has been mentioned previously. The longitudinal extent is described differently by various authors, with the rostral limit generally given as the thalamus; according to Herrick (1956), however, the substance extends from the corpus striatum throughout the brain stem and spinal cord. In his monograph on "The Reticular Formation of the Brain Stem," Brodal (1958) did not include the diencephalon in his definition of the brain stem and commented on the uncertainty of whether the so-called reticular nuclei of the diencephalon are homologous to the reticular formation of the caudal three divisions of the brain stem. The reticular formation of the spinal cord is most prominent in the upper cervical levels, where it may easily be confused with the fasciculus proprius exterior to the base of the dorsal horn.

Phylogenetically, the reticular formation is very old, and in lower animals it contributes a very high percentage of the brain substance. In higher animals a large portion of the brain stem is reticular formation, which serves as a matrix in which the nuclei and tracts are embedded. A retention of the primitive neuronal linkage is evident in higher forms by the predominance of short-axon multisynaptic intrinsic reticular connections. The complexity of the formation is indicated by the mixture of gray and white matter, ascending, descending, and transverse fibers, and sensory and motor neurons with axons passing ipsilaterally and contralaterally. In addition, a single reticular axon gives off collateral fibers, some of which may synapse with motor and some with sensory neurons.

The aggregations of reticular neuron cell bodies are of two types: (a) some of the cell bodies (gray matter) are scattered throughout a localized area in a diffuse manner, with no sharp anatomic borders, whereas (b) other reticular cell bodies coalesce into concentrated, localized, well-defined nuclei. The nomenclature, topography, and cytoarchitecture of various reticular nuclei in a number of species are presented in most of the previously cited references and by Taber *et al.* (1960), O'Leary and Coben (1958), and Valverde (1961). A distinction should be made between the specific nuclei *of* the reticular formation and other important nuclei that are merely located *in* the reticular formation (*e.g.,* the cranial nerve nuclei and the red nucleus).

MAJOR AFFERENT FIBERS TO RETICULAR FORMATION

Fibers come to the reticular formation from three main areas: the cerebellum, spinal cord, and higher levels of the brain. It is generally accepted that the *lateral* portions of the reticular formation are the *afferent* areas in terms of receiving fibers from extrareticular sources, whereas the *medial* reticular areas give origin to the *efferent* fibers that transmit impulses to higher and lower levels of the central nervous system.

The spinoreticular fibers enter the reticular formation to terminate in the lateral reticular nucleus and areas rostral to the trapezoid body. The details of the spinoreticular spinal cord tract are controversial, but according to Brodal (1958) spinoreticular fibers originate from as far caudal as the lumbosacral levels. In addition to the direct ascending reticular fibers, the ascending tracts which pass through the reticular formation send collaterals that synapse with the reticular nuclei. A few examples of such pathways are: spinothalamic, auditory, vestibular, and trigeminal.

Corticoreticular fibers descend to the reticular formation from wide cortical areas, but especially from the motor area. Many authors believe that the corticoreticular fibers follow the pyramidal tract, which they leave in the medulla. Other areas that are thought to send fibers to the reticular formation are the tectum of the midbrain and certain areas of the hypothalamus.

MAJOR EFFERENT FIBERS FROM RETICULAR FORMATION

The same three main areas considered above as the source for the afferent fibers may be listed as the projectile areas for the efferent axons from the reticular formation, that is, the cerebellum, spinal cord, and higher levels of the brain. The reticulocerebellar fibers are well documented with regard to the lateral reticular nucleus, as discussed earlier in this Chapter. The details of the reticulospinal tracts are discussed in Chapter 10. Brodal (1958) estimated that approximately one-half of all the cell bodies within the medullary reticular formation send their axons to the spinal cord. The major higher levels of the central nervous system that receive reticular fibers are: thalamus, hypothalamus, and caudate and lenticular nuclei. For a detailed report of the long ascending reticular fibers, the reader is referred to Brodal and Rossi (1955).

Functions of Reticular Formation

Many of the functions of the reticular formation are based on the interaction of it and the cerebral cortex. The evaluation of reticular-formation activity and overall influence on bodily functions is complicated by the phenomenon of encephalization. The range of this is critical in that research data obtained from one species of mammal cannot be applied to other mammals when the cerebrum is involved directly or indirectly. Even among mammals, the higher the animal phylogenetically, the more dependent it is on cerebral control and hence the greater the incapacitation that results from cerebral deficiency. This is often due to the loss of reticular-formation functions which are still present in lower mammals.

The cerebral cortex does not function autonomously but is greatly dependent on the reticular formation even in the highest mammals, including man. Based on the anatomic characteristics discussed previously, the great ascending (sensory) pathways which pass through the reticular formation are influenced by it. The reticular formation is non-specific in that it responds to all sensory systems which give collaterals to it. The non-specific ascending reticular fibers in turn activate the cerebral cortex, again non-specifically and not localized, by alerting the whole cortex to be attentive and receptive to the impulses of specific modalities mediated by the long sensory pathways. In this context, the reticular formation has been likened to an alarm for the cerebral cortex. Perhaps the analogy of a 'general alarm' would be better, in that the whole cortex is 'awakened.'

Because of this phenomenon, the title *reticular activating system* (RAS) has been coined. The clinical significance of this is that the RAS apparently determines the state of wakefulness, attentiveness, alertness, and arousal of the animal or human. The cat has been used experimentally to obtain a great portion of the physiologic data. Somnolence, stupor, and coma reflect different degrees of cerebral cortical suppression. A destructive lesion of the reticular formation may result in a state of coma from which an animal, or human, may never recover. In addition to the ascending reticular activating system discussed above, the fact that muscle-spindle activity may be altered by stimulation of the reticular formation suggests a descending reticular influence on peripherally initiated sensory impulses.

The medullary reticular formation contains many scattered cell bodies which physiologically function synergistically in a rhythmic coordinated manner for control of vital reflexes (*e.g.*, regulation of respiration and heart beat). In cat, dog, and other mammals diffuse 'centers' for respiration are located in the medullary reticular formation. Some researchers claim to have localized distinct expiratory and inspiratory medullary control areas in many animals, whereas others believe that there is no definite segregation of these areas in cat, dog, or monkey (Carregal *et al.*, 1967).

In addition to the vital rhythmic reflexes, the medullary reticular formation contains 'centers' for protective reflexes, such as coughing and vomiting.

The reticular formation also influences motor systems and reflexes of various forms. The primary influence of the formation is that it synchronizes and regulates the timing of rhythmic activities. Different areas of the formation function synergistically but with opposite effects. One area has an inhibitory effect on cell bodies of lower motor neurons and axons innervating extensor muscles, but at the same time the flexor muscles are excited. Another area exerts the reverse effect on the same muscle antagonists. The medullary reticular formation strongly influences the spinal righting reflex; for example, a low decerebrate cat may still right itself when placed on its side or dropped from the upside-down position. Normally such reflexes are thought to be primarily under vestibular control, as discussed in Chapter 19.

DECEREBRATE RIGIDITY

The physiologic synergism of the rhombencephalic reticular formation and the vestibulospinal tract may be demonstrated experimentally in an animal such as the cat by transection of the mesencephalic brain stem between the rostral and the caudal colliculi. The resulting tremendously exaggerated increase in tone of the antigravity (postural and extensor) muscles of the trunk and limbs characterizes a condition known as decerebrate rigidity. Briefly, a commonly accepted explanation is based on the removal of the inhibitory areas of the reticular formation which are dependent on higher centers (*e.g.*, cerebral cortex, cerebellum, and basal ganglia). The facilitatory reticular regions are left intact with their input from the ascending pathways, and the descending vestibular system is untouched and able to function normally.

Combined with these facilitatory influences primarily on the lower motor neurons of extensor reflexes, the gamma efferent neurons in the spinal cord are made hyperactive, which increases the rate of firing of the muscle spindles, which in turn increases the impulse rate of the skeletal muscles, causing a constant state of contraction in the increased tonic state. A dramatic demonstration of the above principles is based on the fact that decerebrate rigidity can be abolished by specific lesions of the vestibular nuclei and the ventral funiculus of the spinal cord. For a more comprehensive discussion of this and related postural reflexes see Chapter 9 in Ruch *et al.* (1965). Reticulospinal functions are very difficult to evaluate and are undoubtedly much more complicated than we now understand them to be.

Bibliography

Brodal, A.: *The Reticular Formation of the Brain Stem, Anatomical Aspects and Functional Correlations.* Edinburgh, Oliver and Boyd, 1958.

Brodal, A., and Rossi, G. F.: Ascending fibers in brain stem reticular formation of cat. Arch. Neurol. Psychiat. *74:* 68–87, 1955.

Carregal, E. J. A., Williams, B., and Birzis, L.: Respiratory centers in the dog and squirrel monkey. A comparative study. Resp. Physiol. *3:* 333–348, 1967.

Clark, C. H.: Basic Concepts in Neurologic Diagnostics. Chapter 2 in *Canine Neurology—Diagnosis and Treatment* (Hoerlein, B. F.). Philadelphia, W. B. Saunders Co., 1965. Pp. 7–24.

Herrick, C. J.: *The Evolution of Human Nature.* Austin, Tex., University of Texas Press, 1956.

O'Leary, J. L., and Coben, L. A.: The reticular core—1957. Physiol. Rev. *38:* 243–276, 1958.

Olszewski, J.: Cytoarchitecture of the Human Reticular Formation. In *Brain Mechanisms and Consciousness.* Oxford, Blackwell, 1954. Pp. 54–80.

Romer, A. S.: *The Vertebrate Body,* 4th ed. Philadelphia, W. B. Saunders Co., 1970.

Ruch, T. C., Patton, H. D., Woodbury, J. W., and Towe, A. L.: *Neurophysiology,* 2nd ed. Philadelphia, W. B. Saunders Co., 1965.

Taber, E.: The cytoarchitecture of the brain stem of the cat. I. Brain stem nuclei of cat. J. Comp. Neurol. *116:* 27–69, 1961.

Taber, E., Brodal, A., and Walberg, F.: The raphe nuclei of the brain stem in the cat. I. Normal topography and cytoarchitecture and general discussion. J. Comp. Neurol. *114:* 161–187, 1960.

Valverde, F.: Reticular formation of the pons and medulla oblongata. J. Comp. Neurol. *116:* 71–99, 1961.

Walberg, F.: Descending connection to the inferior olive. An experimental study in the cat. J. Comp. Neurol. *104:* 77–173, 1956.

Walberg, F.: Further studies on the descending connections to the inferior olive—reticulo-olivary fibers. An experimental study in the cat. J. Comp. Neurol. *114:* 79–87, 1960.

Yoshikawa, T.: *Atlas of the Brains of Domestic Animals.* Tokyo, University of Tokyo Press, and University Park, Penn., Pennsylvania State University Press, 1968.

chapter 12

Upper Medulla Oblongata and Pons

The internal structures of the upper medulla oblongata appear to meld with those of the pons. It should be remembered that, in the embryo, the pons and medulla were anatomically considered the rhombencephalon (hindbrain). In the adult, the fourth ventricle is retained as a common vesicle for both the pons and the medulla. At the junction of the medulla and pons the most conspicuous structures are: the trapezoid body, the sixth, seventh, and eighth cranial nerves and nuclei, and the dorsal nucleus of the trapezoid body (superior olive).

In the dog brain, unlike that of man, the trapezoid body is visible on the ventral surface (see Fig. 2-5, p. 20). The same figure illustrates that the trapezoid body may be used grossly as a landmark for the locations of cranial nerves VI, VII, and VIII, with the abducens nerve being medial to the latter two.

Based primarily on the differences in the trapezoid, the above-named structures are appropriately considered upper medullary structures of the dog brain (Meyer, 1964), but pontine structures in the human brain (Truex and Carpenter, 1969). The location of the attachment of the trigeminal nerve and its relationship with the middle cerebellar peduncle are similar in man and dog, and therefore they both are considered

184

pontine structures in both mammals. Plates 16, 17, and 18 show the middle cerebellar peduncle (brachium pontis) and the attachment of the trigeminal nerve as they appear in the dog.

In the discussion below, the major functional anatomic properties are described individually for the internal structures. Further details of the physiologic systems to which these structures belong are given in other chapters, to which references are made.

Relationship of Internal Structures of Upper Medulla Oblongata

Eighth Cranial (Vestibulocochlear) Nerve

A relatively comprehensive coverage of the functional significance of the vestibulocochlear nerve and the internal structures involved is contained in Chapter 19. Plates 13, 14, and 15 in the Atlas are transverse sections of the upper medulla–pons junction and show the dorsal and ventral cochlear nuclei (Plate 14), the trapezoid body, and dorsal nucleus of the trapezoid body (superior olive) (Plate 15), all of which function as components of the auditory system.

Without discussing the auditory pathway detailed later in Chapter 19, a few cardinal points seem worthy of mention here.

1. Figure 12-1 is a section through the dog brain, showing the trapezoid fibers and medial lemniscus covered on the ventral surface only by the pyramids medially but with the trapezoid fibers appearing laterally (see Fig. 2-5, p. 20). In comparison, in the human brain the horizontal trapezoid fibers and the medial lemniscus in the pons are very deep and separate the tegmental (upper) from the basilar (lower) area (Fig. 12-2). At a higher level in the dog brain the medial lemnisci are deeper (more dorsal), and there is a basilar portion of the pons (Plate 16), but it is much smaller than in man. The great size of the human basilar

pons, which contains cell bodies (gray matter) and many transverse fibers, is a product of phylogenetic development, as discussed with the middle cerebellar peduncle.

2. The dorsal nucleus of the trapezoid body (superior, or rostral, olive) in the dog is relatively very much larger than that in man. In the dog the nucleus consists of a convoluted platelike lamina which is basically divisible into medial and lateral portions (Figs. 12-1 and 12-3). Study of the topography and cytologic details reveals that in the cat the nuclear divisions are combined with the trapezoid nuclei to form a complex which contains six nuclei (Taber, 1961). Histologically, the dorsal nucleus of the trapezoid body contains bipolar cell bodies which closely resemble those of the dorsal cochlear nucleus (Papez, 1929). Some of the major structures surrounding the dorsal nucleus of the trapezoid body are labeled in Figure 12-3.

Vestibular Nuclei

Beneath the floor of the fourth ventricle in the general area of the upper medulla–pons junction on each side there are four vestibular nuclei: the lateral (Deiters'), medial (Schwalbe's), rostral (superior, Bechterew's), and caudal (spinal, inferior, or descending). As illustrated in Figure 12-4 their relative positions are such that not all four nuclei can be cut in the same transverse section, but Plate 13 shows three, namely, medial, caudal, and lateral. The same Plate shows the vestibular nerve entering the brain stem between the descending (spinal) trigeminal tract and nucleus on its ventral side and the caudal cerebellar peduncle dorsally.

The brief discussion of the vestibular nuclei and system given below is based on the correlation of structure and function. Further functional details are found in Chapter 19 and discussion of cytoarchitectural details will be found in Taber (1961).

The vestibular nerve is classified func-

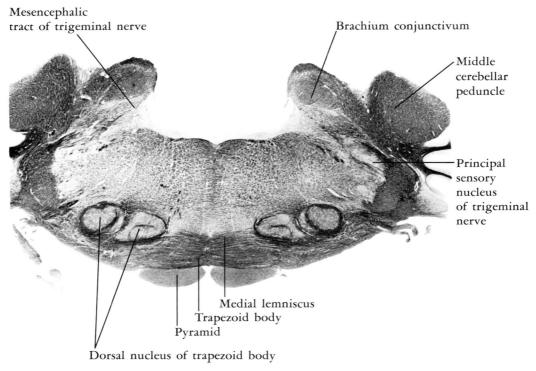

Mesencephalic
tract of trigeminal nerve

Brachium conjunctivum

Middle
cerebellar
peduncle

Principal
sensory
nucleus
of trigeminal
nerve

Medial lemniscus
Trapezoid body
Pyramid

Dorsal nucleus of trapezoid body

Fig. 12-1. Low-power photomicrograph of transverse section through dorsal nucleus of trapezoid body (rostral, or superior, olive), trapezoid body, and medial lemniscus in the dog. Note that in the dog (as in the cat) the trapezoid fibers are covered ventrally only by the pyramids bordering the midline, and are visible on the ventral surface more laterally. Compare this Figure with Figure 12-2, which shows the position of the trapezoid and medial lemniscus fibers in man.

tionally as special somatic afferent (SSA). As its axons grossly enter the brain stem at the lateral end of the trapezoid body from the vestibular (Scarpa's) ganglion, they are dispersed to more structures than directly to the four vestibular ganglia. There are *primary vestibulocerebellar fibers,* which bypass the vestibular nuclei to project directly to the cerebellar cortex. In addition, the cerebellum is an integral part of the vestibular system by means of *secondary vestibulocerebellar fibers,* which are relayed from the vestibular nuclei. To complete the functional anatomic bond, efferent fibers from certain cerebellar cortical areas and the fastigial nuclei project back to the vestibular nuclei (Carpenter, 1959).

The functional importance of the cere-

bellovestibular fibers is evident in the following brief example, as we focus our attention on the *lateral vestibular nucleus.* This nucleus gives rise to descending fibers which enter the ventral funiculus on the same side of the spinal cord as the *lateral vestibulospinal tract* (Fig. 12-4; see Fig. 10-7, p. 157). The lateral vestibular nucleus receives impulses from the vestibular nerve and from the cerebellum and projects fibers to the spinal cord levels. Such anatomic connections permit not only the direct influence of the vestibular nuclei on the lower motor neurons in the spinal cord, but also cerebellar influence on muscle tone and spinal reflexes. The medial vestibular nucleus projects axons caudally in the medial longitudinal fasciculus as the *medial vestib-*

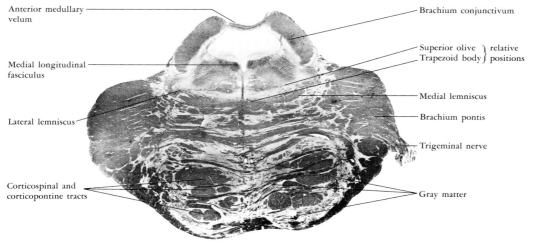

Anterior medullary velum

Medial longitudinal fasciculus

Lateral lemniscus

Corticospinal and corticopontine tracts

Brachium conjunctivum

Superior olive ⎫ relative
Trapezoid body ⎭ positions

Medial lemniscus

Brachium pontis

Trigeminal nerve

Gray matter

Fig. 12-2. Low-power photomicrograph of a transverse section through pons of human brain stem. Compare this with Figure 12-1, taken from dog brain stem. Note particularly the extremely large basilar portion of the pons in man, which has been added ventrally to the level of the trapezoid body, medial lemniscus, and superior olive (dorsal nucleus of trapezoid body in dog). The specific locations of the trapezoid fibers and superior olive are immediately caudal to the relative positions indicated above. The significance of the large basilar pons is in correlation with the corticoponto-cerebellar system, which places the cerebellum under more direct influence of the cerebral cortex in man when compared with carnivores. Additional structures are labeled in this figure of human pons cross section, which may be compared with the same structures as they appear in dog as labeled in Plate 15.

ulospinal tract (Fig. 12-4). This minor tract extends in the ventral funiculus only as far as the mid-thoracic cord levels.

The overall major physiologic effects of the above connections are the exertion of a facilitatory influence on spinal reflexes which control muscle tone, especially to maintain appropriate posture or strength of supporting and balancing movements. Animals with a lesion of the vestibulospinal tracts show atonia, or muscular weakness, and rolling toward the side of the lesion. As discussed in Chapter 14, these symptoms are similar in part to those of certain cerebellar deficiencies, and consequently often make differential diagnosis extremely difficult.

The synergistic facilitatory effect of the vestibulospinal and reticulospinal tracts on the antigravity (postural and extensor) muscles is dramatically demonstrated experimentally in an animal, such as the cat,

by a mesencephalic brain-stem transection to produce decerebrate rigidity, as discussed in Chapter 11.

Vestibular stimuli are initiated chiefly by head movements or positions. Consequently, the proprioceptive impulses from the neck muscles are important contributors to the vestibular mechanism. The vestibular righting reflex is based primarily on the principle that the orientation of the body in space is dictated by head movements.

The highly complicated functional aspects of the vestibular system are evident when one considers some of the symptoms of over-stimulation, disease, and irritative or destructive lesions of the labyrinth. The disequilibrium, staggering, postural changes, falling, and possibly rolling to the side of the lesion have been alluded to previously. In addition, clinical signs such as deviation of the eyes and nystagmus commonly occur. The possible symptoms

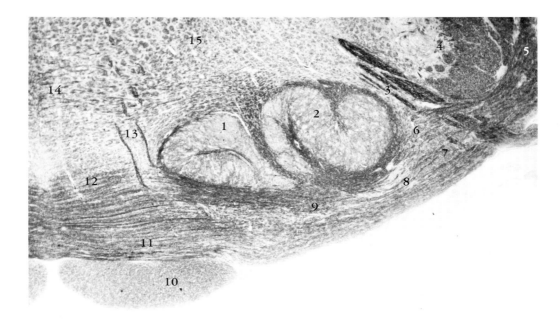

Fig. 12-3. An enlargement of transverse section of dorsal nucleus of trapezoid body and surrounding structures in ventrolateral segment of dog brain stem. Note particularly that the nerve fibers (*13*) of the abducens (VI) pass medially to the dorsal nucleus of the trapezoid body (superior olive) and the fibers (*3*) of the facial nerve (VII) pass on the lateral side between the nucleus and the descending trigeminal tract and nucleus. 1, medial division of dorsal nucleus of trapezoid body (rostral, or superior, olive); 2, lateral division of dorsal nucleus of trapezoid body; 3, facial nerve; 4, nucleus and descending root of trigeminal nerve; 5, lateral lemniscus; 6, rubrospinal tract; 7, ventral spinocerebellar tract; 8, preolivary nucleus; 9, spinothalamic tract; 10, pyramid; 11, trapezoid body; 12, medial lemniscus; 13, abducens nerve fibers; 14, dorsal acoustic striae; 15, reticular formation.

of nausea, vomiting, and vasomotor reactions are discussed later in relation to the importance of the medial longitudinal fasciculus. The involuntary rhythmic typewriter-like movements of the eyes, called nystagmus, are discussed with other effects of labyrinthectomy in Chapter 19. It should be noted that nystagmus is not pathognomonic for any specific deficiency. Similarly, a differential diagnosis should never be based entirely on any one of the above-listed symptoms.

Medial Longitudinal Fasciculus

In the medulla and pons areas the medial longitudinal fasciculus (MLF) borders the median raphe immediately deep to the dorsal median sulcus of the rhomboid fossa and dorsal to the tectospinal tract (Plate 15). The origin, extent, and contents of this fasciculus in the dog are discussed in Chapter 10. Only the vestibular connections are considered here.

All of the vestibular nuclei project axons medially to enter the MLF. On entering the MLF most of the vestibular fibers bifurcate into ascending and descending divisions, some remaining on the same side and some crossing to the other. The rostral (superior) vestibular nucleus, however, projects rostrally a direct *vestibulomesencephalic tract* which enters only the MLF of the same side to innervate the abducens nucleus on its way to termination in the trochlear and oculomotor nuclei in the tegmentum of the

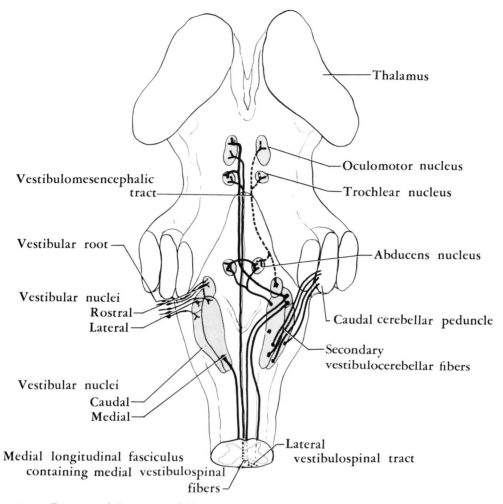

Thalamus

Vestibulomesencephalic tract

Oculomotor nucleus

Trochlear nucleus

Vestibular root

Abducens nucleus

Vestibular nuclei
Rostral
Lateral

Caudal cerebellar peduncle

Secondary vestibulocerebellar fibers

Vestibular nuclei
Caudal
Medial

Lateral vestibulospinal tract

Medial longitudinal fasciculus containing medial vestibulospinal fibers

Fig. 12-4. Diagram of the principal fiber connections of the vestibular nuclei. On the left side of the drawing are shown primary vestibular fibers of the vestibular ganglion (with bipolar cell bodies) terminating in all four vestibular nuclei. The medial vestibular nucleus sends fibers medially to the medial longitudinal fasciculus (MLF) on both sides, where descending vestibular fibers form the *medial vestibulospinal tract* in the ventral funiculus of the spinal cord as far caudad as the mid-thoracic levels. The lateral vestibular nucleus (shown on the right side of the drawing) gives rise to ipsilateral descending fibers to the ventral funiculus of the entire spinal cord as the *lateral vestibulospinal tract.* The rostral (superior) vestibular nucleus (shown on the right side of the drawing) gives rise to ipsilateral ascending fibers directly to the midbrain as the *vestibulomesencephalic tract.* On its way to the trochlear and oculomotor nuclei of the midbrain, this tract sends collaterals to the abducens nucleus.

midbrain (Fig. 12-4). In addition, all of the vestibular nuclei send axons rostrally via the MLF to the above-named three nuclei of origin for neurons innervating the extraocular muscles. These neuroanatomic connec-

tions place the eye muscles under influence of the vestibular nuclei, which in turn receive an input from the cerebellum. This explains, in part, the eye deviation and nystagmus which result from abnormal irritation or destruction of the cerebellum or the true vestibular mechanism.

Descending vestibular fibers which enter the MLF arise almost exclusively from the medial vestibular nucleus and probably extend only to the mid-thoracic levels of the spinal cord as the medial vestibulospinal tract. It is thought that branches of the descending fibers are dispersed to the reticular formation and reach visceral motor and autonomic nuclei and thereby contribute to the reflex symptoms that occur with excessive labyrinthine stimulation (*e.g.*, nausea, vomiting, and vasomotor reactions).

Facial Nerve and Nuclei

As shown in Table 11-1 (p. 177), there are many functional components of the facial nerve, and consequently numerous nuclei are associated with its motor and sensory fibers. Each of the seventh (facial), ninth (glossopharyngeal), and tenth (vagus) cranial nerves has the maximum number of functional components (six) composing any nerve. Chapter 17 contains a relatively comprehensive and complete coverage of the cranial nerve components and the associated nuclei.

The discussion below is focused on only the facial nucleus and its functional component, special visceral efferent (SVE). As general relevant information, the other five functional components and the associated nuclei are: general visceral efferent (GVE)—the rostral salivatory nucleus; general visceral afferent (GVA) and special visceral afferent (SVA)—the solitarius nucleus and tract; general somatic afferent (GSA) exteroceptive—spinal (descending) nucleus and tract of the trigeminal nerve; and GSA proprioceptive—probably the mesencephalic nucleus of the trigeminal nerve. The solitarius nucleus plus its tract and the spinal nucleus of V plus its tract are conspicuous and easily identified (Plates 11, 12, and 13). The mesencephalic nucleus of the trigeminal nerve is at a higher level

(more rostral), and the rostral salivatory nucleus is extremely difficult to identify.

Notice that the only nucleus labeled 'facial' is the SVE nucleus which gives origin to peripheral axons that primarily innervate the muscles of facial expression. Recall that embryologically the facial nerve was within the second branchial (pharyngeal) arch and the adult muscles innervated by this nerve were derived from that arch. One reason SVE is very special is that in this situation there are *visceral* motor impulses conveyed by axons which innervate *skeletal* muscle with typical motor end-plates at the myoneural junction. This is a great contrast to GVE (autonomic) impulses transmitted to visceral muscle with no motor end-plates.

In cross section of the brain stem through the upper medullary levels, the facial nucleus can be identified in the ventrolateral area of the reticular formation (Plate 14). The internal course of the facial root fibers is indirect, in that the axons pass dorsomedially to loop around the abducens nucleus and descend to exit from the brain stem between the superior olive and the descending root and nucleus of the trigeminal nerve (Fig. 12-1). The loop formed by the facial root fibers around the abducens nucleus is the *internal genu of the facial nerve.*

The internal genu has an interesting phylogenetic history. Originally the facial nucleus was near the dorsal median sulcus of the rhomboid fossa close to the abducens nucleus. During phylogenetic development the facial nucleus migrated ventrolaterally and slightly caudally, to align itself with the masticator nucleus of the trigeminal nerve rostrally and the ambiguus nucleus caudally. The emerging facial fibers were established before the migration of the nucleus took place. As the nucleus shifted, the root fibers, which were already attached to the cell bodies in the nucleus, elongated and formed the characteristic internal genu of the facial nerve (Papez, 1929).

The facial nucleus receives afferent axons from a number of cranial nerves to function as the efferent portion of certain reflex arcs, for example, from the spinal trigeminal nucleus for execution of the corneal reflex, and from the cochlear nuclei for the mediation of acousticofacial reflexes, such as closing the eyes as a result of a sudden loud noise.

The most common lesion of the facial nerve is *Bell's palsy,* a sagging of the ipsilateral facial muscles due to destruction of either the facial nucleus or its roots internally or the facial fibers externally. This is a typical lower-motor-neuron lesion with the salient symptom of flaccid paralysis, comparable to the lower-motor-neuron lesions at the spinal cord levels. The normal drooping of the face of certain breeds of dogs (*e.g.,* bloodhound and basset) makes diagnosis of this syndrome difficult in such breeds.

Abducens Nucleus and Nerve

The nucleus of the abducens nerve was mentioned previously in reference to the internal genu of the facial nerve. The nucleus is near the dorsal median sulcus and the rhomboid fossa, from which it is separated by the nerve fibers of the genu. Functionally this nucleus is classified as somatic efferent (SE); therefore it is in direct longitudinal alignment rostrally with the oculomotor and trochlear nuclei and caudally with the hypoglossal nucleus, all of which have the same functional component.

The internal roots of the abducens nerve descend ventrally and slightly laterally to pass medial to the superior olive and emerge lateral to the pyramids (Fig. 12-3). From this point the fibers pass to the extrinsic eye muscles to innervate only one, the lateral (external) rectus. Normal contraction of this muscle holds the eye in proper counterbalance to the pull of the opposing medial rectus muscle innervated by the oculomotor nerve.

A destructive lesion of the nucleus or nerve produces *internal strabismus,* or crossed eye, due to the normal pull of the medial rectus with no balancing force of the lateral rectus muscle. An irritative lesion or hyperstimulation of the abducens nerve or nucleus may cause *external strabismus.*

There is little agreement about the details of the GSA proprioceptive impulses that arise from the muscle spindles in the extrinsic muscles of the eye. One popular belief is that the mesencephalic nucleus of the trigeminal nerve is the location of the cell bodies for such fibers (Peele, 1961).

Trigeminal Nerve and Nuclei

The trigeminal nerve (V) is the largest of the cranial nerves and contains both sensory and motor fibers. All three divisions of the nerve—the ophthalmic, maxillary, and mandibular—contain sensory fibers, but only the mandibular division carries motor axons. Although the motor fibers are fewer in number, they are the very important special visceral efferent (SVE) axons which innervate the muscles of mastication. The nucleus of origin for the axons is the *motor* (*masticator*) *nucleus* of the trigeminal nerve.

The sensory nuclei of the trigeminal nerve constitute a major GSA representation in the brain stem. As shown in Table 11-1 (p. 177), the spinal nucleus and tract are terminals of peripheral GSA exteroceptives for cranial nerves VII, IX, X, and XI, in addition to V. The only sensory component of the trigeminal is GSA of both exteroceptive and proprioceptive types. Recall that the GSA nuclei form a continuous longitudinal, irregularly shaped column from the midbrain caudally to the spinal cord. The respective portions of the sensory column from rostral to caudal are: the mesencephalic nucleus and tract, principal (main), and the spinal (descending) nucleus and tract of the trigeminal nerve which in turn is continuous with the substantia gelatinosa of the spinal cord. There are slight overlaps as indicated in Figure 12-1 with the mesencephalic tract and the principal (chief) nucleus cut in the same transverse section.

The *mesencephalic nucleus of the trigeminal nerve* contains large unipolar cell bodies presumably of the peripheral proprioceptive neurons from the muscles of mastication, temporomandibular joint, teeth, mouth, and extraocular muscles. This nucleus functions as a typical sensory ganglion and is an exception to the rule that sensory neurons of peripheral nerves have their cell bodies located in a ganglion. The unipolar arrangement in this nucleus permits two-neuron reflexes, that is, axons (distal processes) from the cell bodies pass directly to the masticator nucleus, which in turn sends axons back to the muscles of mastication. This two-neuron reflex arising and terminating in the muscles of mastication may control the force of biting in the regulation of mastication (Elliott, 1969).

In monkey and cat there is a bilateral projection to the trigeminal mesencephalic nucleus from the neuromuscular spindles, Golgi tendon organs, and joint receptors which contributes to the integration of function of the muscles of mastication (Smith *et al.*, 1967). The site of the cell bodies of neurons from the muscle spindles in the extrinsic ocular muscles is not universally accepted as being in the trigeminal mesencephalic nucleus. Experimental studies in lambs suggest that a localized portion of the semilunar ganglion contains cell bodies of neurons which transmit impulses from the extraocular muscles (Manni *et al.*, 1966).

The *principal (main, pontine, chief) sensory nucleus of the trigeminal nerve* (Fig. 12-1) is lateral to the masticator nucleus and longitudinally between the mesencephalic and descending nuclei. It is generally agreed that this nucleus of termination receives impulses initiated by touch and pressure. This nucleus has multipolar cell bodies of internuncial neurons for transmitting GSA impulses entering via the trigeminal nerve. The peripheral neurons have their cell bodies within the semilunar ganglion. A major efferent path is the *trigeminal lemniscus,* com-

posed of axons from this nucleus plus the spinal nucleus which decussate to join axons from the contralateral medial lemniscus and ascend to the thalamus. Axons of the third order arise in the thalamus to enter the internal capsule to participate in the thalamic radiation to reach the cerebral cortex. In addition to the direct trigeminal lemniscus, collaterals are given off to synapse with motor nuclei for completion of important reflexes, for example, corneal (efferent via VII), tongue (efferent via XII), and salivary (efferent via VII and IX).

The *spinal (descending) trigeminal tract and its nucleus* subserving pain and thermal impulses, especially from the skin and mucous membranes of the entire head, form a long column from the level of the pons (Fig. 12-1) caudally to the upper cervical spinal cord. The spinal tract and nucleus of cranial nerve V form a conspicuous landmark throughout the lateral area of the medullary reticular formation. The nucleus is continuous rostrally with the principal sensory nucleus and caudally with the substantia gelatinosa of the spinal cord. This nucleus contains multipolar cell bodies of internuncial neurons connecting the trigeminal and other nuclei of the brain stem and cord.

Middle Cerebellar Peduncle

The middle cerebellar peduncle (brachium pontis) is the most lateral of all three cerebellar peduncles as observed in cross section (see Fig. 11-1). It is the simplest in that it is composed of only fibers from the pontine nuclei as afferent axons to the cerebellum as a final portion of the *corticopontocerebellar system.* In the dog and cat the diameter of this peduncle is approximately the same as that of the other two peduncles, but in man this peduncle is by far the largest and the most highly developed. This is because the corticopontocerebellar system is correlated both phylogenetically and ontogenetically with the development of the neocortex and the neocerebellum, that is, the cerebellar hemispheres (Everett, 1965). The functional significance of this phenomenon is that the cerebellum in man is dominated by the cerebral cortex to a much greater extent. The lesser development of the basilar portion of the pons in dog and cat, when compared with that of man, has been mentioned previously, and is illustrated in Figures 12-1 and 12-2.

Rostral Cerebellar Peduncle

During the first portion of its course the rostral (superior) cerebellar peduncle (brachium conjunctivum) is within the dorsal tegmental portion of the pons, bordering the rostral part of the fourth ventricle (Fig. 12-1). As it proceeds toward the mesencephalon, the peduncle passes progressively deeper, and in the midbrain it forms a complete decussation (Plate 19). The rostral peduncle is both afferent and efferent to the cerebellum. The most prominent afferent tract is the ventral spinocerebellar tract which forms a dorsal cap on the peduncle (Fig. 12-1). The more numerous efferent fibers arise mostly from the dentate and interpositus cerebellar nuclei and pass mainly to the midbrain. In addition to fibers to the red nucleus in the mesencephalon, other efferent fibers extend to the thalamus and possibly the globus pallidus, reticular formation, and motor nuclei of the brain stem.

It should be remembered that grossly the *rostral (anterior) medullary velum* stretches between the two rostral peduncles to form the roof of the pontine portion of the fourth ventricle.

Bibliography

Carpenter, M. B.: Lesions of the fastigial nuclei in the rhesus monkey. Am. J. Anat. *104:* 1–33, 1959.

Elliott, H. C.: *Textbook of Neuroanatomy,* 2nd ed. Philadelphia, J. B. Lippincott Co., 1969.

Everett, N. B.: *Functional Neuroanatomy,* 5th ed. Philadelphia, Lea & Febiger, 1965.

Manni, E., Bortolami, R., and Desole, C.: Eye muscle proprioception and the semilunar ganglion. Exp. Neurol. *16:* 226–236, 1966.

Meyer, H.: The Brain. Chapter 8 in *Anatomy of the Dog* (Miller, M. E., Christensen, G. C., and Evans, H. E.). Philadelphia, W. B. Saunders Co., 1964. Pp. 480–512.

Papez, J. W.: *Comparative Neurology.* New York, Thomas Y. Crowell Co., 1929.

Peele, T. L.: *The Neuroanatomic Basis for Clinical Neurology,* 2nd ed. New York, McGraw-Hill Book Co., Blakiston Division, 1961.

Smith, R. D., Marcarian, H. Q., and Niemer, W. T.: Bilateral relationships of the trigeminal mesencephalic nuclei and mastication. J. Comp. Neurol. *131:* 79–91, 1967.

Taber, E.: The cytoarchitecture of the brain stem of the cat. I. Brain stem nuclei of cat. J. Comp. Neurol. *116:* 27–69, 1961.

Truex, R. C., and Carpenter, M. B.: *Human Neuroanatomy,* 6th ed. Baltimore, Williams & Wilkins Co., 1969.

chapter 13

Mesencephalon and Metathalamus

The mesencephalon (midbrain) and the metathalamus (medial and lateral geniculate bodies plus the pretectal area) are anatomically and physiologically related, but the two are derived from separate embryonic divisions of the neural tube (see Chapter 2). The adult midbrain evolves directly from the embryonic division of the same name (mesencephalon), and the adult metathalamus is considered a portion of the diencephalon, from which it developed. It should be recalled that the *pretectal area,* or *pretectum,* is the area immediately rostral to the colliculi at the diencephalon-mesencephalon junction.

As stated previously, the mesencephalon is the simplest brain division in that it is the only segment of the original three cephalic divisions of the neural tube that does not give rise to superstructures. The midbrain connects the diencephalon rostrally and the pons caudally. It retains its original cylindrical shape to a large extent, with four rounded eminences, the *corpora quadrigemina,* on the dorsal surface, and the ventral surface is modified to form two 'legs,' the *crura cerebri,* or *cerebral peduncles.* The nomenclature for the designated areas as seen in a transverse section of midbrain varies among authors. According to the Nomina Anatomica Veterinaria (1968), the four areas of the mesencephalon in a dorso-ventral sequence are tectum, tegmentum, substantia nigra, and crus (pedunculus) cerebri. The tectum and tegmentum are separated indistinctly at approximately the horizontal level of the dorsal border of the *cerebral aqueduct,* or *iter.* This is the mesencephalic simple tubular portion of the ventricular system that connects the pontine portion of the fourth ventricle caudally and the third ventricle rostrally. The homogeneous-looking gray matter which surrounds the cerebral aqueduct is referred to as the *periaqueductal,* or *central, gray matter* (Figs. 13-1 and 13-2).

Tectum

The tectum (L., roof), or *quadrigeminal plate,* is composed primarily of two pairs of rounded eminences: the rostral (superior) and the caudal (inferior) colliculi. Collectively, the four elevated bodies on the dorsal surface are referred to as the *corpora quadrigemina.* Each colliculus has a brachium (arm) between it and a geniculate body. The rostral colliculi and the lateral geniculate bodies are joined by brachia of the rostral colliculi, and the caudal colliculi and the medial geniculate bodies by the brachia of the caudal colliculi. The brachium of the rostral colliculus is less conspicuous than that of the caudal colliculus and is very difficult to identify grossly. In addition to these connections, the two rostral colliculi are connected by fibers forming the commissure of the rostral colliculi, and the caudal colliculi are joined similarly by the commissure of the caudal colliculi (Fig. 11-1,

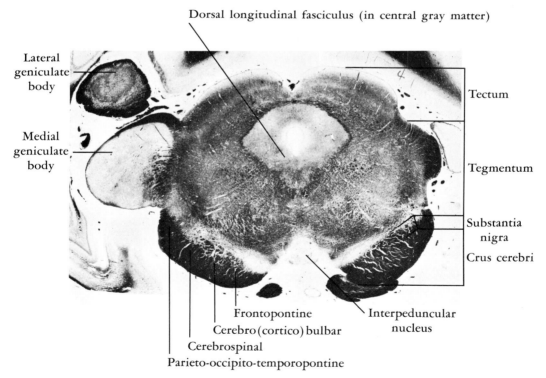

Dorsal longitudinal fasciculus (in central gray matter)

Lateral geniculate body

Medial geniculate body

Tectum

Tegmentum

Substantia nigra

Crus cerebri

Frontopontine
Cerebro (cortico) bulbar
Cerebrospinal
Parieto-occipito-temporopontine

Interpeduncular nucleus

Fig. 13-1. Photomicrograph of a cross section through the rostral colliculus of the midbrain. The four areas are labeled from dorsal to ventral: tectum, tegmentum, substantia nigra, and crus cerebri (cerebral peduncle). The organization of the corticifugal fibers in the crus cerebri is in agreement with Singer (1962). Additional structures are labeled for orientation.

p. 172). The latter is much larger and longer than the rostral commissure, so that the caudal colliculi lie caudal and lateral to their rostral counterparts.

Prior to a discussion of specific fiber connections, a brief generalized introduction may aid in understanding the functional relationships between the colliculi (mesencephalon) and the geniculate bodies plus the pretectum (metathalamus). Functionally, the rostral colliculi and pretectum are centers for visual reflexes, whereas the lateral geniculate bodies are relay centers for fibers transmitting visual impulses to optic cerebral cortex. The caudal colliculi serve as centers for auditory reflexes and also as relay stations for auditory impulses to the medial geniculate bodies, which in turn relay impulses to the auditory cerebral cortex.

The Caudal Colliculus

The caudal (inferior) colliculus is a correlation center, but is not as complex as the rostral colliculus. Microscopically, the caudal colliculus appears homogeneous, whereas the rostral colliculus is laminated, as may be observed by comparison of Plates 17 and 19.

The greatest source of afferent impulses to the caudal colliculus is the lateral lemniscus, which transmits impulses primarily from the cochlear nuclei and the superior olive (see Chapter 19). Most of the lemniscal fibers that transmit auditory impulses synapse in the caudal colliculus with neurons which send fibers through the brachium to terminate in the medial geniculate body. Experimental studies in the cat indicate that the cells of the caudal colliculus are arranged to respond to different frequencies to present a tonotopic localization (Rose et al., 1963). A few lemniscal fibers bypass the colliculus to pass directly into the brachium and terminate in the medial geniculate body. Some lemniscal fibers terminate in the caudal colliculus for auditory reflexes. The large commissure interconnecting the two caudal colliculi is one of many decussations that account for the auditory pathway and reflexes being functionally bilateral. Physiologic studies such as those of Goldberg and Neff (1961) indicate that the brachium of the caudal colliculus in the cat may send fibers to centers other than the medial geniculate body (e.g., the rostral colliculus and certain nuclei of the thalamus).

The Rostral Colliculus and Pretectal Area

A majority of the thinly myelinated smaller fibers of the optic tract which bypass the lateral geniculate body go to the rostral (superior) colliculus or the pretectal area, or both. These fibers mediate impulses which are concerned with reflex connections such as pupillary adjustments, accommodation, convergence, and blinking (see Chapter 18).

The optic tectum of submammalian forms, which is homologous to the rostral colliculus of mammals, has a complex laminated cellular appearance similar to that of the cerebral cortex. The physiologic significance of the optic tectum (rostral colliculus) diminishes in mammals as its connections increase, particularly with the thalamus and cerebral cortex. As mentioned previously, the mammalian rostral colliculus retains the laminated appearance of alternating gray (cell bodies) and white (fibers) layers. Evidence of the anatomic similarity between the rostral collicular areas among mammals is that the layers are named the same in the dog (Singer, 1962) as in man (Crosby et al., 1962). The latter authors indicate that there are some differences in size of the individual layers among mammals. As an example, the corticotectal systems are proportionately larger in man than in other mammals and this should be considered in interpreting functions as based on experimental animals. In summary of the detailed information available, the following brief statements should suffice for basic appreciation of the complicated lamination of the rostral colliculus. (1) The superficial layers are associated with the external corticotectal fibers. (2) The intermediate gray and white layer is the chief receptive layer of the multiple inputs of various modalities such as: vision, pain and temperature sense, touch, and possibly hearing. (3) The deep white layer is the greatest efferent layer, in that nearly all of the efferent fibers pass through this layer.

Based on the report of Altman and Carpenter (1961), some of the efferent projections from the rostral colliculus in the cat are: (1) *Descending efferents.* Tectoreticular fibers project diffusely to the mesencephalic reticular formation bilaterally (the contralateral fibers cross in the commissure of the rostral colliculus). Crossing fibers in the dorsal tegmental decussation descend to the pontine and medullary reticular nuclei near the midline. Some descending fibers enter the medial longitudinal fasciculus in the medulla and some continue to the spinal cord as a tectospinal tract. Specific tectospinal tract fibers were identified in the lateral portion of the ventral funiculus of the cervical cord levels. (2) *Ascending efferents* were identified to: the ventral portion of the lateral geniculate body, the medial geniculate body, the pretectum, dorsal thalamus, and pulvinar. Discussion of additional functional aspects of the rostral colliculus and the pretectal area is found in Chapter 18.

Tegmentum

The tegmentum is the central core of the mesencephalon as a rostral continuation of the pontine tegmentum, with similar characteristics; for example, both contain nuclei and pathways embedded in the reticular formation.

The Red Nucleus (Figs. 13-1 and 13-2)

The red nucleus lies in the tegmentum at the level of the rostral colliculus, embedded within the reticular formation ventral to the oculomotor nucleus. The principal *afferent fibers* to this nucleus are those from the opposite side of the cerebellum via the rostral cerebellar peduncle (brachium conjunctivum) which enters the tegmentum and crosses caudal to the red nucleus as the *decussation of the brachium conjunctivum* (Plate 19). These fibers from the dentate nucleus of the cerebellum, the *dentatorubral tract*, constitute the main mass of the brachium conjunctivum. In turn there are fibers which originate in the red nucleus and project to the thalamus as the rubrothalamic tract. Therefore, the two above-named pathways are portions of the dentatorubrothalamic system. Finally, the thalamic efferent fibers project to the cerebral cortex so that one main function of the red nucleus is as a way station for proprioceptive impulses from the cerebellum to the cerebral cortex. This major discharge from the cerebellum provides the feed-back route to the cerebrum to complete the circuit with the cerebropontocerebellar system.

The two principal descending *efferent tracts* from the red nucleus are the rubrospinal and rubrobulbar. The *rubrospinal* fibers emerge from their cell bodies in the red nucleus to cross the midline in the *ventral tegmental decussation* and descend through the reticular formation to enter the lateral funiculus of the spinal cord. The *rubrobulbar* fibers are mingled with the rubrospinal fibers but leave the common bundle to reach the branchiomeric motor nuclei of cranial nerves V, VII, IX, X, and XI. Examples of the functional significance are: the fibers reaching the masticator nucleus (V) may represent the final link in a non-pyramidal pathway concerned with sucking movements and semi-automatic chewing. Those fibers reaching the facial nucleus (VII) may influence involuntary or emotional facial expressions, and the input into nucleus ambiguus (IX, X, and XI) may be involved in involuntary phases of swallowing (Carpenter and Pines, 1957).

The Interpeduncular Nucleus (Fig. 13-1)

The interpeduncular (intercrural) nucleus (of Gudden) is phylogenetically old and therefore prominent in submammalian forms. As the name indicates, this nucleus is located between the cerebral peduncles at the dorsal apex of the interpeduncular fossa. This nucleus is the termination of the habenulopeduncular tract (fasciculus retroflexus), which is a segment of olfactory pathways discussed under Rhinencephalon in Chapter 16. In addition to afferents from the habenulae, this nucleus receives fibers from the mammillary bodies. The major efferents from this nucleus are to the dorsal tegmental nucleus. Functionally the interpeduncular nucleus is a way station in discharges from the hypothalamic olfactovisceral and epithalamic olfacto-somatic correlation centers to the midbrain tegmentum (Ariëns Kappers *et al.,* 1967). To emphasize the fact that anatomic data do not always support logical physiological expectations, the work of Bailey and Davis (1942b) is cited in which lesions involving the interpeduncular nucleus were made in the cat with the aid of a stereotaxic instrument. Based on the known anatomic connections, the authors expected some alteration in behavior relating to food. Instead of the expected response, the cats propelled forward in an obstinate manner, made low cries, and did not turn for any object. In a cage the cat would push against the corner with his head until completely exhausted. The hair was rubbed off and the scalp macerated. If on a table it would walk straight off the edge and fall to the floor. The animal showed no interest in its environment and merely propelled himself straight forward until meeting an obstacle and then would keep pushing. This behavior lasted as long as the animal lived, usually about three days. Because of this reaction, the authors labeled the behavior 'the syndrome of obstinate progression.' They chose not to call the syndrome a result of interpeduncular nuclear lesion because the destruction was not confined to this nucleus, although the behavior did not occur if the interpeduncular nucleus was missed.

Nucleus Pigmentosus. (Plate 18)

The nucleus pigmentosus (nucleus of the locus ceruleus) consists of pigmented cell bodies which appear scattered along the ventrolateral angle of the central gray matter medially, adjacent to the mesencephalic tract of the trigeminal nerve. As in the substantia nigra, the intracellular melanin is virtually absent in young dogs and cats but increases with age (Brown, 1943). The functions of this nucleus are not definite, but electrical stimulation of the area containing the locus ceruleus in cat has a definite effect on the respiratory pattern (Johnson and Russell, 1952; Baxter and Olszewski, 1955). Johnson and Russell suggest the possibility that the locus ceruleus is the pneumotaxic respiratory center.

Mesencephalic Nucleus (Plate 18)

The mesencephalic nucleus and tract of the trigeminal nerve is immediately lateral to the locus ceruleus. Some anatomic and physiologic properties of these structures are considered with the other trigeminal nuclei in Chapter 12. Recall that this nucleus has predominantly unipolar cell bodies resembling those of sensory ganglia outside of the central nervous system. The chief neuronal processes of the nucleus form the mesencephalic tract of the trigeminal nerve, which passes caudally into the pons to the level of the trigeminal motor nucleus, with which its fibers are intimately associated.

Reports rather consistently designate the trigeminal mesencephalic nucleus as the site of cell bodies for proprioceptive fibers from the muscles of mastication, temporomandibular joint capsule, teeth, and areas of the mouth. The site of the cell bodies of neurons from the muscle spindles in the extraocular muscles is not universally accepted as being in the trigeminal mesencephalic nucleus.

Experimental studies in the lamb suggest that a localized portion of the semilunar ganglion contains cell bodies of neurons which transmit impulses from the muscle spindles in the extrinsic ocular muscles (Manni *et al.*, 1966). In monkey and cat there is a bilateral projection to the trigeminal mesencephalic nuclei from the neuromuscular spindles, Golgi tendon organs, and joint receptors which contributes to the functional integration of the muscles of mastication (Smith *et al.*, 1967).

As a third alternative location for cell bodies of proprioceptive neurons concerned with extraocular muscles, cell bodies have been reported to occur along the course of the oculomotor, trochlear, and abducens nerves (Crosby *et al.*, 1962).

Periaqueductal Gray Matter
(Figs. 13-1 and 13-2)

The periaqueductal (central) gray matter is the homogeneous-looking gray area surrounding the cerebral aqueduct of Sylvius. The principal afferent fibers to the area are components of the *dorsal longitudinal fasciculus*. This conveys impulses from the preoptic and hypothalamic areas caudally to the nuclei of the periaqueductal gray matter, rostral colliculus, reticular formation, parasympathetic, and other motor nuclei of the cranial nerves.

The physiologic importance of the periaqueductal gray matter is suggested by the experimental studies in the cat by Bailey and Davis (1942a). Slight destruction of the periaqueductal gray matter resulted in the animals acting wildly and at the same time they stared into space without signs of interest in their surroundings. They exhibited widely dilated pupils, and frequently spit or snarled and struck at what seemed to be imaginary enemies. This state would last for a day or two and then disappear, leaving a cat which appeared to be normal.

More extensive lesions resulted in the above-mentioned characteristics for a short time (a few hours), followed by complete unresponsiveness and silence. After some days the animal might right itself and walk if stimulated, but it still showed no interest in its surroundings, showed no spontaneous activity, and would never feed itself. It seems evident that lesions in the periaqueductal gray matter may involve fibers of the dorsal longitudinal fasciculus, and that this could account for the emotional and feeding reactions.

Trochlear Nucleus and Nerve
(Plate 19)

The nucleus of the trochlear nerve (IV) is directly caudal to the oculomotor complex at the ventral border of the periaqueductal gray matter. The trochlear nerve passes dorsolaterocaudally in the lateral border of the central gray matter (Plate 18) to enter the rostral medullary velum where it decussates to emerge from the dorsolateral surface of the brain stem (Plate 17). From here the slender nerve follows the contour of the brain stem as it passes ventrally within the transverse fissure between the cerebral and cerebellar hemispheres. On the ventral surface the nerve passes rostrally through the cavernous sinus and the orbital fissure to enter the orbit, where it innervates only one extrinsic muscle of the eye, the dorsal (superior) oblique. The motor functional component of this nerve is somatic efferent (SE). The trochlear fibers conveying general somatic afferent proprioceptive impulses probably have their cell bodies in the mesencephalic nucleus of the trigeminal nerve, as discussed previously.

Oculomotor Nuclear Complex and Nerve (Fig. 13-2)

The nuclear complex of the oculomotor nerve (III) is at the level of the rostral colliculus, bordered dorsally by the central gray matter and lateroventrally by the diverging fibers of the medial longitudinal fasciculus, as shown in Figure 13-1. There are three functional components of the oculomotor nerve: somatic efferent, general somatic afferent proprioceptive and general visceral efferent (parasympathetic). The *somatic efferent* fibers arise from the principal portion of the oculomotor nucleus and pass ventrally across the medial border of the red nucleus to emerge as an inseparable segment of the nerve at the interpeduncular fossa as schematically illustrated in Figure 13-2. From this location the thick oculomotor nerve passes forward through the cavernous sinus and orbital fissure to the orbit, where the somatic efferent fibers innervate all of the extraocular muscles except the lateral rectus, retractor bulbi, and the dorsal oblique. Researchers agree that there is a localization pattern for the oculomotor nucleus correlated with specific extraocular muscles, but various patterns have been presented by different investigators.

The *general somatic afferent proprioceptive* component of the oculomotor nerve is similar in characteristics to that of the trochlear nerve in that proprioceptive cell bodies probably are located in the trigeminal mesencephalic nucleus, although other locations have been reported, as discussed above.

The *general visceral efferent (parasympathetic)* component of the oculomotor nerve originates from a subdivision of the oculomotor nuclear complex named the *Edinger-Westphal nucleus.* Fibers that convey preganglionic parasympathetic impulses arise in this nucleus and terminate within the ciliary ganglion where they synapse with postganglionic fibers which penetrate the sclera of the eyeball and pass to the ciliary body and the sphincter of the iris. The Edinger-Westphal nucleus and its efferent fibers are functionally concerned with light reflexes and accommodation.

It should be remembered that the medial

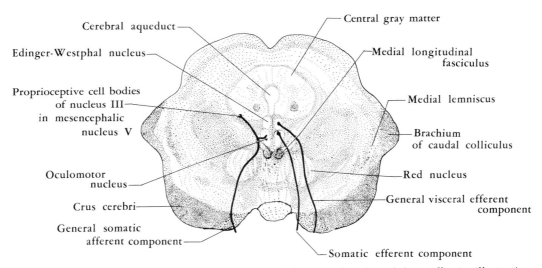

Fig. 13-2. Diagram of a transverse section through the rostral region of the midbrain, illustrating the oculomotor nuclear complex as correlated with functional components.

longitudinal fasciculus is a prominent structure of the brain stem (Fig. 13-1) which as one major function connects the vestibular nuclei with the motor nuclei of origin concerned with the innervation of the extraocular muscles. This and other functions of the medial longitudinal fasciculus are discussed in Chapter 12. The reticular formation of the midbrain is also a major constituent of the tegmentum. This is considered in Chapter 11 in a general discussion of the reticular formation.

The Subcommissural Organ (Plate 21)

Because of its location at the diencephalon-mesencephalon junction, this organ may logically be discussed with either brain division. Its main points of reference are the posterior commissure (of the diencephalon) and the cerebral aqueduct (of the mesencephalon).

Immediately beneath the posterior commissure on the dorsal surface of the cerebral aqueduct the ependymal cells are modified into a pendulous structure, the subcommissural organ. In the dog the modification of the ependymal cells is much greater than in primates where the subcommissural organ is merely a cellular plate (Truex and Carpenter, 1969). The same authors state that this organ is regarded as the only neurosecretory structure in the midbrain, and is one of the few brain areas not included in the blood-brain barrier. Functionally this organ is implicated in the secretion of aldosterone. Electrocoagulative ablation of the subcommissural organ in rats results in an immediate and drastic fall in water consumption, and it is suggested that this organ may be the so-called volume receptor of the body, regulating the total body fluid content (Gilbert, 1960).

Substantia Nigra

In contradistinction to the progressive phylogenetic regression of the interpe-duncular nucleus, the substantia nigra, which is rudimentary in lower vertebrates, is best identified in mammals, and reaches its greatest development in man (Truex and Carpenter, 1969).

The substantia nigra is located in the basal portion of the mesencephalon, where it separates the tegmentum from the crus cerebri. It is best seen in the rostral collicular level, where it can be followed into the diencephalon (Plates 20 and 21). In the mesencephalon the substantia nigra is an outstanding lamina of gray matter on the dorsal side of the fibrous crus cerebri. The name refers to the intracellular melanin pigmentation, which is virtually absent at birth but increases with age. The amount of pigment is greatest in man, where it is an outstanding gross landmark, but it also may be seen grossly in old dogs in transverse sections of the midbrain.

The functional importance of the substantia nigra is poorly understood and there are many controversies in the literature. It appears to be functionally related to the basal ganglia and thereby indirectly to the cerebral cortex as a segment of the reflex arcs important in regulating muscle tonus and stabilizing voluntary movements (see Chapter 16).

Crus Cerebri

The crus (pedunculus) cerebri (cerebral peduncle) is the base of the midbrain which is represented by a massive band of corticifugal fibers (Fig. 13-1). The term 'cerebral peduncle' is used differently by various authors. Frequently it is used in a general sense in reference to the entire 'leg' on each ventral side of the mesencephalon. In this context the cerebral peduncle includes the substantia nigra in the dorsal portion of the peduncle between the tegmentum and the *basis pedunculi* at the ventral surface, which is named *crus cerebri* in this text, as given in Nomina Anatomica Veterinaria (1968).

In a cross section through the midbrain

one can observe a somatotopic organization of the corticifugal fibers which descend through the crus cerebri (Fig. 13-1). *Cortico-pontine fibers* are located in the extreme medial and lateral portions of the crus cerebri. *Frontopontine fibers* are at the medial end, and the *corticopontine fibers* from the *temporal, occipital,* and *parietal cortical lobes* form the lateral border. The central area of the crus cerebri contains corticobulbar fibers medial to the corticospinal fibers. The most laterally placed fibers in this central segment are to the hindlimb levels of the spinal cord; the middle region contains fibers going to the lower motor neurons for innervation of the forelimbs; and the medial fibers of the central region are to the motor nuclei containing the cell bodies of the lower motor neurons for innervation of the musculature of the face, larynx, and pharynx. The corticifugal fibers in the crus cerebri are organized similarly in dog (Fig. 13-1) and man (Elliott, 1969).

Bibliography

Altman, J., and Carpenter, M. B.: Fiber projections of the superior colliculus in the cat. J. Comp. Neurol. *116:* 157–177, 1961.

Ariëns Kappers, C. U., Huber, G. C., and Crosby, E. C.: *The Comparative Anatomy of the Nervous System of Vertebrates, Including Man.* New York, Hafner Publishing Co., 1967. Vol. 2.

Bailey, P., and Davis, E. W.: Effects of lesions of the periaqueductal gray matter in the cat. Proc. Soc. Exp. Biol. Med. *51:* 305–306, 1942a.

Bailey, P., and Davis, E. W.: The syndrome of obstinate progression in the cat. Proc. Soc. Exp. Biol. Med. *51:* 307, 1942b.

Baxter, D. W., and Olszewski, J.: Respiratory responses evoked by electrical stimulation of pons and mesencephalon. J. Neurophysiol. *18:* 276–287, 1955.

Brown, J. O.: Pigmentation of substantia nigra and locus coeruleus in certain carnivores. J. Comp. Neurol. *79:* 393–405, 1943.

Carpenter, M. B., and Pines, J.: The rubro-bulbar tract—anatomical relationships, course, and terminations in the rhesus monkey. Anat. Rec. *128:* 171–185, 1957.

Crosby, E. C., Humphrey, T., and Lauer, E. W.: *Correlative Anatomy of the Nervous System.* New York, Macmillan Co., 1962.

Elliott, H. C.: *Textbook of Neuroanatomy,* 2nd ed. Philadelphia, J. B. Lippincott Co., 1969.

Gilbert, G. J.: The subcommissural organ. Neurology *10:* 138–142, 1960.

Goldberg, J. M., and Neff, W. D.: Frequency discrimination after bilateral section of the brachium of the inferior colliculus. J. Comp. Neurol. *116:* 265–289, 1961.

Johnson, F. H., and Russell, G. V.: The locus coeruleus as a pneumotaxic center. Anat. Rec. *112:* 348, 1952 (abstract).

Manni, E., Bartolami, R., and Desole, C.: Eye muscle proprioception and the semilunar ganglion. Exp. Neurol. *16:* 226–236, 1966.

Nomina Anatomica Veterinaria, Adopted by the General Assembly of the World Association of Veterinary Anatomists, Paris, 1967. Vienna, Adolf Holzhausen's Successors, 1968.

Rose, J. E., Greenwood, D. D., Goldberg, J. M., and Hind, J. E.: Some discharge characteristics of single neurons in the inferior colliculus of the cat. I. Tonotopical organization, relation of spike-counts to tone intensity, and firing patterns of single elements. J. Neurophysiol. *26:* 294–320, 1963.

Singer, M.: *The Brain of the Dog in Section.* Philadelphia, W. B. Saunders Co., 1962.

Smith, R. D., Marcarian, H. Q., and Niemer, W. T.: Bilateral relationships of the trigeminal mesencephalic nucleus and mastication. J. Comp. Neurol. *131:* 79–91, 1967.

Truex, R. C., and Carpenter, M. B.: *Human Neuroanatomy,* 6th ed. Baltimore, Williams & Wilkins Co., 1969.

chapter 14

Cerebellum

The cerebellum is the caudal superstructure which is derived from the dorsal portion of the embryonic metencephalon. It lies caudal to the occipital lobes of the cerebral hemispheres, from which it is separated *in vivo* by the tentorium cerebelli within the transverse fissure. The gross configuration of the cerebellum is analogous to a globular mushroom with a hilus to which are attached three pairs of 'stems,' the cerebellar peduncles, which are firmly planted in the brain stem. The three pairs of peduncles, separated by the fourth ventricle, appear in a medio-lateral orientation in cross section (see Fig. 14-6), but the fiber connections are basically in a rostro-caudal relationship. The caudal (inferior) cerebellar peduncle (restiform body) is attached to the medulla, the middle cerebellar peduncle (brachium pontis) to the pons, and the rostral (superior) cerebellar peduncle (brachium conjunctivum) to the midbrain.

Although the cerebellum has no primary motor nuclei nor any direct projection fibers to the lower motor neurons of the cerebrospinal nerves, it is primarily concerned with the synergy of motor activity, equilibrium, and muscle tone. As a superstructure it functions in an indirect manner to send impulses to the brain stem which in turn are relayed to the upper and lower levels of the central nervous system. The smoothness of muscular activity depends on the proper coordination of muscle groups contracting and relaxing. These activities, plus the essential muscle, tendon, and joint sense, have been referred to previously as *proprioception*. The major influence of the cerebellum on such complex physiologic mechanisms was indicated by Charles Sherrington (1947) in labeling the cerebellum as the "head ganglion of the proprioceptive system."

Phylogenetic Functional-Anatomic Correlations

The relative size and shape of the cerebellum are correlated with an animal's type of limb movement, center of gravity, and species posture. Reptiles and birds, which have predominantly trunk-muscular or symmetrical limb movements as a means of locomotion, generally have a well-developed middle cerebellar portion corresponding to the vermis. This area is more highly developed in birds which fly than in the flightless ones (Ranson and Clark, 1959). In mammals with well-developed limbs, and especially with independent limb movements, the cerebellar hemispheres are better developed. In primates, with the progressively upright posture and independent limb movements, the cerebellar hemispheres and the corticopontocerebellar system are best developed (Crosby *et al.*, 1966). Also illustrating the correlation of cerebellar differences with animal types are the following facts: (1) The cerebellar lingula is better developed in an animal with a relatively large tail, such as a rat, than in an animal with a relatively insignificant tail, such as a pig (Larsell, 1952). (2) The paraflocculus is very well developed in aquatic mammals which possess well-synchronized movements of axial and appendicular muscles.

Structure of Cerebellum

In order to understand the functional and dysfunctional aspects of the cerebellum it is important for one to become acquainted with the commonly presented subdivisions based on its phylogenetic development and correlated with species variations.

The following discussion of phylogenetic divisions is based on subhuman mammals as presented by Brodal (1969). The cerebellum can be separated into three divisions, based on phylogeny (Fig. 14-1). (1) The *archicerebellum* consists of the flocculonodular lobe, that is, the nodulus of the vermis and its lateral floccular appendages. Phylogenetically this is the oldest portion, which is separated from the body of the cerebellum by the *caudo(postero)lateral fissure*, the first to develop embryonically. (2) The *paleocerebellum* is represented by the vermis of the rostral (anterior) lobe plus the pyramis, uvula, and paraflocculus. The rostral and caudal lobes are separated by the *primary fissure*, the second fissure to develop embryonically. (3) The *neocerebellum* consists of the lateral portions of the cerebellum and the middle parts of the vermis. This division is better developed in higher forms, especially the hemispheric portions, which are greatly enlarged in primates.

Grossly the relationship of the gray and white matter in the cerebellum appears to be the reverse of that in the spinal cord, that is, the outside cerebellar cortex is gray matter which encloses the central white matter, the medullary body. Embedded within the central white matter, however, are three cerebellar nuclei. The outside surface of the lobes is composed of elongated gyri called *folia*. Each folium has a white medullary core, an extension of the central body, which contains the afferent and efferent fibers of the cortex. In a sagittal section of cerebellum the general appearance of the folia, sulci, and fissures is likened to that of an evergreen shrub, and hence, as stated previously, it is referred to as *arbor vitae*. Figure 14-2 reviews some cerebellar relationships. Further discussion of basic gross anatomy of the cerebellum is contained in Chapter 2 (pp. 26–28) and will not be repeated here.

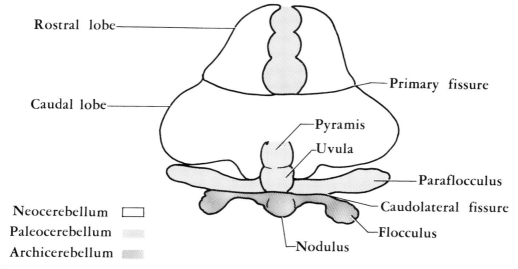

Fig. 14-1. Diagrammatic illustration of the mammalian cerebellum to indicate the phylogenetic divisions. The archicerebellum is the oldest, and the neocerebellum is the youngest, phylogenetic division. (Modified from Brodal, 1969.)

Histology of the Cerebellar Cortex

The histologic appearance of the cerebellar cortex is relatively simple and extremely uniform throughout all folia. Each folium has a central white core of myelinated afferent and efferent fibers. The peripheral gray area (cortex) is composed of three distinct layers: (1) an outer molecular layer, (2) a middle Purkinje cell layer, and (3) an inner granular layer (Fig. 14-3).

Peripheral Gray Area

The *molecular layer* has a low cellular density and appears as a rather homogeneous peripheral layer. This layer contains the dendritic ramifications of the Purkinje cells and the axons of the granular cells from intrinsic sources as well as terminal arborizations of certain afferent (climbing) fibers from the white core of the folium. The few cell bodies which are present are of two types: the outer small stellate cells located peripherally, and the deeply placed basket cells. The processes of both types tend to run parallel to the surface of the folium. The axons of the basket cells pass immediately peripheral to the bodies of the Purkinje cells and send collaterals deep to form basket-like networks as one axon synapses with many Purkinje cell bodies (Fig. 14-3).

The *Purkinje cell layer* is outstanding, in that the single layer of flask-shaped cell bodies separates the other two layers. The extremely large dendritic arborizations of these cells are in the molecular layer, whereas the axons pass deeply through the granular layer and the medullary core to terminate by synapses within the cerebellar nuclei.

The *granular layer* is an extremely cellular layer deep to the Purkinje cells. As the name indicates, the very small cell bodies are packed extremely densely to give a granular appearance. The cells have little cytoplasm and resemble lymphocytes, a resemblance which on casual observation gives the granular layer the general appearance of lymphoid tissue.

Nerve Fibers (Fig. 14-3)

There is only one type of *efferent* fiber from the cerebellar cortex: the axons of the

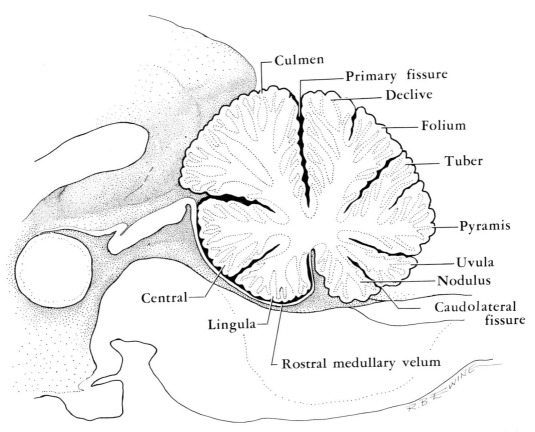

Fig. 14-2. Sagittal section of the caudal portion of the brain to show the nine lobules of the cerebellar vermis in relation to surrounding structures. The nomenclature used here for the dog agrees with human terminology, but it should be realized that there are other commonly accepted methods of identification which differ from the classical 'nine-lobule plan' given here.

Purkinje cells which pass through the granular layer and medullary core of the folium to the cerebellar nuclei for relay of impulses out of the cerebellum. The *afferent* fibers of the cortex are of two anatomic types. These are terminals of the tracts entering the cerebellum via the peduncles. The exact origins of both types are questionable. The *climbing fibers* pass from the white matter through the granular and Purkinje cell layers to reach the molecular layer, where they synapse chiefly with the abundant dendrites of the Purkinje cells. The *mossy fibers* are so named because of their structural resemblance to moss. The profuse number of branches enter the granular layer, where they are confined and terminate by synapses with the granular cells.

Nuclei of Cerebellum (Fig. 14-4)

The cerebellar nuclei are embedded in the white matter core dorsal to the fourth ventricle. The nuclei are different from those of man in two respects: there are only three principal nuclei in the dog and cat, whereas in man there are four; also the nuclei are anatomically not as discrete in the dog and cat.

(1) The *fastigial (medial,* or *roof) nucleus*

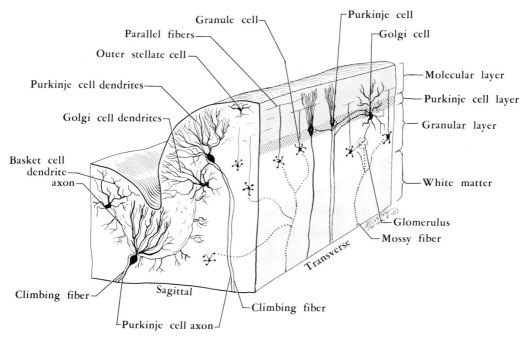

Molecular layer
Purkinje cell layer
Granular layer
White matter
Glomerulus
Mossy fiber

Granule cell
Purkinje cell
Golgi cell
Parallel fibers
Outer stellate cell
Purkinje cell dendrites
Golgi cell dendrites
Basket cell
dendrite
axon
Climbing fiber
Sagittal
Purkinje cell axon
Climbing fiber
Transverse

Fig. 14-3. Schematic diagram of the cerebellar cortex in sagittal and transverse planes, showing the basic histologic relationships of cells and fibers. The three histologic cerebellar cortical layers are labeled on the right side. (After Truex and Carpenter, 1969.)

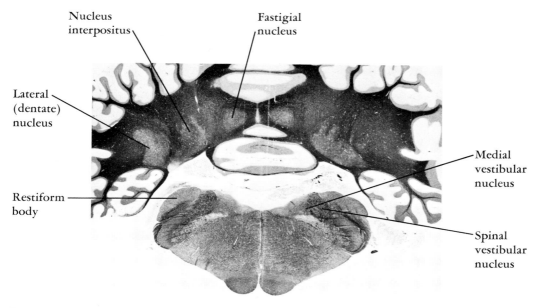

Nucleus interpositus
Fastigial nucleus
Lateral (dentate) nucleus
Restiform body
Medial vestibular nucleus
Spinal vestibular nucleus

Fig. 14-4. Low-power photomicrograph of a transverse section through the lateral recess of the fourth ventricle of the brain stem and cerebellum of the dog. The three principal cerebellar nuclei are labeled and a few brain-stem structures are identified for orientation. The name *lateral nucleus* is preferred for the *dentate nucleus* in animals, in agreement with Nomina Anatomica Veterinaria (1968). (Cf. Plate 12 for additional identification of brain-stem structures). Luxol fast blue–cresyl violet stain.

borders the midline in the roof of the fourth ventricle. Phylogenetically this is the oldest nucleus. (2) The *nucleus interpositus* is between the other two in a transverse plane (Fig. 14-4). It is a composite of the nuclei globosus and emboliformis of man, which are distinctly developed in only the highest mammals (Crosby *et al.,* 1962). Authors vary in their specific terminology regarding the equivalents of the divisions of nucleus interpositus and the emboliform and globosus nuclei of man (Flood and Jansen, 1961; Nomina Anatomica Veterinaria, 1968). (3) The *lateral (dentate) nucleus* is the most lateral of the three nuclei. In dog the tooth-like appearance of this nucleus is more vague than in man, and it is indistinctly separated from the nucleus interpositus on its medial side. Therefore, the name lateral nucleus is preferred by Nomina Anatomica Veterinaria (1968).

Peduncles and Fiber Connections of Cerebellum

The cerebellum is attached to the brain stem by three pairs of cerebellar peduncles, the caudal, middle, and rostral. Recall that the peduncles as seen in their cross section appear grossly related in a medio-lateral plane (Fig. 14-6), but the peduncular fiber connections are oriented in a rostro-caudal relationship.

Caudal Cerebellar Peduncle

The caudal (inferior) cerebellar peduncle (restiform body) is attached to the medulla and is composed primarily of afferent cerebellar fibers. Grossly, the central portion of this peduncle is between the other two, with the rostral peduncle medially and the middle peduncle laterally. In a cross section of the medulla through the area of the descending and medial vestibular nuclei, the caudal cerebellar peduncle is lateral to the descending vestibular nucleus and dorsal to the tract of the descending root of the tri-

geminal nerve (Plate 12). The most significant *afferent* fibers within this peduncle are: (1) the olivocerebellar fibers from the (inferior) olive, (2) the dorsal spinocerebellar fibers from the same side of the spinal cord, (3) the external or superficial arcuate fibers on the lateral surface of the medulla superficial to the descending tract of the trigeminal nerve, (4) the reticulocerebellar fibers from the reticular nuclei of the brain stem, and (5) the vestibulocerebellar fibers from the vestibular nuclei. The principal *efferent* fibers from the fastigial nuclei through this peduncle are the cerebellovestibular and the cerebelloreticular fibers which complete the possible feedback circuit between the brain stem and the cerebellum. Based on the phylogenetic correlations given previously, it should not be surprising that the cerebellar flocculus and nodulus project fibers to the vestibular nuclei and establish the flocculonodular lobe as the classical 'vestibulocerebellum' (Truex and Carpenter, 1969).

The fastigial-vestibular nuclear connections in the cat are somatotopically organized among the cerebellar vermis, fastigial nucleus, and the lateral vestibular nucleus (Walberg *et al.,* 1962). Figure 14-5 illustrates the representative areas for forelimb and hindlimb in the cat according to the above-cited authors. In addition, the vestibulospinal projection from the lateral vestibular nucleus shows a somatotopic organization in the cat (Pompeiano and Brodal, 1957).

Middle Cerebellar Peduncle

The middle cerebellar peduncle (brachium pontis) is the most lateral peduncle and is composed primarily of the afferent fibers from the pontine nuclei. As a portion of the corticopontocerebellar system the middle cerebellar peduncle aids in placing the cerebellum under the influence of the cerebral cortex. This functional relationship is best developed in primates and is correlated anatomically with the extremely large pontine protuberance that covers the

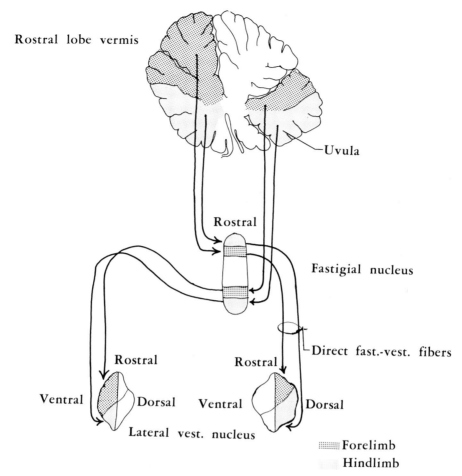

Rostral lobe vermis

Uvula

Rostral

Fastigial nucleus

Direct fast.-vest. fibers

Rostral

Rostral

Ventral

Dorsal

Ventral

Dorsal

Lateral vest. nucleus

▓ Forelimb
░ Hindlimb

Fig. 14-5. Diagram illustrating the somatotopical organization of projections from the hindlimb and forelimb areas in cerebellar vermis in the cat to a fastigial nucleus and its relay to the lateral vestibular nuclei. (After Walberg *et al.,* 1962.)

trapezoid body (see Fig. 12-2, p. 187), which in the dog is seen exposed on the ventral surface caudal to the pontine protuberance (see Fig. 2-5, p. 20).

Rostral Cerebellar Peduncle

The rostral (superior) cerebellar peduncle (brachium conjunctivum) is the most medial of the three peduncles as seen in a dorsal view of the brain stem with the cerebellum removed (Fig. 14-6). In contradistinction to the other two peduncles, the principal fibers of the rostral cerebellar peduncle are efferent from the cerebellum. One outstanding afferent tract, however, is the ventral spinocerebellar pathway. Grossly the peduncle attaches to the mesencephalon to distribute fibers to the following nuclei: red nucleus, reticular formation, and thalamus. The brachium contains fibers from all the cerebellar nuclei (Cohen *et al.,* 1958). Efferent messages from the cerebellum via the rostral cerebellar peduncle are sent in duplicate to permit synchronized coordination of motor mech-

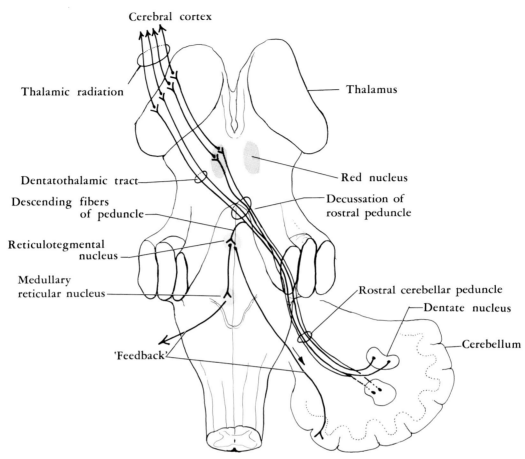

Fig. 14-6. Diagram illustrating the major efferent fibers in the rostral (superior) cerebellar peduncle (brachium conjunctivum). The cerebellar efferents to the red nucleus cross in the midbrain tegmentum as the decussation of the brachium conjunctivum. From the red nucleus these fibers are relayed rostrally to the thalamus and then to the cerebral cortex as illustrated. In addition, the descending fibers of the rubrospinal and rubrobulbar tracts to the lower motor neurons are influenced by the cerebellum via this pathway relay (not shown in the figure). Some fibers, especially from the lateral (dentate) nucleus, bypass the red nucleus and pass directly to the thalamus as the dentatothalamic tract. Other descending fibers and a feedback mechanism to the cerebellum are illustrated.

anisms at all levels. The cerebellum influences the rubrospinal tract via the nucleus interpositus, which sends efferents through the rostral cerebellar peduncle to the contralateral red nucleus. This circuit is completed by the rubrospinal fibers to chiefly the flexor motor neurons (Pompeiano, 1967). The dentate nucleus also sends fibers via the rostral cerebellar peduncle to the contralateral red nucleus which are then relayed to the thalamus (dentatorubrothalamic fibers) and directly from the lateral (dentate) nucleus to the thalamus (dentatothalamic fibers) (Fig. 14-6). As the figure illustrates, terminal axons enter the thalamic radiation (thalamocortical fibers) and extend to the cerebral cortex. Note that within the mesencephalic tegmentum the fibers of the rostral cerebellar peduncle cross to the opposite side as the *decussation of the brachium conjunctivum,* or *decussation of the rostral cerebellar peduncle* (Fig. 14-6; Plate 19).

Spinocerebellar Tracts; Summary of Functions

Recall that the nucleus of Clarke in the spinal cord gives rise to the dorsal spinocerebellar tract. This tract and the ventral spinocerebellar tract convey impulses to the cerebellum from only the hindlimb and lower trunk in the cat. The cervical cord equivalent to the nucleus (column) of Clarke is the lateral (accessory) cuneate nucleus, located in the lower medulla lateral and a little rostral to the cuneate nucleus (Brodal, 1969; Plate 10). Fibers from this nucleus enter the cerebellum via the caudal cerebellar peduncle as the *cuneocerebellar tract.* Anatomically this is for the forelimb, neck, and upper trunk what the dorsal spinocerebellar tract is for the hindlimb and lower trunk.

Somatotopical Localization

Numerous experimental procedures have been performed in the cat to study the functional localization in the cerebellum. Somatotopical mappings of fiber connections between the cerebellum and brain stem have been drawn in detail (Brodal, 1967). The vestibular fastigial nuclear somatotopic organization has been referred to and illustrated in Figure 14-5. Pompeiano (1967) presented an illustrated report of the functional organization of cerebellar projections to the spinal cord in the cat.

The concept that the cerebellum is topologically patterned in a single complete outline of the body has been conclusively discarded (Elliott, 1969). As illustrated in Figure 14-7, the body parts are represented in two cerebellar cortical areas; one with the head pointed rostrally and the other with

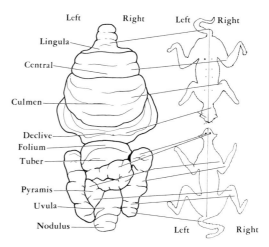

Fig. 14-7. Diagrammatic illustration of somatotopical localization in the cerebellum of cat determined by observation of movements produced by stimulation of cerebellar cortex in the decerebrate animal. (After Hampson *et al.*, 1952.)

the head pointed caudally (Hampson *et al.*, 1952).

Functions and Dysfunctions of the Cerebellum

As alluded to in the introduction to this Chapter the cerebellum has no primary motor nuclei and no direct fiber connections to the lower motor neurons of the cerebrospinal nerves, but it exerts its important influence on the synergy of motor activity, equilibrium, and muscle tone. The proprioceptive qualities of the cerebellar action are not at the conscious level. Defects resulting from lesions of the cerebellum apparently cannot be voluntarily controlled or modified to any extent (Gardner, 1968).

It is generally believed that the stretch receptors (*i.e.*, muscle spindles and Golgi tendon organs) do not participate in the circuit which reaches the level of position sense. They serve to provide an input of information to the cerebellum, primarily via the spinocerebellar tracts, concerning the state of muscular tone for proper posture, equilibrium, and synergy. It should be understood that in order for these complex phenomena to be executed properly there must be a delicate reciprocating action between muscle groups. The *agonist* is the muscle which actually provides the desirable movement, and the *antagonist* is the opposing muscle which must simultaneously relax to permit the smooth action; for example, while the flexors of the limb are contracting, the opposing extensors are relaxing. To add to the complexity of the process, it also is necessary for certain muscle groups to fix the proper joints for the desired synergistic actions of the agonist and antagonist. It is the cerebellum which acts in an involuntary fashion, not to initiate, but to modulate the muscular action required for proper sitting, standing, walking, running, and jumping.

The correlation of cerebellar lesions with the symptomatology may in general be based on the three phylogenetic developmental segments. Lesions involving the caudal cerebellar vermis and the flocculus result in the *archicerebellar syndrome.* To mention two possible causes: (1) a tumor in the roof of the fourth ventricle may press on the nodulus and peduncles of the flocculi, or (2) internal hydrocephalus may produce pressure on the archicerebellum. The closely associated vestibular mechanism also may be affected, to produce nystagmus and bizarre positions. Muscle tone is not altered, but the animal walks with an incoordinated, broad-based, staggering gait with a tendency to fall to either side or backward when held up on its hind legs. Experimentally, the emetic effects of motion in susceptible dogs have been abolished by ablation of the nodulus (Tyler and Bard, 1949).

Experimental data suggest that the paleocerebellum (chiefly the anterior lobe) exerts an inhibitory effect on muscle tone. Anterior cerebellar lobe lesions, therefore, result in a form of extensor rigidity plus asynergy as the outstanding symptoms, whereas stimulation of the lobe decreases extensor

rigidity from other sources (Elliott, 1969). Elliott suggested that this effect is greater in subprimates, and related the paleocerebellum with the quadruped form of locomotion where extension is the basic body posture. Therefore, synergistic action involving the cerebellar influence would require that extension be inhibited.

Phylogenetically the neocerebellum is the newest acquisition and therefore the cerebellar hemispheres are functionally correlated with modulation of the fine movements of the extremities, best utilized in man. Hemispheric lesions, therefore, produce asynergy with related symptoms, some of which are tested easily in man but with difficulty in animals. Many of the pathologic symptoms may be interpreted as various forms of asynergic disturbances. *Dysmetria,* the inability to judge distances, probably is best observed when the animal is attempting to ascend or descend a flight of stairs and during feeding. *Ataxia,* or incoordination, is characterized by contractions of muscle groups which are irregular in force and direction. This is observed when the animal staggers and demonstrates jerky movements while walking. When standing, the animal will show a wide-based stance to overcome the ataxia. Intention tremor is the uncontrollable involuntary shaking which takes place mostly in the extremities as the animal attempts to perform purposeful movements, such as placing the paw to hold a bone for chewing. In addition, the animal may show muscle weakness or flabbiness and cerebellar nystagmus, which is nystagmus on fixation. It should be kept in mind that unilateral cerebellar lesions manifest their effects ipsilaterally.

Experimental Procedures for Study of Cerebellar Functions

The relative ease with which the cerebellum can be approached surgically and the possibility of observing definite subdivisions grossly or with the aid of a surgical dissecting microscope are factors favoring study of cerebellar function by electrical stimulation and lesions, or by surgical ablation techniques. Sprague and Chambers (1959) reported the effects of cerebellar and cortical lesions in the cat. The discussion below is based on their work to present the results of the nervous system deficiencies *per se,* but also to include some of the important ancillary problems of postoperative animal care, which must be handled properly in chronic experiments. The discussion also elucidates the progressive behavioral improvement to a certain point in time (two weeks in the reported experiment), after which the syndrome becomes stabilized and the deficiencies permanent. Truly, these considerations are obvious to the researcher who has performed such experiments, but to the uninitiated reader such factors usually are not realized.

Total Cerebellectomy

Total cerebellectomy included removal of the total cerebellar cortex plus the deep nuclei within the medullary core. The major symptoms during the first three or four days postoperatively were: Frequent spasms of opisthotonus occurred with marked extension of all limbs, especially the forelimbs. Spasticity occurred in the limb and masticatory muscles. Vestibular symptoms of head tilt, horizontal resting nystagmus, and rolling were absent. The placing, righting, and hopping reflexes were absent, and the cat had no coordination. The bladder was atonic and urine had to be expressed manually. Because of spasticity in muscles of mastication and incoordination of the tongue plus malocclusion, forced feeding was necessary.

By seven days after operation the opisthotonus had disappeared and the spasticity was reduced. Righting of the forebody and visual placing occurred, but the latter was clumsy and hypermetric.

During the second week there was im-

provement, but muscular coordination was poor and there was no coordination between forelimbs and hindlimbs. The animals could support themselves when placed upright, but stepping and walking were never initiated. The cats seemed to show more interest in their surroundings and some would attempt to bat and chew a moving string, but with very little coordination or accuracy.

At approximately two weeks postoperatively the cerebellar syndrome became stabilized, with very slight improvement up to a survival time of over six months. The cardinal persistent symptoms of the syndrome were: lack of coordination between forelimbs and hindlimbs, presence of tremor and ataxia in all movements, lack of total bodily righting, inability to stand without support or eat without help. As expected, there were variations among the animals as to degrees of improvement.

Cerebral Lesions in Totally Cerebellectomized Cats

In an attempt to learn more about the influence of the cerebral cortex on the cerebellum, Sprague and Chambers (1959) placed cerebral lesions in the chronic cerebellectomized cats used to obtain the data given above. As a second group they cerebellectomized cats which had previously received cerebral cortical lesions. They were cognizant that cats with total neocortical ablation retain essentially normal locomotion (Bard and Rioch, 1937). From their experiments Sprague and Chambers concluded that ablation of only the sigmoid (cruciate) gyri or most of the neocortex greatly reduces or abolishes cerebellar ataxia and tremor. Placing and hopping reflexes are extremely reduced or abolished by loss of either cerebral cortex or cerebellum, whereas the righting reflex survives destruction of either the cerebral cortex or the cerebellum, but not of both.

Bibliography

Bard, P., and Rioch, D. McK.: A study of four cats deprived of neocortex and additional portions of the forebrain. Bull. Johns Hopkins Hosp. *60:* 73–147, 1937.

Brodal, A.: Anatomical Studies of Cerebellar Fibre Connections and Functional Localization. In Vol. 25, *Progress in Brain Research* (Fox, C. A., and Snider, R. S., Eds.). Amsterdam, Elsevier Publ. Co., 1967.

Brodal, A.: *Neurological Anatomy in Relation to Clinical Medicine,* 2nd ed. New York, Oxford University Press, 1969.

Cohen, D., Chambers, W. W., and Sprague, J. M.: Experimental study of the efferent projections from the cerebellar nuclei to the brainstem of the cat. J. Comp. Neurol. *109:* 233–259, 1958.

Crosby, E. C., Humphrey, T., and Lauer, E. W.: *Correlative Anatomy of the Nervous System.* New York, Macmillan Co., 1962.

Crosby, E. C., Schneider, R. C., De Jonge, B. R., and Szonyi, P.: The alterations of tonus and movements through the interplay between the cerebral hemispheres and the cerebellum. J. Comp. Neurol. *127* (Suppl. 1): 1–91, 1966.

Elliott, H. C.: *Textbook of Neuroanatomy,* 2nd ed. Philadelphia, J. B. Lippincott Co., 1969.

Flood, S., and Jansen, J.: On the cerebellar nuclei in the cat. Acta anat. (Basel) *46:* 52–72, 1961.

Gardner, E.: *Fundamentals of Neurology,* 5th ed. Philadelphia, W. B. Saunders Co., 1968.

Hampson, J. L., Harrison, C. R., and Woolsey, C. N.: Cerebro-cerebellar projections and the somatotopic localization of motor function in the cerebellum. Res. Publ. Ass. Nerv. Ment. Dis. *30:* 299–316, 1952.

Larsell, O.: The morphogenesis and adult pattern of the lobules and fissures of the cerebellum of the white rat. J. Comp. Neurol. *97:* 281–356, 1952.

Nomina Anatomica Veterinaria, Adopted by the General Assembly of the World Association of Veterinary Anatomists, Paris, 1967. Vienna, Adolf Holzhausen's Successors, 1968.

Pompeiano, O.: Functional Organization of the Cerebellar Projections to the Spinal Cord. In Vol. 25, *Progress in Brain Research* (Fox, C. A., and Snider, R. S., Eds.). Amsterdam, Elsevier Publ. Co., 1967.

Pompeiano, O., and Brodal, A.: Spino-vestibular fibers in the cat—An experimental study. J. Comp. Neurol. *108:* 353–381, 1957.

Ranson, S. W., and Clark, S. L.: *The Anatomy of the Nervous System—Its Development and Function,* 10th ed. Philadelphia, W. B. Saunders Co., 1959.

Sherrington, C.: *The Integrative Action of the Nervous System,* New Haven, Yale University Press, 1947; Issued as a Yale Paperbound, Forge Village, Mass., The Murray Printing Co., 1961.

Sprague, J. M., and Chambers, W. W.: An analysis of cerebellar function in the cat, as revealed by its partial and complete destruction, and its interaction with the cerebral cortex. Arch. ital. Biol. *97:* 68–88, 1959.

Truex, R. C., and Carpenter, M. B.: *Human Neuroanatomy,* 6th ed. Baltimore, Williams & Wilkins Co., 1969.

Tyler, D. B., and Bard, P.: Motion sickness. Physiol. Rev. *29:* 311–369, 1949.

Walberg, F., Pompeiano, O., Brodal, A., and Jansen, J.: The fastigiovestibular projection in the cat. An experimental study with silver impregnation methods. J. Comp. Neurol. *118:* 49–75, 1962.

chapter 15

Diencephalon

The diencephalon is the most rostral segment of the brain stem. Because of the
intimate association of the telencephalon
and diencephalon, many authors prefer to
exclude the diencephalon from the brain
stem. The German term for diencephalon
is *Zwischenhirn,* which is translated as 'between brain' (Skinner, 1961).

The segments of the brain stem caudal
to this area also are bilateral, but as a solid
core with the right and left sides as almost
completely attached mirror images. The
sides of the diencephalon, however, are
almost completely separated by the vertically oriented midline third ventricle. The
two thalami are joined by a cylindrical connection, the *massa intermedia,* or *interthalamic
adhesion* (Fig. 15-1), which is surrounded by
the third ventricle.

In human brain the interthalamic connection is much smaller, being very little
larger in diameter than the anterior commissure. The functional significance of this
structure in man seems questionable, since
approximately only 80 per cent of human
brains have an interthalamic adhesion
(Truex and Carpenter, 1969).

The complicated diencephalon may be
subdivided into: (*a*) epithalamus, (*b*) thalamus, or dorsal thalamus, (*c*) hypothalamus,
(*d*) subthalamus, or ventral thalamus, and
(*e*) metathalamus. All of the divisions except
the subthalamus may be identified grossly—
the first three in a sagittal section (Fig. 15-1),

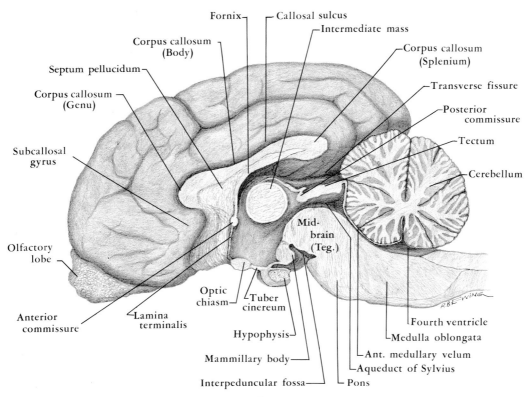

Fig. 15-1. Sagittal section of the brain of the dog.

and the last in a whole brain. A portion of the hypothalamus may be identified grossly on the ventral surface of the whole brain (see Fig. 2-5, p. 20), as well as in a sagittal view of the hemisected brain. The subthalamus is caudal and lateral to the hypothalamus and is not visible either in a whole brain or on sagittal section. The metathalamus (medial and lateral geniculate bodies) is discussed in Chapter 13 because of its structural and functional intimacy with the tectum of the midbrain.

Epithalamus

The epithalamus (Gk. *epi*—upon), the most dorsal portion of the diencephalon, forms a caplike covering of the thalamus. The anatomic components of the epithalamus are: the pineal body (gland), or epiph-

ysis; the habenulae with their commissure; the striae medullares; and the posterior commissure, which some authors exclude as a component of the epithalamus. All of these structures either border or cross or are exactly in the midline.

Pineal Body

The pineal body (gland), or epiphysis (Fig. 15-1), is extremely small in the dog. According to Meyer (1964), it is only about 1 mm long. It is attached at the dorsocaudal midline of the diencephalon and points caudally between the rostral colliculi of the midbrain. There are many unsolved mysteries regarding the anatomic, physiologic, and pathologic aspects of the pineal body. The literature contains numerous interesting reports of various types of studies on this organ, but there is little agreement among researchers regarding interpretation of the data. A species-specific characteristic of the pineal body in the ox is the presence of a few smooth and striated muscle fibers (Trautmann and Fiebiger, 1957) in addition to the normally expected brain sand (calcareous particles) and possibly amyloid bodies. There are many conflicting reports regarding the endocrinologic functions of this organ, which seem unnecessary to cite here.

Habenula

Each habenula (Plate 22) is an oval-shaped eminence of the epithalamus, which is divisible into a medial and a lateral nucleus. The smaller medial nucleus is more homogeneous in appearance, whereas the lateral nucleus has fiber bundles of the *stria medullaris* within it. The more definite connections of each habenula are the *stria medullaris* as the major afferent and the *habenulopeduncular tract (fasciculus retroflexus of Meynert)* as the major efferent tract.

The stria medullaris is a complex bundle of fibers conveying impulses to the habenula possibly from the hippocampus, septal area, amygdaloid area, and the preoptic area. Some fibers of the stria medullaris are contralateral, in that they cross to the opposite side within the habenular commissure. The efferent habenular fibers of the habenulopeduncular tract (Plate 22) arise from both the medial and lateral habenular nuclei and form a definite bundle which passes ventrocaudally through the dorsomedial area of the thalamus and the red nucleus before its primary termination in the interpeduncular nucleus of the mesencephalon.

Posterior Commissure

The posterior commissure (Plate 21) is located in the pretectal area immediately rostral to the mesencephalic tectum as an inconspicuous structure. Many authors do not include it as a component of the epithalamus. The functions of the posterior commissure are obscure. Crosby *et al.* (1962) cite evidence that stimulation in and lateral to this commissure results in lowering of the head and trunk in the cat. They mention species variations with a reduction in size in primates as compared with that in other animals. Some of the components of the posterior commissure among mammals are listed as: (1) interconnections between the whole diencephalic-mesencephalic transitional area; (2) tectohabenular and habenulotectal fibers, some of which decussate in the commissure; (3) fibers from caudal areas of the brain and even the spinal cord; and (4) fibers connecting the right and left medial longitudinal fasciculi.

Thalamus
(Plates 21, 22, 23, 24, 25)

The *thalamus*, or *dorsal thalamus*, is the largest mass of gray matter in the diencephalon. It is subdivided by medullary laminae into gray areas, which in turn are composed of numerous nuclei. The two thalami have the third ventricle as their *medial border*, except where they are joined by the massa intermedia (thalamic adhesion). *Laterally* the thalamus is bordered by the internal capsule, which separates it from the lenticular nucleus (putamen and globus pallidus). *Dorsolaterally* the caudate nucleus lies medial to the internal capsule. The caudate nucleus, with the large head rostrally and its smaller body, can be followed caudally in its dorsolateral relationship with the thalamus (Plates 23, 24, 25).

The *dorsal* surface of the thalamus is mostly free, in that it contributes to the floor of the lateral ventricle. The *ventral* boundary is anatomically marked medially by the hypothalamic sulcus (Plate 25) in the wall of the third ventricle and *ventrolaterally* the thalamus is in contact with the subthalamus. *Rostrally* the dorsal surface approaches and contributes to the walls of the *interventricular foramen* between the lateral and third ventricles. The head of the caudate nucleus extends rostrally beyond the thalamus (Plates 26 and 27). *Caudally* the thalamus is connected with the midbrain by the pretectal area at the diencephalic-mesencephalic junction (Plate 21). In this area the dorsocaudal thalamus (pulvinar) is lateral and more dorsal than the rostral colliculi of the mesencephalon.

Nuclei of the Thalamus
(Plates 22, 23, 24; Fig. 15-2)

There are different methods for classification of the thalamic nuclei, and there are different interpretations of specific areas. The following discussion of the thalamic nuclei is based primarily on the reports of Rioch (1929) and Ingram *et al.* (1932), which are based on brains of the dog and cat. In general, these reports coincide with the more recent three-dimensional model of the thalamus of the dog (Sychowa, 1961; Fig. 15-2). To emphasize the close similarities of basic anatomy of brain stems among various

mammals, in his classification of the nuclei in the dog thalamus Rioch stated that he followed particularly the description of the thalamus of the albino rat given by Gurdjian (1927). It should be understood that the following discussion is not intended to be complete, but it is hoped that it is sufficiently adequate so that, when combined with study of the plates, it will give the reader a basic understanding of the thalamus. For further details based on the presented classification, the reader is referred to the reports cited above.

The thalamus is composed of five nuclear groups based on their relative topography: nuclei of the midline, rostral (anterior), medial, lateral, and ventral groups.

The *internal medullary lamina* is a rather poorly defined longitudinal plate of fibers which passes through the thalamus to divide it into three of the basic nuclear groups: rostral (anterior), medial, and lateral. Toward the rostral end of the thalamus the internal medullary lamina bifurcates to form two borders of the rostral nuclear group.

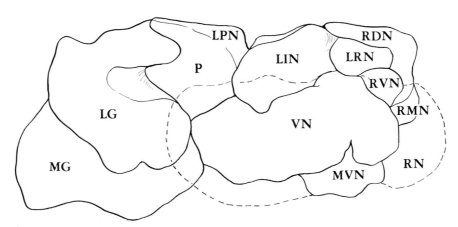

Fig. 15-2. Diagram representing lateral view of thalamus in dog, showing location of various nuclei. The reticular nucleus, indicated by dotted lines, has been removed. *LG*, lateral geniculate body; *LIN*, laterointermediate nucleus; *LPN*, lateroposterior nucleus; *LRN*, laterorostral nucleus; *MG*, medial geniculate body; *MVN*, medioventral nucleus; *P*, pulvinar; *RDN*, rostrodorsal nucleus; *RMN*, rostromedial nucleus; *RN*, reticular nucleus; *RVN*, rostroventral nucleus; *VN*, ventral nucleus. (After Sychowa, 1961.)

Midline Nuclei

The midline nuclei are so named because most of them form the gray matter of the massa intermedia between the two thalami, but some border the lateral walls of the third ventricle. Phylogenetically these nuclei are among the oldest and therefore they are classified as archethalamus, or primitive thalamus (Elliott, 1969). The midline nuclei in the dog are small and poorly defined, except for the *nucleus centralis medialis,* which lies in the midline between the right and the left internal medullary lamina (Plates 22 and 23). There are several other less distinct nuclei in the midline area, including the commissural (associated with the interthalamic adhesion), paraventricular (Gk. *para*—alongside) and periventricular (Gk. *peri*—around), referring to the third ventricle.

As might be expected from their phylogenetic age, the midline nuclei connect chiefly with other phylogenetically ancient structures: afferents largely from sensory relays through the reticular formation and rhinencephalon, and efferents mostly to the hypothalamus, basal ganglia, and amygdala. Their connections with the hypothalamus, primarily via the periventricular system, suggest that they may be involved with visceral activity. Other functions of these nuclei are obscure; it is thought they are more active in lower animals but have been retained possibly to activate the cortex in a vague general way.

Rostral Nuclei

The rostral (anterior) group of nuclei (Plate 24) compose the rostral, or anterior, tuberculum which can be observed grossly at the extreme dorsorostrolateral thalamic pole separated from the caudate nucleus by the *stria terminalis.* This group consists of three distinct nuclei: the rostro(antero)dorsalis, rostro(antero)ventralis, and the rostro(antero)medialis. The stems for the synonyms 'rostral' and 'anterior' are both

used here because the latter term is firmly established in the literature and more commonly used; however, the Nomina Anatomica Veterinaria (1968) stipulates 'rostral,' with no alternative.

The *rostrodorsal nucleus* (*n. anterodorsalis*) is the most dorsal of the rostral nuclei and appears as a dense cellular cap dorsal to the other two nuclei of the group lateral to the stria medullaris (Plate 24). The *rostroventral nucleus* (*n. anteroventralis*) lies in the ventrolateral portion of the rostral tuberculum deep to and extending laterally beyond the rostrodorsal nucleus. This is the largest of the three nuclei in the group at the level of the section shown in Plate 24. The rostral portion of the rostroventral nucleus receives a majority of the fibers of the mammillothalamic tract (Sychowa, 1961). The *rostromedial nucleus* (*n. anteromedialis*) is in the ventromedial area of the rostral tuberculum, deep to the other two nuclei (Plate 24).

Medial Nuclei

The medial group of nuclei, according to Rioch (1929), consists of nine nuclei which extend the entire length of the thalamus. The *medial dorsal nucleus,* the largest of the medial group, includes most of the area between the internal medullary lamina and the nuclei of the midline (Plates 22, 23, 24). In Plate 22 immediately ventral to the habenula is a small but distinct *nucleus of the habenulopeduncular tract,* within the tract.

Lateral Nuclei

The lateral group of nuclei form the dorsolateral area of the thalamus throughout its length between the internal and external medullary laminae. The boundaries of the individual nuclei are indistinct. In this area, too, many nuclei are described in detailed studies of the thalamus that are unnecessary for our consideration. Due to the above-stated anatomic vagueness and the lack of conclusive evidence for definite functional significance of the subdivisions, it seems adequate to consider this entire region as the *lateral nuclear area* (Plate 22), with the most caudal portion designated as the *pulvinar* (Plates 21 and 22). Comparison of slides of human and animal brains cut through this area, or gross observation of the brains, reveals the poorly differentiated pulvinar in the dog and cat when compared with that in man.

Lateral to the external medullary lamina of the thalamus is the thin, elongated *reticular nucleus* which follows the contour of the external medullary lamina and the parallel concave medial surface of the internal capsule on its lateral side (Plates 22, 23, 24).

Ventral Nuclei

The ventral group of thalamic nuclei occupies the ventro(lateral) area of the thalamus (Plates 22, 23, 24). The general boundaries of the ventral nuclear group are: *rostrally*, at the level of the optic chiasm, the nucleus reticularis, from which it is separated by the external medullary lamina; *dorsally*, the lateral group of nuclei; *medially*, the midline nuclei and the ventromedial portion of the internal medullary lamina; *laterally*, the external medullary lamina; *caudally* and *laterocaudally*, the medial geniculate body; and *ventrally*, the ventral portion of the external medullary lamina, which curves somewhat horizontally to separate the ventral thalamic nuclear group from the subthalamus and hypothalamus.

The ventral group of thalamic nuclei is subdivided into rostral (anterior) and caudal (posterior) segments in the parasagittal plane, plus medial and lateral segments in the horizontal plane. Each quadrant is referred to as a specific nucleus, for example, the ventral posterior (caudal) medial nucleus and the ventral posterior (caudal) lateral nucleus, as identified in Plates 22 and 23. Such specific details are mentioned for this nuclear group because of their functional significance, considered below.

Functions of the Thalamus

In a general way the thalamic nuclei may be classified according to their fiber connections. Thalamocortical axons from thalamic nuclei enter the internal capsule to be distributed to the cerebral cortex. Such fibers form thalamic radiations as the last neurons in the long ascending (sensory) pathways. It is generally agreed that all sensory impulses, with the single exception of those of olfaction, enter the thalamus, from which they are relayed to specific cerebral cortical areas via the thalamocortical radiations.

In addition to the specific relay by thalamocortical fibers, that is, from one single nucleus directly to a specific cortical area, there are diffuse cortical or subcortical connections, or both, from dispersed nuclei. The thalamus is complex and should not be considered solely as a relay center; for example, there is substantial reason to conclude that the thalamus is the chief sensory-integrating mechanism of the neuraxis (Truex and Carpenter, 1969). Those authors maintain that there is abundant evidence that certain thalamic areas play a dominant role in the maintenance and regulation of states of consciousness, awareness, attention, and alertness.

The thalamus also participates in the emotional factors that are associated with most sensory inputs to the cerebral cortex. The thalamic afferents from the cerebellum and the basal ganglia indicate that the thalamus also functions as an integrative center for motor activity.

It should be remembered that there are two types of sensation: discriminative and affective. Both types are undoubtedly present to some extent in dogs and cats, and the basic general neuronal patterns are similar to those in man. The more highly differentiated cerebral cortex in man suggests that discriminative sensory mechanisms would be much more prominent in man than in the dog; and the relatively 'simple life' of lower animals, even cats and dogs, leads one to question the significance of affective sensation in subhuman animals.

As alluded to above, the discriminative sensations are related to the specific thalamocortical relays from a single nucleus to a specific cortical area. A highly complicated cerebrum is necessary for the degree of sophistication that enables man to identify precisely and compare stimuli with regard to such relative properties as intensity, location, position, texture, and time.

Specific nuclei which are noted for such precise relays are the ventral caudo-(postero)lateral and ventral caudo(postero)-medial nuclei. The *ventral caudolateral nucleus* is a relay center for GSA exteroceptive and proprioceptive impulses from the body. This nucleus receives the spinothalamic tracts plus the medial lemniscus and relays their impulses to the sensory cerebral cortex. The *ventral caudomedial nucleus* receives fibers conveying GSA exteroceptive and proprioceptive impulses from the head area and relays them to the sensory cerebral cortex. The medial and lateral geniculate bodies of the metathalamus also belong to this specific relay system for terminal cortical representation of hearing and vision, respectively.

These are very complex anatomic and functional interrelationships of structures of the brain and there are many unanswered questions and controversies that arise when comparisons are attempted between the brains of humans and subhumans. It may be argued that certain areas of the dog cerebral cortex may be more highly differentiated than the comparable area in man, for example, the auditory radiations from the medial geniculate body to the auditory cortex (see Chapter 19).

The affective quality of sensation results from the subjective interpretation of stimuli which depends on the physiologic and psychologic state of the individual. Elliott (1969) uses the term 'affect' as a synonym for 'emotion,' and refers to the affective influence of the thalamus on the cerebral cortex as an 'emotionally motivating influence.' Even in man, this is difficult to define qualitatively and identify, and it obviously differs among individuals. I realize the possible risk involved in suggesting that dogs possess an affective quality of sensation; nevertheless it seems obvious that different breeds of dogs or those raised in different types of environment demonstrate different temperaments and 'personalities.' This is the general non-specific or non-discriminatory quality that is credited primarily as thalamic-cortical interaction.

Hypothalamus
(Plates 22, 23, 24)

The hypothalamus is the most ventral portion of the diencephalon, being exposed on the ventral surface of the brain. The hypothalamic structures on the ventral surface, from rostral to caudal, are: optic chiasm, infundibulum and neurohypophysis, tuber cinereum, and the mammillary bodies. The neurohypophysis (posterior lobe of the pituitary gland) frequently is not included as a component of the hypothalamus. When it is included, one is reminded that the neurohypophysis is composed of nervous tissue and that there are specific tracts to it from the rostral area of the hypothalamus. The hypophysis is attached to the brain by the infundibulum (stalk of the pituitary). In routine removal of the brain the pituitary is most frequently separated from the brain by severance through the infundibulum. Grossly, one can observe the hollow center of the infundibulum, the ventral extension of the third ventricle. Since the neurohypophysis is attached to the infundibulum, which in turn is attached to the tuber cinereum, it seems logical to consider a simple plan of hypothalamic organization for nuclear content based on only three basic areas with reference to the optic chiasm, tuber cinereum, and mammillary bodies.

Before the specific hypothalamic nuclei are discussed the anatomic boundaries of the hypothalamus should be reviewed. *Medially,* the right and left hypothalami are separated by the narrow, vertically oriented slitlike third ventricle. *Dorsally,* the hypothalamic sulcus, a groove in the wall of the third ventricle, separates the hypothalamus from the (dorsal) thalamus. *Laterally,* the hypothalamus is indistinctly separated from the subthalamus. *Rostrally,* the hypothalamus is indistinct from the preoptic area, but for convenience the optic chiasm is generally considered the rostral border. *Rostrolaterally,* the first segment of the optic tract borders the hypothalamus. *Caudally,* the mammillary bodies are considered as the limiting border, but there is a melding into the caudal (posterior) perforated substance and tegmentum of the midbrain. *Caudolaterally,* the cerebral peduncles form a border on the ventral surface.

Hypothalamic Nuclei

In a rostro-caudal sequence there are three gray areas, which are subdivided into specific nuclei. The areas are: **(1)** Supraoptic (rostral); **(2)** tuberal (intermediate); and **(3)** mammillary (caudal).

The specific nuclei of the hypothalamus are not as well defined anatomically as those of the thalamus. Since frequently there are no exact boundaries between nuclei, identification is based on the relative positions of cellular condensations and sub-areas within the three major areas. The anatomic identification and hypothalamic connections of pathways combined with the functional and dysfunctional specificities aid in positive identification of localized nuclei or sub-areas.

Supraoptic Area

The supraoptic, or rostral, area lies above the optic chiasm and fuses rostrally with the preoptic area which extends into the telencephalon. The supraoptic area contains two functionally well-known nuclei, the *supraoptic* and the *paraventricular.* The former tends to stay associated with the dorsal side of the optic tract and therefore mostly lateral to the latter, which lies dorsomedially in close proximity to the wall of the third ventricle. Both of these nuclei give rise to axons which pass to the neurohypophysis, or posterior lobe of the pituitary gland. The supraoptic nucleus is referred to as the *nucleus tangentialis* in the cat by Ingram *et al.* (1932) and in the dog by Rioch (1931), who does not state the synonyms but illustrates the position of the nucleus and has labeled it 'nucleus tangentialis.'

The Nomina Anatomica Veterinaria (1968) and reports more recent than those cited above refer to the nucleus as supraoptic in many animals: rat (deGroot, 1959), dog (Singer, 1962), cat (Snider and Niemer, 1961), and man (House and Pansky, 1967). The paraventricular nucleus also is referred to in the older literature as the *nucleus filiformis principalis.*

A distinction should be made between the paraventricular and the periventricular nucleus. The larger portion of the *paraventricular nucleus* is a functionally important region in the preoptic area best demonstrated in cross sections near the optic chiasm where it is in the dorsal area of the hypothalamus as illustrated in Plates 24 and 25. The *periventricular nucleus* (Plate 24) is in the border of the third ventricle and not restricted to any rostro-caudal area, nor is it restricted to the hypothalamus.

Bordering the third ventricle in the dorsal thalamic area are cells and fibers of the periventricular system. The location of the periventricular nucleus and fibers in the wall of the third ventricle indicates the probable association with the central group of thalamic nuclei. In lower mammals (below carnivores), the central group of nuclei and fibers form a commissural system between the right and left dorsolateral thalamic areas. In carnivores the periventricular system possesses vertically oriented fibers connecting the thalamus and hypothalamus (Rioch, 1931).

Tuberal Area

The tuberal, or intermediate, area of the hypothalamus refers to the area dorsal to the tuber cinereum located on the ventral surface of the brain between the optic chiasm and the mammillary bodies. The *infundibulum,* or stalk of the pituitary gland, attaches to the *median eminence* of the tuber cinereum. Near the infundibular attachment the ventral portion of the *periventricular nucleus* is indistinguishable from the *arcuate nucleus,* and both names are used collectively for identification of the gray area enveloping the base of the third ventricle (Plate 24).

The *dorsomedial* and *ventromedial nuclei* (Plate 24) are indistinct areas immediately lateral to the periventricular nuclear border of the third ventricle. Their relative positions are indicated by their names, but there is no distinct separation between the two nuclear areas. Lateral to these nuclei is the *lateral hypothalamic nucleus,* which includes the area bordered medially by the dorsomedial and ventromedial nuclei and laterally by the hippocampal fissure, optic tract, and subthalamus.

Mammillary Area

The mammillary, or *caudal hypothalamic,* area is named for the mammillary bodies, which are the most prominent structures in this area (Plates 22 and 23). Plate 23 shows fibers of the fornix terminating in the lateral portion of the mammillary body. From the medial portion of the body heavily myelinated fibers of the *mammillothalamic tract (of Vicq d'Azyr)* are easily identified as they pass dorsolaterally. Directly dorsal to the mammillary bodies between the diverging mammillothalamic fibers is the *caudal hypothalamic nucleus.*

Major Afferent and Efferent Fibers of the Hypothalamus

Some of the fiber connections of the hypothalamus are conspicuous tracts, whereas others are rather diffuse. Examples of hypothalamic afferents and efferents indicate the complexity of the interconnections of the hypothalamus with the rest of the brain.

Afferent Pathways

FORNIX

The hippocampal–fornix relations are considered in Chapter 16, in the discussion of the rhinencephalic–limbic systems. The hippocampus originates in the piriform area and arches caudodorsally to continue rostrally beneath the corpus callosum as the fornix. The fornix arches rostroventrally where fibers pass rostrally and caudally to the anterior commissure (see Fig. 16-12, p. 257). The caudal fibers pass ventrally as the postcommissural fibers of the fornix and extend ventrocaudally to terminate in the lateral portion of the mammillary body (Plate 23).

MEDIAL FOREBRAIN BUNDLE

At least a portion of the medial forebrain bundle (olfactohypothalamic tract) originates in the septal nuclear area and the olfactory stria area of the rhinencephalon. Fibers are dispersed throughout the hypothalamic nuclei, including the lateral portion of the mammillary body. Caudal to the hypothalamus this bundle continues to the midbrain tegmentum.

THALAMOHYPOTHALAMIC FIBERS

The thalamohypothalamic (periventricular) fibers have been mentioned previously as a connection between the medial thalamic and hypothalamic nuclei.

MAMMILLARY PEDUNCLE

The mammillary peduncle arises caudally in the midbrain tegmentum and is thought to transmit impulses to the hypothalamus indirectly from the ascending tracts of the brain stem and spinal cord.

STRIA TERMINALIS

The stria terminalis also begins in the piriform area, more specifically in the amygdaloid complex. The fibers arch in a course similar to the first half of the hippocampus–fornix fibers, but axons of the stria terminalis terminate in the rostral hypothalamic area.

RETINAL FIBERS

The existence of fibers from the retina to the hypothalamus in carnivores is controversial.

Efferent Pathways (Fig. 15-3)

MAMMILLARY BODY EFFERENTS

From the mammillary body there is a short vertical tract, the *fasciculus mammillaris princeps*, which projects dorsally and divides into two: the *mammillothalamic* and the *mammillotegmental tract*. By careful gross dissection with the aid of a dissecting microscope these tracts may be identified. The mammillothalamic tract passes rostrodorsally to terminate in the rostroventral nucleus of the thalamus. This nucleus probably interchanges fibers with the cingulate gyrus, and therefore interaction between the hypothalamus and the cortical region is possible. The mammillotegmental tract courses caudally to reach the tegmentum of the midbrain and thereby establishes a caudal brain-stem connection with the hypothalamus.

PERIVENTRICULAR FIBERS

The lightly myelinated and nonmyelinated periventricular fibers pass to and from the dorsomedial and midline thalamic nuclear areas. These areas are in general connected with the frontal cortex and therefore put the hypothalamus under the influence of the cerebral cortex. In addition, there are periventricular fibers that pass caudally through the ventral periaqueductal gray matter of the mesencephalon and in the subependymal area in the floor of the fourth ventricle as the *dorsal longitudinal bundle (of Schütz)* (Fig. 15-3; Plate 18).

HYPOTHALAMOHYPOPHYSEAL TRACT

The hypothalamohypophyseal tract is of special interest to neuroendocrinologists. From the supraoptic and paraventricular nuclei nonmyelinated axons pass toward the neurohypophysis as the *supraopticohypophyseal tract*. As this tract passes through the median eminence area at the base of the infundibulum it is joined by a smaller bundle of fibers, the *tuberohypophyseal tract*, from the tuber cinereum. The combination of the two tracts forms the *hypothalamohypophyseal tract*, which extends to the posterior lobe of the pituitary gland (hypophysis) (Fig. 15-3).

Functions of Hypothalamus

An indication of the important functions of the hypothalamus is that this small area of the brain directs the following bodily functions: appetite, water balance (*i.e.,* between water consumption and urinary output), sexual performance and behavior, sleep-wake cycle, body temperature, blood pressure regulation, and emotions.

Two important regulatory mechanisms are under the physiologic jurisdiction of the hypothalamus: (1) the autonomic nervous system, and (2) the pituitary gland.

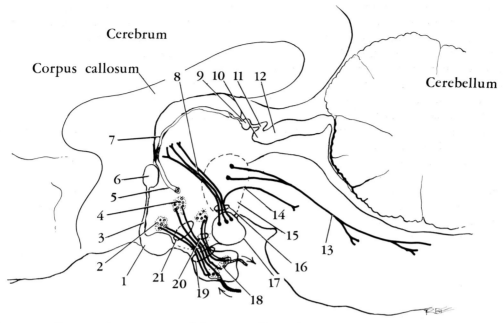

Fig. 15-3. Diagram showing some of the major efferent hypothalamic pathways. 1, optic chiasm; 2, supraoptic nucleus; 3, lamina terminalis; 4, paraventricular nucleus; 5, column of fornix; 6, anterior (rostral) commissure; 7, stria medullaris thalami; 8, mammillothalamic tract; 9, habenula; 10, posterior (caudal) commissure; 11, pineal body; 12, tectum of midbrain; 13, dorsal longitudinal fasciculus; 14, mammillotegmental tract; 15, caudal nuclear area; 16, fasciculus mammillaris princeps; 17, mammillary body; 18, capillary plexus in neurohypophysis; 19, hypothalamohypophyseal tract; 20, tuberohypophyseal tract; 21, supraopticohypophyseal tract.

Hypothalamic–Autonomic Nervous System Relations

The hypothalamic internuclear connections serve as a mechanism for balancing the sympathetic and parasympathetic activities. There are many hypothalamic areas other than the single example used above which contribute to the dorsal longitudinal bundle (of Schütz). This fasciculus plus other multisynaptic relays give off fibers which connect with the parasympathetic nuclei of the brain stem.

Some of the hypothalamic areas have anatomic and physiologic relations with both the sympathetic and parasympathetic systems. This is merely one example of the extreme complexity of hypothalamic activity. In general, however, it may be stated

that the rostromedial hypothalamic area is parasympathetic dominant whereas the caudolateral region is sympathetic dominant.

Chapter 9 discusses specific characteristics of the sympathetic and parasympathetic systems correlated with their physiologic effects. The following discussion includes the hypothalamic influence on autonomic functions. It should be kept in mind that autonomic effects are far more widespread than are the sympathetic and parasympathetic efferent neurons from the neuraxis to the target organs. There is constant interplay between the autonomic nervous system and the endocrine system. The hypothalamus and other brain areas (*e.g.,* certain cortical areas) are reciprocating their influences, which in turn produce an impact on the emotional and personality factors involved in autonomic function.

When the hypothalamus is released from cortical control, by either decortication or a lower interruption of the cortical connections with the hypothalamus, an animal demonstrates an exaggerated state of emotional stress and anger which is termed 'sham rage.' The name 'sham rage' is given to the syndrome because it is reasoned that without cerebral cortical function the animal undoubtedly does not feel angry, but merely looks (and acts) as if it were angry (Crosby *et al.*, 1962). The cat is commonly the animal of choice for experimentally producing sham rage for demonstration in medical teaching laboratories. Extreme caution must be used in handling and caring for such an animal.

Based on the caudolateral hypothalamic area as the sympathetic dominant region as discussed above, it should not be surprising that stimulation of this area results in sympathomimetic responses characteristic of emotional stress and anger similar to those in sham rage. Examples of the specific responses are: erection of the hair, dilation of the pupil, elevation of blood pressure, increase in rate and amplitude of respiration, inhibition of activity of the gut and bladder, and overt somatic struggling movement (Truex and Carpenter, 1969). The same authors state what is expected, namely, destruction of the caudal hypothalamus results in emotional lethargy, abnormal sleepiness, and fall in body temperature due to the reduced visceral and somatic activity.

Experimental hypothalamic stimulations and lesions furnish additional evidence that the hypothalamus is a principal center concerned with emotional behavior. Bilateral lesions of the ventromedial nucleus in the tuberal region produce an animal that is hyperphagic (possesses a voracious appetite) and demonstrates savage, ragelike behavior. An animal may, for other reasons, exhibit rage and savage-like behavior without developing excessive appetite, but fol-

lowing hypothalamic lesions it is very rare for an animal to have an excessive appetite without exhibiting savageness (Ingram, 1956). These animals are not hungry but will eat heartily as long as food is available. Hunger is correlated with the need for food and is associated with an empty stomach, a low blood sugar, and the like, but these animals possess none of these symptoms and become excessively obese (Fig. 15-4). Ingram also states that bilateral lesions destroying portions of the lateral hypothalamic area reduce or abolish the desire for food in hyperphagic and normal animals. This obviously results in severe emaciation (Fig. 15-4). Based on the above data, the ventromedial hypothalamic nucleus is concerned with *satiety*, whereas the lateral hypothalamic area is considered to be a *feeding center.*

Evidence of the hypothalamic internuclear physiologic synergism is illustrated by Netter (1957). He indicates that destruction of the ventromedial hypothalamic nucleus produces rage, as discussed above, but the same behavioral change may be produced by stimulation of the dorsomedial hypothalamic nucleus.

Sympathetic and parasympathetic interaction participates in the regulation of body temperature. For many of the factors involved in body-temperature regulation the anatomic-physiologic correlations of rostromedial-parasympathetic and caudolateral-sympathetic seem apparent.

The caudal hypothalamus is stimulated by environmental changes which threaten a decrease in body temperature. Therefore this area is actively concerned with the conservation and the increased production of body heat. The cutaneous blood vessels are constricted and sweat secretion ceases in an attempt to prevent heat loss. In addition, there is an increase in visceral and somatic muscular activity for the advantageous change in body hair movement (piloerection in subhumans and goose pimples in man) and shivering. Bilateral lesions in the caudal

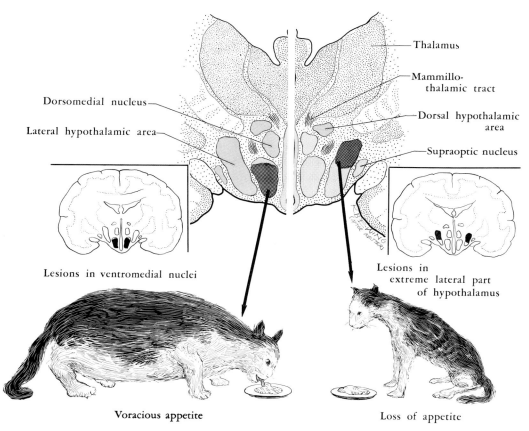

Thalamus

Mammillo-
thalamic tract

Dorsomedial nucleus

Lateral hypothalamic area

Dorsal hypothalamic
area

Supraoptic nucleus

Lesions in ventromedial nuclei

Lesions in
extreme lateral part
of hypothalamus

Voracious appetite

Loss of appetite

Fig. 15-4. The effects on appetite of discrete bilateral lesions in specific hypothalamic areas in the cat. Bilateral lesions in the ventromedial nuclei produce hyperphagia (voracious appetite), accompanied by rage. Bilateral lesions in the extreme lateral part of the hypothalamus produce loss of appetite (anorexia), which leads to emaciation. (After Netter, 1957.)

hypothalamic area destroy descending pathways concerned with both the conservation and dissipation of body heat and therefore produce a poikilothermic animal (*i.e.*, one in which the body temperature tends to vary with the temperature of the environment).

The rostral hypothalamus, especially the preoptic area, is stimulated by environmental changes which threaten to increase body temperature. Therefore this area is functionally concerned with initiating mechanisms that will dissipate excess body heat. Such required changes are: vasodilation of cutaneous blood vessels, profuse sweating, and increased rate and decreased amplitude of the respiration (panting). Destructive lesions of the rostral hypothalamus result in hyperthermia (hyperpyrexia) because of the inability to dissipate body heat.

A thorough search of the literature reveals that not all authors fully agree on the exact loci and mechanisms involved in body-temperature regulation, but in general the above statements are accepted.

Hypothalamic–Pituitary Gland Relations

The pituitary gland is generally referred to as the 'master endocrine gland' because it regulates the functional activity of the other endocrine glands. It should be emphasized, however, that the hypothalamus is largely responsible for the regulation of the activity of the pituitary gland. These relationships are extremely complex and our knowledge is incomplete in many respects, especially concerning the circuits of feedback of specific hormones to both the hypothalamus and the anterior pituitary. Such details are among the main concerns of neuroendocrinologists.

POSTERIOR LOBE (NEUROHYPOPHYSIS)

Hypothalamic connection with the neurohypophysis is via the hypothalamohypophyseal tract discussed previously. The posterior lobe is composed largely of glial-like cells called *pituicytes*. The pituicytes do not secrete hormones but act simply as supporting structures for large numbers of terminal neuronal elements originating in the supraoptic and paraventricular nuclei of the hypothalamus. These neuronal elements are concerned with the production and secretion of two posterior pituitary hormones.

1. *Antidiuretic hormone (ADH)* is the preferred name for the first, although the names vasopressin and Pitressin are also frequently used. The absence of ADH results in *diabetes insipidus,* characterized primarily by polydipsia (excessive thirst) and polyuria (the excretion of excessive amounts of urine with a low specific gravity). The supraoptic nuclei receive one of the richest blood supplies in the brain. The activity of the nuclei seems to be dependent on the osmotic pressure of the blood in the hypothalamic vessels. As an example, when the blood supplying these nuclei becomes hyperosmotic, as during dehydration, the nuclei are stimulated to increase the production and release of ADH. This hormone, which is in the systemic blood, acts on the distal convoluted tubules of the kidney to conserve water. The opposite action occurs during excessive hydration, when the blood

supplying the nuclei is hypo-osmotic.

2. *Oxytocin* is the second hormone released by the neurohypophysis. This hormone is produced by the paraventricular nuclei and is especially important in the process of lactation. Essentially, the stimulus of suckling on the nipple of the mammary gland causes impulses to be transmitted through sensory nerves to the hypothalamus, influencing the activity of the paraventricular nuclei which results in release of oxytocin from the posterior lobe. The hormone causes contraction of the myoepithelial cells surrounding the alveoli of the mammary gland, resulting in ejection of the milk from the alveoli.

Increased secretion of oxytocin is also stimulated by sensory impulses from the cervix which results in contraction of uterine muscle believed to be important in fertilization of the ovum and parturition.

There are many unknown facts concerned with the specific actions of these two hormones, but it is generally accepted that they are produced and secreted by neurons and subsequently discharged into the capillaries of the posterior pituitary lobe.

ANTERIOR LOBE (ADENOHYPOPHYSIS)

The hypothalamic control over the anterior pituitary lobe and its hormones has received much attention from researchers in recent years. Rather than direct neural connections, as for the posterior pituitary lobe, there is a neurovascular mechanism composed of neurons from the hypothalamus which terminate on numerous hypophyseal portal vessels in the median eminence which transport blood and the absorbed neurogenic factors directly to the anterior lobe of the pituitary. The hypothalamus produces an activating agent, a *releasing factor (neurohormone)*, for each of the anterior pituitary hormones, with the exception of prolactin, which has an *inhibitory factor* (PIF).

The five releasing factors that have been extracted are: thyrotropin-releasing factor

(TRF), growth hormone–releasing factor (GHRF), follicle-stimulating hormone-releasing factor (FSHRF), luteinizing hormone–releasing factor (LRF), and corticotropin-releasing factor (CRF) (Martini and Ganong, 1966). These factors are similar to hormones except that they are produced by neurons prior to being absorbed by capillaries.

Subthalamus

This diencephalic division, also referred to as the ventral thalamus, is not visible either in a whole brain or from a sagittal view. It is bordered dorsally by the thalamus (dorsal thalamus), ventrally and laterally by the internal capsule as it approaches the basis pedunculi, and medially and rostrally by the hypothalamus; caudally the subthalamus is continuous with the tegmental and basilar portions of the mesencephalon. The melding of the caudal subthalamic area and the rostral midbrain is evident by observation of mesencephalic structures which extend into the subthalamus (*i.e.,* the red nucleus and substantia nigra).

The major gray subthalamic areas are: the zona incerta, nucleus of the field of Forel, the entopeduncular nucleus, and the subthalamic nucleus, plus the mesencephalic rostral extensions of the red nucleus and the substantia nigra.

The following discussion is a general summary of the known subthalamic connections in carnivores, based primarily on the work of Rioch (1929). The reader is referred to the original source for further details of the subthalamic connections in carnivores. All of the nuclear areas of the subthalamus have principal connections with the striatal areas via fibers of the *ansa lenticularis.* Therefore emphasis is placed below on additional connections which are not common to all.

The *zona incerta* (Plate 23) is a somewhat circumscribed nucleus lying on the dorsal surface of the internal capsular fibers as they begin to form the basis pedunculi. The zona extends rostrally to the chiasm and caudally to the level of the subthalamic nucleus. Laterally the zona incerta appears to blend with the reticular nucleus of the thalamus. Dorsally it is bordered by the external medullary lamina and the dorsal thalamus. Medially are the three *'H' fields of Forel.* Forel used the German word *Haube* to designate the 'cap' of fibers rostral to the red nucleus. The H field of Forel designates the scattered cell bodies (gray matter) in the prerubral area among the lenticular fasciculus and the ansa lenticularis, which is therefore referred to as the nucleus of the field of Forel. The H_1 and H_2 fields are fasciculi that for the most part arise in the striatum and pass through the area: The H_1 field refers to the *thalamic fasciculus,* and the H_2 field is the *lenticular fasciculus.* The retrolenticular portion of the internal capsule contributes fibers which connect with the zona incerta. Projections to the midbrain tectum and tegmentum are the main efferents of the zona incerta. They give off fibers to the dorsal supraoptic decussation and the lateral geniculate body. The zona incerta is connected to the ventral nucleus of the thalamus, and to the ventral and lateral hypothalamus. The caudal portion receives fibers from the striatum via the lenticular fasciculus.

The *entopeduncular nucleus* (Plates 24 and 25) is a relatively large gray area with fibers coursing through it. This nucleus is compressed between the optic tract ventrally and the genu of the internal capsule dorsolaterally. This nucleus in the cat is directly continuous with the globus pallidus, but in the dog there is an indistinct separation between the two. There are connections between this nucleus and the dorsal thalamus.

The *subthalamic nucleus* (Plate 22) is a biconvex lens–shaped mass at the dorsomedial surface of the peduncle. Ventromedially the subthalamic nucleus fuses with the lateral hypothalamus. There are commissural connections between the right and left nuclei via the commissure of Forel. Internuclear fibers connect the subthalamic nucleus with the hypothalamus and the zona incerta.

The *red nucleus* (Plates 20 and 21) is a prominent mesencephalic tegmental structure which is best known for its caudal efferents, the rubrospinal and rubrobulbar tracts, as discussed in Chapter 13. From its rostral projections, mainly to the diencephalon, the red nucleus is connected with the striatal areas via the lenticular fasciculus.

The *substantia nigra* (Plates 20 and 21) also is considered mainly as a midbrain structure. It extends rostrally into the diencephalic area to the level of approximately the middle of the mammillary bodies. The substantia nigra is connected with many brain areas with various functions, for example, the cerebral cortex, corpus striatum, tectum and tegmentum of the midbrain, cerebellum, subthalamus, hypothalamus, and secondary sensory centers of the brain stem.

Metathalamus

This division of the diencephalon, the metathalamus, consists of the *medial* and *lateral geniculate bodies.* Because of the anatomic and physiologic associations of these bodies with the mesencephalon, discussion of the metathalamus is included in Chapter 13. Functionally the medial geniculate body is a relay center to the auditory cortex (see Chapter 19), and the lateral geniculate body is a relay center to the optic cortex (see Chapter 18).

Bibliography

Crosby, E. C., Humphrey, T., and Lauer, E. W.: *Correlative Anatomy of the Nervous System.* New York, Macmillan Co., 1962.

De Groot, J.: *The Rat Forebrain in Stereotaxic Co-ordinates.* Amsterdam, N. V. Noord-Hollandsche Uitgevers Maatschappij, 1959.

Elliott, H. C.: *Textbook of Neuroanatomy,* 2nd ed. Philadelphia, J. B. Lippincott Co., 1969.

Gurdjian, E. S.: Studies on the brain of the rat. II. The diencephalon of the albino rat. J. Comp. Neurol. *43:* 1–114, 1927.

House, E. L., and Pansky, B.: *A Functional Approach to Neuroanatomy,* 2nd ed. New York, McGraw-Hill Book Co., Blakiston Div., 1967.

Ingram, W. R.: The Hypothalamus. Clin. Symp. *8:* 117–156, 1956.

Ingram, W. R., Hannett, F. I., and Ranson, S. W.: The topography of the nuclei of the diencephalon of the cat. J. Comp. Neurol. *55:* 333–394, 1932.

Martini, L., and Ganong, W. F., Eds.: *Neuroendocrinology.* New York, Academic Press, 1966. Vol. 1.

Meyer, H.: The Brain. Chapter 8 in *Anatomy of the Dog* (Miller, M. E., Christensen, G. C., and Evans, H. E.). Philadelphia, W. B. Saunders Co., 1964. Pp. 480–532.

Netter, F. H.: *Hypothalamus,* supp. to Vol. I, *Nervous System, The Ciba Collection of Medical Illustrations.* Summit, N.J., Ciba Pharmaceutical Products, Inc., 1957.

Nomina Anatomica Veterinaria, Adopted by the General Assembly of the World Association of Veterinary Anatomists, Paris, 1967. Vienna, Adolf Holzhausen's Successors, 1968.

Rioch, D. McK.: Studies on the diencephalon of carnivora. I. The nuclear configuration of the thalamus, epithalamus, and hypothalamus of the dog and cat. J. Comp. Neurol. *49:* 1–119, 1929; II. Certain nuclear configurations and fiber connections of the subthalamus and midbrain of the dog and cat. *idem. 49:* 121–153, 1929.

Rioch, D. McK.: Studies on the diencephalon of carnivora, III. Certain myelinated-fiber connections of the diencephalon of the dog (Canis familiaris), cat (Felis domestica) and aevisa (Crossarchus obscurus). J. Comp. Neurol. *53:* 319–388, 1931.

Singer, M.: *The Brain of the Dog in Section.* Philadelphia, W. B. Saunders Co., 1962.

Skinner, H. A.: *The Origin of Medical Terms.* Baltimore, Williams & Wilkins Co., 1961.

Snider, R. S., and Niemer, W. T.: *A Stereotaxic Atlas of the Cat Brain.* Chicago, The University of Chicago Press, 1961.

Sychowa, B.: The morphology and topography of the thalamic nuclei of the dog. Acta Biol. Exp. *21:* 101–120, 1961.

Trautmann, A., and Fiebiger, J.: *Fundamentals of the Histology of Domestic Animals* (translated and revised from 8th and 9th German edition by R. E. Habel and E. L. Biberstein). Ithaca, N.Y., Comstock Publishing Associates, Division of Cornell University Press, 1957.

Truex, R. C., and Carpenter, M. B.: *Human Neuroanatomy,* 6th ed. Baltimore, Williams & Wilkins Co., 1969.

chapter 16

Telencephalon

The telencephalon represents the ultimate in phylogenetic development of the brain. The phenomenon of encephalization, which has been mentioned in earlier chapters, is directly referable to the dominance of the telencephalon. This applies particularly to the neocortex (as discussed in the section on the Cerebral Cortex later in this Chapter). Recall that 'encephalization' also occurs in the ontogeny (development of the individual) of higher animals. Commensurate with the anatomic development of this area of the brain is its progressive functional dominance exemplified by the ascent of behavioral control of the animal into the telencephalon.

In passing from lower to higher animals, and especially to man, an outstanding result of the increased anatomic complexity is that the higher animals possess neuronal circuits which permit greater versatility in response to stimuli, and greater learning, memory, conceptualization, and syntheses of input, rather than the limited species-stereotyped reactions characteristic of lower animals. Again it is the cerebral cortex that is largely responsible for such characteristics, but one should realize that parallel with the higher development of the cortex is the greater sophistication of its connections with subcortical segments of the central nervous system.

Without accompanying afferent and efferent connections with subcortical segments, the cerebral cortex could not function at its highest level of neuronal activity. The clinical significance of these connections is appreciated when one studies the functional deficiencies resulting from subcortical lesions. Much of the pertinent information pertaining to cortical functions has been obtained from data derived from research on experimental animals.

Basal Ganglia (Nuclei)

Location and Contents

The basal ganglia are subcortical nuclei located deep within the basal portion of the telencephalon in close relationship to the diencephalon, but for the most part separated from it by the internal capsule. The original term 'ganglia' is not acceptable in the pure terminology of the present day, but the term is established in the literature and is more commonly used than the proper term, 'nuclei.'

Authors differ in what they include as basal ganglia. Originally the term was used to refer to all of the large areas of gray matter deep within the interior of the brain, especially the forebrain. This, of course, included the thalamus. But as more details became known about the phylogeny, embryology, and fiber connections, the thalamus was excluded, and such gray areas as the claustrum and amygdala were added by some authors.

Differences of opinion still exist regarding the contents of the basal ganglia, but the caudate nucleus and the lentiform nucleus are considered to be the principal components. Recall that the *lentiform,* or *lenticular, nucleus* is composed of the laterally placed *putamen* and the medially placed *globus pallidus,* or *pallidum.* In carnivores the inner segment of the pallidum is represented by the entopeduncular nucleus (Fox *et al.,* 1966).

In this text the *basal ganglia* are considered to include the least number of generally accepted structures, namely, the caudate nucleus, lenticular nucleus (putamen and globus pallidus), and amygdala. Frequently the claustrum and sometimes the red nucleus, subthalamus, and substantia nigra are considered basal ganglia. Occasionally the amygdala is excluded.

Frequently the *corpus striatum* is named as a component of the basal ganglia. The anatomic structures which compose the corpus striatum are the caudate nucleus, putamen, and globus pallidus (Fig. 16-1). In more general usage, corpus striatum is a collective term used to describe the gray and white striations formed by the rostral (anterior) limb of the internal capsule partially separating the caudate nucleus and the lentiform nucleus.

The lentiform nucleus is subdivided by a thin sheet of white matter, the *external medullary lamina,* into the lateral putamen and the medial globus pallidus. In addition, the globus pallidus is separated from the entopeduncular nucleus by a medullated partition, the *internal medullary lamina,* which adds to the alternating gray and white areas and thereby contributes to the striated appearance (Plate 24).

The reference above to a partial separation of the caudate and the lentiform nucleus is based on the transverse cellular bands that may be seen crossing the internal capsule and connect the caudate and lentiform nuclei. The connection is primarily between the caudate nucleus and putamen, which are similar phylogenetically and developmentally and differ from the globus pallidus.

The globus pallidus is phylogenetically older than the putamen and caudate nucleus and develops from the diencephalon, whereas the putamen and caudate nucleus are phylogenetically younger and develop from the telencephalon. The putamen and caudate nucleus increase in size proportionately with the development of the cerebral cortex and increase in size to a greater

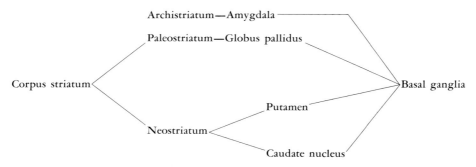

Fig. 16-1. Outline indicating phylogenetic age and relationships of corpus striatum and basal ganglia.

extent than does the globus pallidus (Brodal, 1969).

In keeping with the accepted terminology for phylogenetic age, the older globus pallidus is considered as *paleostriatum,* and the caudate nucleus plus the putamen as *neostriatum.* Since the amygdala is phylogenetically oldest, it is considered as *archistriatum.* An outline of the anatomic components of the basal ganglia, incorporating this nomenclature, is presented in Figure 16-1.

Fiber Connections of Basal Ganglia

Our knowledge of the intrinsic and extrinsic fiber connections of the basal ganglia is obscure. Some of the possible reasons for incomplete and controversial data are: (1) the difficulty of access to the basal ganglia; (2) species differences in anatomic and physiologic relationships; (3) the abundant interconnections, which have no totally acceptable functional significance; (4) the presence of intrinsic and extrinsic feedback loops and reciprocating circuits which are extremely difficult to evaluate.

Much precise information has been obtained from studies using silver impregnation methods in normal brains and in those with lesions experimentally placed by stereotaxic techniques. Such investigative methods have revealed that there are more connections than were previously known, and the internuclear relationships are more

specific than was previously assumed. Although it is not always true, it may be said that most of the afferents to the basal ganglia are received by the putamen and caudate nucleus (neostriatum), whereas most of the external efferent fibers originate in the globus pallidus (paleostriatum). Most of the intrinsic efferents from the caudate nucleus and putamen pass only to the globus pallidus (Brodal, 1969).

Afferents to Basal Ganglia

The input to the basal ganglia comes from many sources. The known afferents to basal ganglia that are generally considered come from: (1) the thalamus, (2) subthalamus, (3) cerebral cortex, (4) brain stem nuclei, and (5) substantia nigra.

The cortical projections to the corpus striatum are definite in that all cerebral cortical areas send efferents to the caudate nucleus and putamen in such varied mammals as the rat (Webster, 1961) and the cat (Webster, 1965). The cortical projection is predominantly ipsilateral and is topically organized, that is, a specific cortical area projects to a specific area in the neostriatum (caudate–putamen). The lesser contralateral projection consists of fibers that cross in the corpus callosum and reach the caudal area of the head of the caudate nucleus via the subcallosal fasciculus and the putamen via the external capsule (Brodal, 1969).

Although specific details are known for corticostriatal projections such as presented above, Brodal reports that it is not known whether the corticostriate fibers are from separate cortical cell bodies or collaterals of other corticifugal fibers (*e.g.,* corticospinal). He also mentions that, although the neostriatum must be functionally dependent on the cortex, there is a lack of conclusive evidence that the corpus striatum sends fibers directly back to the cortex.

Efferents of Basal Ganglia

The efferent fibers from the basal ganglia are dispersed with many of the projections composed of fibers which connect the sources of input to the basal ganglia, as mentioned previously. This forms the anatomic basis for feedback systems and reciprocating circuitry. It should be emphasized again that, although the cerebral cortex strongly influences the functions of the corpus striatum, probably there are no direct feedback striate–cortical fibers. This does not mean that the cerebral cortex receives no feedback from the striate body, but the 'information' is relayed via the thalamus.

The complete circuit for these relationships is: cerebral cortex → putamen → globus pallidus → thalamus → cerebral cortex. An alternative route is: cerebral cortex → caudate → (putamen) → globus pallidus → thalamus → cerebral cortex (Fig. 16-2). Since the cerebral cortex has so many corticifugal fibers, the basal ganglia may influence such descending pathways as the corticospinal, corticobulbar, and corticoreticular. As one can imagine, these types of functional relays are extremely difficult to evaluate.

The major efferent projections of the basal ganglia are to: (1) the thalamus, (2) subthalamus, (3) hypothalamus, (4) brain stem nuclei, that is, the red nucleus and caudal (inferior) olivary nucleus, (5) reticular formation, and (6) substantia nigra.

The intrinsic efferent projections from the basal ganglia are mostly from the globus pallidus. These are not functionally independent, but are influenced largely by intrinsic inputs to the pallidum (*i.e.,* from the caudate nucleus and putamen) (Fig. 16-2). In addition, the pallidum receives inputs from a number of different subcortical nuclei, including the thalamus, red nucleus, and substantia nigra. Also there are large reciprocal connections between the globus pallidus and the subthalamus.

In reference to the caudal projections of the basal ganglia, it should be emphasized that there are no direct descending pathways from the basal ganglia to the spinal cord. This, of course, does not mean that the basal ganglia have no influence on the lower motor neurons of the spinal cord. The following facts show how the basal ganglia exert indirect influence on the lower motor neurons of the spinal cord (Fig. 16-2): (1) The *central (medial) tegmental tract* contains axons from cell bodies located largely in the periaqueductal gray matter of the rostral midbrain level and the red nucleus which terminate chiefly in the caudal (inferior) olive. This nucleus is the origin of the olivospinal tract, a minor, obscure, and questionable pathway to the lower motor neurons of the cervical spinal cord. (2) The *rubrospinal tract* (see Chapter 10) to the lower motor neurons of the spinal cord is influenced by the basal ganglia through the red nucleus in the tegmentum of the midbrain. It should be remembered that the red nucleus also receives a major input from the cerebellum (see Chapter 14). (3) The *reticulospinal tracts* are indirectly associated with the basal ganglia via the red and tegmental nuclei which send fibers to the reticular nuclei.

It should be realized that the above are not the only known connections, but they indicate that the basal ganglia are connected with a large number of structures either directly or by circuitous routes with many reciprocating connections. The difficulties that such complex relationships create for

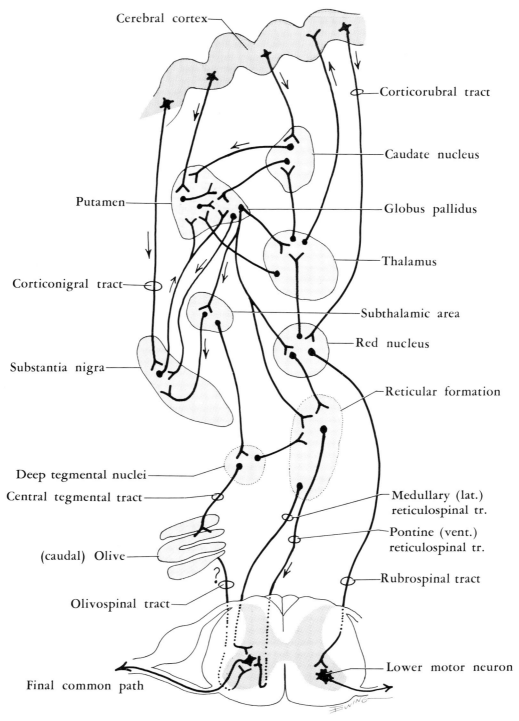

Cerebral cortex

Corticorubral tract

Caudate nucleus

Putamen

Globus pallidus

Thalamus

Corticonigral tract

Subthalamic area

Red nucleus

Substantia nigra

Reticular formation

Deep tegmental nuclei

Central tegmental tract

Medullary (lat.) reticulospinal tr.

Pontine (vent.) reticulospinal tr.

(caudal) Olive

Rubrospinal tract

Olivospinal tract

Lower motor neuron

Final common path

E. WING

Fig. 16-2. Diagram showing some of the interconnections of the corpus striatum (caudate nucleus, putamen, and globus pallidus) with surrounding structures of the brain stem and recurrent 'feedback' fibers to the cerebral cortex. The indirect influence of the basal ganglia on the lower motor neuron as the final common path is illustrated. The olivospinal tract reported by some authors is questionable, according to others.

the researcher in his proper interpretation of experimental data should be appreciated. Compounding the problem are species differences that must be considered when one is critically evaluating basal ganglia and their relations with other systems, such as pyramidal, extrapyramidal, and sensory.

Functions of the Basal Ganglia

As mentioned above, the functions of the basal ganglia are complex, and it is extremely difficult to correlate the functional significance of a single nucleus with the overall basal ganglionic influences on so many systems. McGrath (1960) stated "lesions of the basal ganglia are commonly seen in the dog but we are unable to ascribe specific neurologic signs to them." It seems that our present-day clinical application of knowledge of veterinary neurology in this area has not advanced significantly over what it was a decade ago.

Papez (1929) considered the corpus striatum a primitive mechanism for directing movements into proper channels. He referred to the striate body as being composed of "a large series of motor valves for releasing various groups of purposive movements commonly called instinctive responses." As examples he cited securing and selection of food, chewing, swallowing, eye and head movements, and walking.

Experimental Investigations

Functionally the neostriatum (caudate nucleus and putamen) differs from the paleostriatum (globus pallidus). Generally the neostriatum exerts an inhibitory influence on somatic motor activity, whereas the paleostriatum produces an excitatory effect resulting in hypertonus. Prolonged stimulation of the globus pallidus may produce tremor on the opposite side (Noback, 1967). The same author cites results of ablation experiments in animals, which are discussed below.

Experimentally it is necessary to place lesions bilaterally for study of nuclear areas in the brain. Unilateral ablation of the globus pallidus or the putamen results in few or no obvious symptoms. Bilateral ablation of the putamen results in a hyperactive animal (cat) that disregards its environment and shows behavioral changes. As an example, the animal may walk off the edge of a table, or it may walk into an object and continue to push against it, even a wall ('obstinate progression'). Such a cat may show no fear, even when placed with a hostile dog.

In contradistinction to these results, bilateral ablation of the globus pallidus produces the opposite behavioral changes; that is, the animal is hypoactive and somnolent, and exhibits hypotonus. Such an animal will retain a set posture for an extended time, even though it may be an unnatural position of a limb or unnatural body posture. The hypoactivity exhibited by the animal is somewhat similar to that in Parkinson's disease in man, but the animals show no rigidity or tremor.

It should be emphasized that attempts to experimentally simulate in animals (cats and monkeys) the total symptoms resulting from disease of the basal ganglia in man have been unsuccessful (Brodal, 1969). Partial success has been achieved, however, as shown by the following: small unilateral lesions in the rostroventral region of the caudate nucleus in the cat produce permanent athetoid and choreiform hyperkinesias which resemble those in man.

Larger lesions with generalized destruction of the caudate nucleus do not produce these symptoms (Liles and Davis, 1969). Those authors defined athetoid movements as incessant alternating flexion and extension movements of the toes and paws of each forelimb. Usually this action alternated between the two forelimbs, similar to the kneading action observed in normal household cats as an expression of pleasure when they are caressed. The intensity of athetoid movements and the hyperextension of the toes increased when the animals lay on their sides, and these movements were absent during locomotion and sleep.

Basal Ganglionic Influence on Motor Activity

Based on numerous reports in the literature and his own work, Brodal (1969) points out that the basal ganglia do not have prominent projections that would make possible their exertion of a marked effect on brain stem nuclei which are origins for descending fibers to the lower motor neurons of the spinal cord. The present-day evidence indicates that "the basal ganglia can scarcely be considered as important motor centers as was previously assumed." He considers the main concern of the basal ganglia to be the collaboration between the thalamus and the cerebral cortex.

In addition to the basal ganglia Brodal includes the subthalamic nucleus and the substantia nigra as contributors to a closed circuit from these nuclei to the thalamus and then to the cortex. By means of this circuit the basal ganglia (and the other nuclei named above) can influence descending fibers to the spinal cord and motor functions to only a limited degree. There appears to be no support for considering that "the motor area in the dog is located in the basal ganglia rather than in the cerebral cortex as it is in man" and that "total destruction of most of the basal ganglion results in total paralysis," as stated by Clark (1965).

The influence of basal ganglia on the red nucleus is probably responsible, at least in part, for the incoordination that results from lesions of the red nucleus and the rubrospinal pathway. This tract is probably a more important motor tract in animals than in man. In animals the red nucleus relays impulses from the cerebral cortex, cerebellum, and basal ganglia, and functionally is concerned with coordination and possible locomotion (Palmer, 1965). The *possible* lo-comotion function of the red nucleus and the rubrospinal tract is the degree of certainty that is commonly accepted. Reports in the literature do not appear to fully support the view of Clark (1965) that the rubrospinal pathway is "the motor tract . . . which if severed in the spinal cord will cause a voluntary motor paralysis."

Most investigators believe that the corpus striatum is dependent on the cortex. This is supported by the inability to elicit movement by stimulating the caudate nucleus and putamen (Ruch *et al.*, 1965). The same authors cite reports of experiments on cats designed to identify and assess the independent functions of the striatum. Decorticate cats were compared with other decorticate cats in which, in addition, the striatum was ablated. Changes in behavior were noted, rather than added paresis. The results of such experimentation led to the conclusion that the striatum appears to be involved in "ordering the component parts of complex movement."

Rhinencephalon, or 'Olfactory Brain'

Olfaction is a very complicated modality which involves much more than just the perception of odors. The degree of olfactory specialization differs among animals, and is reflected in varieties of behavior. Although the olfactory nerve is functionally classified as *special visceral afferent*, the sensory input via this nerve can be a tremendous stimulus to practically all parts of the brain.

Olfaction may produce behavioral changes which definitely are not restricted or correlated with visceral activity. Truly the correlation of olfaction with food is very apparent, as well as the physiological effects of certain aromas from food (*e.g.*, salivation and lip licking). Also in animals olfaction plays a major role in such phenomena as sexual desire and mating, and recognition of friends or foes—including the emotional changes that are associated with these phenomena.

Vertebrates with a well-developed sense of smell are referred to as *macrosmatic,* and those with a poorly developed olfactory system are known as *microsmatic.* Animals without a sense of smell are *anosmatic.* Phylogenetically the rhinencephalon, or 'olfactory brain,' is an old area of the brain. In fishes and amphibians the telencephalon is dominated by the olfactory input to its pallium (cortex), although the latter has no cortical-like structural features as are seen in mammals (Brodal, 1969).

In mammals the first fissure that develops embryonically is the rhinal fissure which separates the olfactory (paleocortex, or piriform area) from the nonolfactory cortex (neocortex, or neopallium). Based on phylogenetic age, the hippocampus (Ammon's horn) develops from the oldest cortex (archicortex), which is pushed inwardly by the developmental growth of the newer cortices. When the corpus callosum appears in mammals, coincident with development of the cerebral hemispheres (neocortex), the hippocampus (archicortex) in this area is drastically reduced, to become the insignificant *indusium griseum* on the dorsal surface of the corpus callosum (Plate 27).

The rostral extension of the original hippocampus is thought to pass beneath the genu of the corpus callosum to become the subcallosal gyrus within the septal area. The important point to remember in summary of the phylogenetic cortical development is that the archicortex and the paleocortex, which represent the entire hemisphere in lower vertebrates, are completely overpowered by the neocortex in mammals. In brief, the functional significance of this phenomenon is that the archicortex and paleocortex, which were almost exclusively concerned with olfaction in lower forms, have acquired many nonolfactory connections and functions in mammals. This is especially true in higher mammals.

Definition of Rhinencephalon

Commonly the components of a system or area of the brain are defined according to their location. However, the rhinencephalon may be defined according to the structures which are thought to have, or perhaps phylogenetically have had, olfactory function. It is not surprising that the utilization of such a broad criterion would result in great differences of opinion among authors regarding the contents of the rhinencephalon, or 'olfactory brain.' In this text the least number of rhinencephalic components that are generally considered will be used, namely, olfactory bulb, olfactory tract, piriform lobe, and hippocampus (Ammon's horn). This is in agreement with McGrath (1960).

It may rightly be argued that inclusion of the hippocampus is incorrect, but it is included because of the popular belief, until recently, that the hippocampus is functionally related to olfaction. Current evidence, however, suggests that the hippocampus probably has no olfactory functions (Brodal, 1969). Dogs in which both hippocampi have been ablated show no deficit in conditioned responses depending on olfaction (Allen, 1940). He reported that after dogs had learned that a paper package might contain meat, removal of both hippocampi did not reduce their ability to select immediately a meat package from three paper packages of identical size and wrapping. Of added interest is that their responses to taste (also classified functionally as *special visceral afferent*) of salt, sugar, quinine, and acetic acid were not altered.

Anatomic support for not including the hippocampus in the rhinencephalon is that it displays its greatest development in microsmatic man and is well developed in anosmatic aquatic mammals, such as the whale and porpoise (Truex and Carpenter, 1969). Based on such physiologic and anatomic evidence, it seems advisable to exclude the hippocampus from the already short list of rhinencephalic structures given above. From a student's standpoint, then, one can find support for either including or excluding the hippocampus, but he should be aware of the reasons for the choice.

In addition to the few structures named above (including the hippocampus), it is not uncommon to find the following structures added as rhinencephalic components: septal cortex, amygdala, rostral (anterior) perforated substance, fornix, and rostral (anterior) commissure (Meyer, 1964). Notice that some of these structures are considered components of the basal ganglia and/or limbic system. Such anatomic intimacy, and in some cases duplication, suggests close functional interrelationships among rhinencephalon (according to the 'long list' of components), basal ganglia, and the limbic system.

In higher mammals, including carnivores, the functions of these structures are not related solely to olfaction, but are concerned also with such traits as personality, behavior, and emotion. Consideration of such complex phenomena may well serve as a basis for understanding that practically the whole brain reacts to olfactory input.

Olfactory Pathways

Olfactory Receptors

The respiratory mucosa within the nasal cavity contains thickened areas of olfactory (neuro)epithelium which varies in area and color among animals. The olfactory epithelium primarily consists of three types of cells: (1) the basal cells; (2) the supporting, or sustentacular, cells; and (3) the olfactory cells, which are bipolar neurons. The free distal surface of the olfactory cell has short olfactory hairs which are thought to be stimulated by odoriferous substances dissolved on the moist surface. The proximal end of the olfactory cell (neuron) is an attenuated elongated process which continues as a nonmyelinated fiber of the *olfactory nerves* (cranial nerves I).

Olfactory Bulb

Grossly the olfactory bulb is tremendously large in dog when compared with that in man (Fig. 16-3). In some dog brains the olfactory bulb contains a cavity which connects with the rostral portion of the lateral ventricle of the same side. The olfac-

DOG   MAN

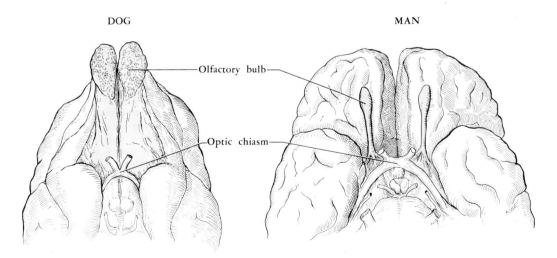

Olfactory bulb

Optic chiasm

Fig. 16-3. Olfactory bulbs and tracts in dog and man, showing comparative size.

tory bulbs lie in the cribriform fossae of the ethmoid bone. In the process of removal of the dog brain the olfactory nerves tear from the bulbs, which gives the rough surface to the bulbs.

Microscopically the olfactory bulbs of the dog have a laminar organization that is much better defined than it is in man. The general microscopic laminar appearance with the important cellular arrangements is illustrated in Figure 16-4. The proximal fibers from the neurons within the olfactory epithelium are grouped into a number of bundles which pass through the many small foramina of the cribriform plate to enter the *olfactory glomeruli* within the bulb. Here they synapse with dendrites from the *brush (tufted)* and *mitral cells* (Fig. 16-4) which project axons caudally, the secondary olfactory fibers, as components of the olfactory tract.

Olfactory Tract

The olfactory tract (see Fig. 2-5, p. 20) appears grossly as a short fibrous caudal extension of the olfactory bulb. It bifurcates to form the *lateral* and *medial olfactory striae.* The area between the diverging striae is the *anterior perforated substance,* so named because of the many small blood vessels that penetrate the surface and when removed leave many small holes, producing the perforated appearance. The lateral olfactory gyrus is the cortical area between the lateral olfactory stria and the rostral rhinal sulcus (see Fig. 2-5, p. 20).

For a detailed discussion of olfaction based on phylogenetic differences, the reader is referred to Ariëns Kappers *et al.* (1967). The following discussion is a general summary based on the basic mammalian

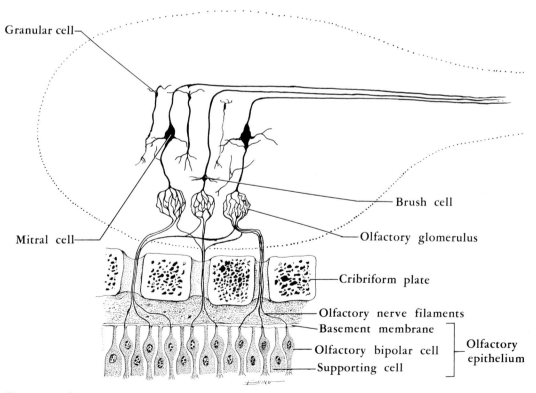

Fig. 16-4. Section through the cribriform plate, illustrating some of the histologic structure of the olfactory epithelium and the olfactory bulb. (After House and Pansky, 1967.)

plan for olfaction derived from many sources.

The lateral olfactory stria and gyrus pass lateral to the anterior perforated substance to enter the rostral portions of the piriform (prepiriform) area. Within this area the fibers of the lateral stria terminate in the cortex and in parts of the amygdaloid complex. This area is frequently referred to as the *primary olfactory cortex.* In this context, it should be noted that olfactory fibers are the only sensory fibers that reach the cortex without being relayed by the thalamus.

Axons of the medial olfactory stria pass to the subcallosal area and the paraterminal gyrus ventral to the genu of the corpus callosum (see Fig. 2-7, p. 25). Some fibers from the medial stria enter the rostral (anterior) portion of the anterior commissure to cross and project rostrally to the olfactory bulb of the opposite side (Fig. 16-5). Other fibers in the rostral (anterior) portion of the anterior commissure are commissural fibers which connect the two piriform areas.

Alternate Pathways

The following three pathways are simplifications of complex reflexes. Much of the discussion below is fundamentally similar to that pertaining to man as found in most textbooks of human neuroanatomy. It should be remembered that the basic anatomic framework of olfactory neuronal patterns is relatively stable among mammals, although the higher forms have acquired more complex properties of nonolfactory function.

All pathways actually originate peripher-

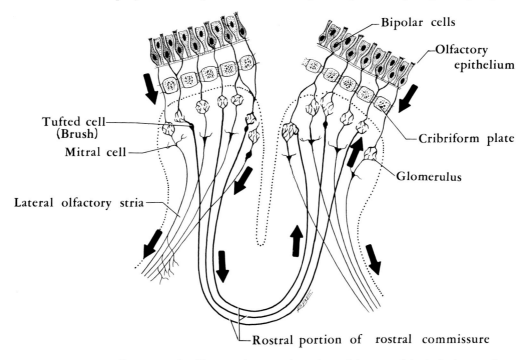

Fig. 16-5. Diagram illustrating the fibers in the rostral portion of the rostral (anterior) commissure connecting the two olfactory bulbs. Many of the larger axons of the mitral cells pass from the olfactory tract into the lateral olfactory stria, whereas most axons of the tufted (brush) cells pass caudally to enter the rostral commissure and continue to the opposite olfactory bulb. (Modified from Everett, 1971.)

ally in the olfactory epithelium, from which impulses are transmitted via the fascicles of the *olfactory nerve* through the cribriform plate to the *olfactory bulb*. From this location three basic pathways are discussed below and schematically illustrated in Figure 16-6.

PATHWAY A (FIG. 16-6)

Pathway A is thought to be the most primitive and direct path. From the olfactory bulb fibers pass via the medial olfactory stria to the subcallosal and septal nuclear areas. Here they synapse with neurons whose axons pass caudally via the *medial forebrain (olfacto-hypothalamic) bundle*. Most of the fibers in this fasciculus terminate in the preoptic and lateral hypothalamic nuclei, plus the mammillary bodies. Other medial forebrain bundle fibers continue caudally into the midbrain tegmentum and extend through the reticular formation of the brain stem. Connecting fibers between the reticular gray matter and various cranial nerve and spinal cord autonomic nuclei permit *olfacto-visceral reflexes,* for example, secretion by salivary and gastric glands (stimulation of parasympathetic nuclei of cranial nerves VII, IX, and X) in response to odors.

PATHWAY B (FIG. 16-6)

Pathway B is a more complicated reflex involving pathway A with the addition of the following: via the lateral stria to the piriform area; relays from here along with the subcallosal area via the *stria medullaris thalami* to synapse with cell bodies in the *habenula*. From this epithalamic nucleus axons pass ventrally in a prominent bundle, the *habenulopeduncular tract (fasciculus retroflexus)* to terminate in the *interpeduncular (intercrural) nucleus*. From here axons pass to the reticular formation and extend rostrally and caudally via the *dorsal longitudinal fasciculus (of Schütz)* to synapse with brain stem nuclei.

This pathway fortifies pathway A by involving visceral efferent (autonomic), somatic efferent, and special visceral efferent nuclei of the brain stem. Impulses to the reticular formation of the brain stem could be relayed caudally to the spinal cord. These connections would permit the complex participation of various types of motor activity as occur in vomiting and retching.

PATHWAY C (FIG. 16-6)

Pathway C presents the most complicated reflex pattern and includes structures of the limbic system which are discussed in the next section of this Chapter. The cerebral cortex is ultimately involved in this pathway, therefore it seems possible that this is the circuit utilized for the differentiation between pleasant and unpleasant odors.

Fibers from within the piriform area caudal to the amygdaloid complex follow the *hippocampus* caudally, dorsally, and rostrally to continue in the *fornix*, which curves ventrally and caudally to terminate in the *mammillary bodies* of the hypothalamus. Efferent fibers from the mammillary bodies form the *mammillothalamic tract (of Vicq d'Azyr)* which projects dorsally to terminate in the *rostral (anterior) nucleus of the thalamus*.

From here impulses are relayed to the *cingulate gyrus*, which in turn sends association fibers to the frontal lobe cerebral cortex. This circuitry permits the correlation of olfaction and psychic factors which play a great role in the emotional behavior of animals, including man. It should be remembered that the mammillary body also sends fibers caudally within the brain stem as the *mammillotegmental tract*. This may also fortify the reticular formation synapses and relays to cranial nerve and spinal cord motor nuclei, as considered above in discussions of pathways A and B. These pathways as discussed include the basic patterns and do not consider the reciprocal connections which are known to exist.

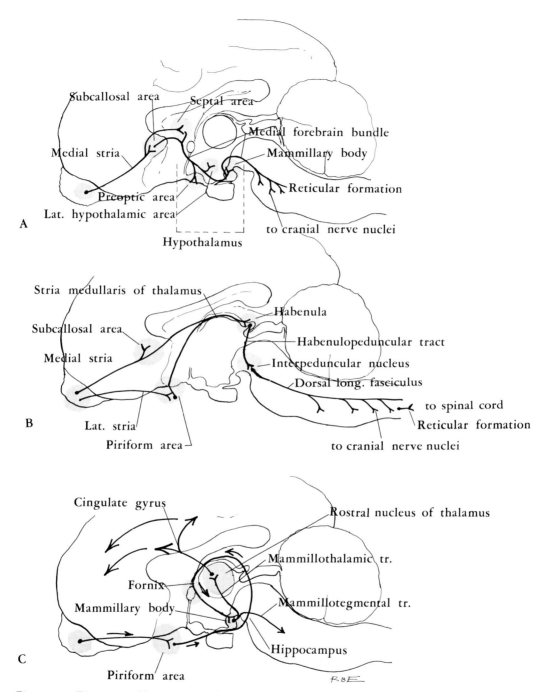

Fig. 16-6. Diagrams of basic neuronal pathways utilized in olfactory reflexes. See text for explanation.

Limbic System

Definitions and Anatomy

The sharing of some anatomic structures and functions of the so-called limbic lobe and the rhinencephalon, defined in broad concepts, has been mentioned. Similar to the difficulties encountered in attempting to define 'rhinencephalon,' there is no generally accepted definition of 'limbic system.' The term 'limbic lobe' seems to be more definite, because there is a point of reference, based on the report of Broca (1878). He found a large horseshoe-shaped cerebral convolution that is constant in the brains of all mammals. The literal meaning of limbic is 'border,' and therefore he called this *le grand lobe limbique,* because it forms a border around the rostral portion of the brain stem. The concept of the limbic lobe has been expanded, with subcortical components being added. From this expansion, including the many controversial additions made by various researchers, the limbic system evolved.

Although there are inconsistencies in what should be included, the following definitions are basic but adequate. The *limbic lobe* consists of the parahippocampal, cingulate, and subcallosal gyri, plus the deeper hippocampus and dentate gyrus. The *limbic system* consists of the limbic lobe as defined here, plus associated subcortical nuclei (*e.g.,* amygdaloid complex, hypothalamus, epithalamus, septal nuclei, and the anterior thalamic nuclear area). This system is in a position to receive and associate olfactory, visceral, oral, sexual, and basic sensory impulses, that is, olfactory (SVA), optic (SSA), auditory (SSA), exteroceptive (GSA), and interoceptive (GVA) impulses, and project them to the hypothalamus.

In addition, it should be emphasized that this phylogenetically old system is connected to the neocortex.

Functions of the Limbic System

The limbic system is involved with emotional and behavioral patterns. The limbic system–hypothalamic relationship is largely responsible for these functions correlated with the hypothalamic influence on the autonomic nervous system. Based on the fundamental phenomenon that emotion involves visceral reaction, the limbic system is frequently referred to as the *'visceral brain.'*

A functional unit which includes many structures of the limbic system is known as the *Papez circuit,* which includes: hippocampus → fornix → mammillary body → anterior thalamic nucleus → cingulate gyrus → cerebral cortex. Papez (1937) proposed this unit as a mechanism of emotion. This phenomenon applies to animals as well as to man. Since the neocortex is far superior in man, when compared with other mammals such as the dog and cat, many of the emotional, psychic, and behavioral changes observed in man are generally thought of as strictly human characteristics.

There is evidence that similar changes occur in animals, but perhaps to a different degree in many instances. Perhaps our inability to adequately measure many changes in animals (*e.g.,* psychic, personality, memory, thinking ability, emotional, etc.) accounts for the common skepticism regarding the existence of these changes.

Obviously it is virtually impossible to adequately apply and evaluate in animals such classic statements of the functional characteristics of the human limbic system as: (1) the limbic system provides the emotional background for intellectual functions, and (2) how an individual 'feels' about something (or his 'gut reaction') is mediated by the visceral brain, but what he 'thinks' or 'knows' is provided by the neocortex (House and Pansky, 1967).

Such limitations imposed by the human-animal gap do not preclude experimental studies on the limbic system in animals as an acceptable procedure for attempting to understand the system. Klüver and Bucy (1939) extirpated large portions of both temporal lobes in monkeys. The major areas destroyed in each lobe were: the amygdaloid complex, hippocampus and parahippocampal gyrus, and much of the temporal neocortex.

After the operation, monkeys which were previously wild became docile and showed no evidence of fear or emotional anger, even to threatening enemies (*e.g.,* dogs, cats, snakes, and strangers). The monkeys exhibited strong oral tendencies; almost all objects were examined tactually, smelled, and mouthed. This compulsion to examine everything and apparent inability to recognize or appreciate objects seen is termed 'psychic blindness.' Dietary changes in monkeys subjected to temporal lobectomy included eating of foods unusual for monkeys (*e.g.,* fish, bacon, beef, and other meats). Sexual aberrations included: attempted interspecies mating and autosexual, homosexual, and excessive heterosexual activities.

After several months most of the bizarre behavior disappeared, with the exception of the tameness and the oral tendencies. This syndrome, referred to as the Klüver-Bucy syndrome, has been reported in man following bilateral removal of large areas of the temporal lobes (Terzian and Ore, 1955).

It should be noted that subtotal temporal lobectomy in the cat is reported by some investigators to produce a docile animal, whereas other investigators have produced cats with savage and easily provoked rage behavior. Experimental gross ablations of specific areas of the brain are difficult to repeat with complete accuracy. Such discrepancies in results may occur if identical structures are not removed in different animals, even in those of the same species.

The reports of experimental ablations and stimulations of structures in the limbic system are very confusing. A brief discussion of the major structures, including some re-

sults of experimentation, illustrates some of the controversy.

The Amygdala
(Plate 25)

The amygdala was named because of its almond nut–like shape (L. *amygdala*—almond nut). The amygdaloid body, or complex, is considered a component of both the basal ganglia and limbic system in this text, and frequently is included elsewhere as a component of the rhinencephalon. The complex is composed of a number of nuclear masses deep within the rostral piriform areas along the ventrolateral and rostral walls of the ventral horn of the lateral ventricle. This gray matter mass is continuous ventrally with the cerebral cortex of the temporal pole.

The amygdaloid complex may be subdivided into two nuclear groups which basically are similar in all mammals. The *corticomedial* group includes the cortical, central, and medial nuclei; the *basolateral* group includes the basal and lateral nuclei. There is evidence that the individual subnuclear groups are rather specific regarding the very complex connections and projections, but there are conflicting reports pertaining to such details.

The *afferent fibers* to the amygdala come from many sources. Olfactory afferents from the olfactory bulb appear to terminate only in the corticomedial group, whereas the basolateral group is thought to receive fibers from the entire piriform cortex. Other less definite afferents have been reported from the thalamus, hypothalamus, and neocortex.

Functions of the Amygdala

There are many controversies and uncertainties concerning the functions of the amygdala. Experimental studies in animals and observations in man with appropriately placed lesions have revealed much information, but it seems apparent that much more sophisticated research methods are needed for study of this area. A variety of visceral, somatic, and behavioral responses have been produced in the cat by stimulation of the amygdala. The following discussion is based primarily on the reports of Brodal (1969), Kaada and Ursin (1957), Shealy and Peele (1957), and others. *Visceral* or autonomic effects are numerous and include both sympathetic and parasympathetic responses. Examples include salivation; increased gastric acidity; changes in gastrointestinal motility, respiration, and blood pressure; pupillary dilation; piloerection; and spontaneous micturition and defecation.

Controversial reports of *somatic* movements include: turning of the head, clonic facial contractions, eye movements, licking, chewing, and swallowing. In addition, both inhibition and facilitation of spinal reflexes have resulted from amygdaloid stimulation. Examples of variable *behavioral changes* include: a general arousal effect on the neocortex similar to that produced by stimulation of the reticular formation of the brain stem, a state of fear or anger, withdrawal, and generalized rage reactions.

It should be realized that, since the amygdala is a component of the limbic system, it is therefore functionally affected by the other limbic components as well as by the neocortex. Evidence of the complexity of such a system is that the literature contains reports of identical or similar effects of stimulation of other regions of the brain. It is not known whether the effects of stimulation of the limbic system are expressed wholly or in part via the hypothalamus (Peele, 1961).

A major reason for such discrepancies in results of amygdaloid stimulation is that many of these reactions are merely components of more complex emotional and behavioral patterns. Such complex patterns tend to vary among individuals.

In summary, some reasons for the resulting inconsistencies are:

1. The amygdaloid complex does not function by itself and is not a 'center' for anything.

2. The amygdala is a component of a functional unit or system and has many intrinsic and extrinsic connections. It should be remembered that the amygdala is considered a component of the basal ganglia and the limbic system, and frequently is included as a rhinencephalic structure.

3. Electrical stimulation of brain areas is a crude and unnatural method of investigation. The current used is usually above normal physiologic limits *in vivo,* and the spreading of current to surrounding areas may give misleading data. The stimulation of fibers passing through the area also may produce effects that are not the result of stimulation of the specific area.

4. There is no precise mapping of the amygdaloid area that is universally accepted. This leads to difficulties in duplication of experiments by different investigators. Species differences should be considered.

5. The fact that the behavioral reactions considered for the amygdala can be produced by stimulation of other brain areas has a definite impact on the credibility of what might be reported as amygdaloid function.

6. Often the investigator's subjective interpretation, particularly of such complex responses as behavioral changes, leads to confusion.

7. The variations, both anatomic and physiologic, among animals or even the same species lead to different behavioral responses to similar stimuli. Perhaps past physiologic and psychologic conditioning, especially in man, account for discrepancies in behavioral results.

The above list could be much longer, but it should be long enough for the intended purpose. Obviously the influence of the factors cited is not limited to investigation of the amygdala, but may also apply to that of other areas of the brain.

Hippocampus (Ammon's Horn) (Fig. 2-15 [p. 40]; Plates 19 through 23)

The origins of both names for this structure have interesting histories. The name hippocampus is derived from its curved form similar to that of a sea horse, which in Greek is 'hippokampos.' Historically, the name Ammon's horn is derived from resemblance of the structure to the ram's horn–like plumes which supposedly were displayed by the mythological deity, Ammon.

Phylogenetically, the hippocampus develops from the oldest cortex (archicortex), which is pushed ventrally and folded inwardly by the newer cortices. Originally the hippocampus extended from the present temporal lobe area in an arch caudally, dorsally, and rostrally to the septal area rostral to the anterior commissure. When the corpus callosum appeared in mammals, the hippocampus dorsal to the callosum became reduced to a rudimentary elongated gray mass, the *indusium griseum,* located on the dorsal surface of the corpus callosum (Plate 27).

In a dog, cat, or human brain one can locate the hippocampus beneath the corpus callosum in the floor of the caudal portion of the lateral ventricle. In a hemisected brain, on opening the lateral ventricle between the fornix and the corpus callosum near the splenium, the hippocampus appears as a rounded prominence in the ventricular floor. The thin lip on the side of the hippocampus is the *fimbria.* The hippocampus extends rostrally to become the *fornix.* There is no line of demarcation between the two, and at this location the fimbria is referred to as the fimbria either of the hippocampus or of the fornix.

Beneath the splenium of the corpus callosum the left and right fornices are joined by the *commissure of the fornix,* which is also called the *hippocampal commissure* (see Chapter 2). From the ventricular floor beneath the splenium, the hippocampus and fimbria can be followed as they arch caudoventrorostrally into the temporal lobe within the wall of the inferior horn of the lateral ventricle.

Histology of the Hippocampus (Fig. 16-7)

Because so much present-day research involves the hippocampus, a brief and very superficial account of its microscopic structure follows. The reader is referred to histology texts for further details. In addition to the embryologic and histologic investigations of the limbic system structures including the hippocampus, the literature reports additional details acquired by special studies in electron microscopy, histochemistry and pharmacology, electrophysiology, and neuroendocrinology.

It should be remembered that the hippocampus is the phylogenetically oldest cortex that has been rolled into the lateral ventricle. In cross section the hippocampal layers appear to turn into the curved *dentate gyrus,* which is deep within the hippocampus as seen in cross section (Fig. 16-7). Essentially in agreement with Trautmann and Fiebiger (1957), seven histologic layers can be identified in a cross section through the hippocampus of the dog, beginning with the ependymal lining of the ventricle: (1) ependyma; (2) alveus; (3) stratum oriens (polymorphous layer), a relatively sparsely cellular layer of polymorphous neurons; (4) layer of large and small pyramidal cells; (5) stratum radiatum; (6) lacunar layer; and (7) molecular layer, which is adjacent to the dentate gyrus, as illustrated in Figure 16-7.

Functions of the Hippocampus

The controversial inclusion of the hippocampus as a component of the rhinencepha-

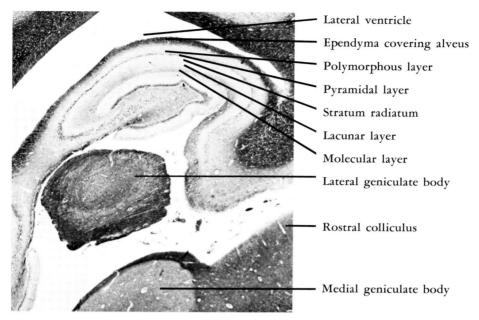

Lateral ventricle
Ependyma covering alveus
Polymorphous layer
Pyramidal layer
Stratum radiatum
Lacunar layer
Molecular layer
Lateral geniculate body

Rostral colliculus

Medial geniculate body

Fig. 16-7. High-power photomicrograph of the hippocampus and dentate gyrus of the dog, showing the histologic layers. This section is through the dorsal portion of the hippocampus as it forms an eminence in the floor of the lateral ventricle in this area. The lateral and medial geniculate bodies and the rostral colliculus are included for orientation.

lon has been discussed. Recent research indicates that the hippocampus has no direct olfactory function. According to Brodal (1969) it probably will never be possible to "define the function of the hippocampus as such, and it will certainly be misleading to consider the hippocampus as a 'center' for any particular function."

Details of functional correlations with the histologic structure and synapses indicate that certain cells are excitatory in function (Andersen *et al.,* 1966), whereas others are inhibitory (Andersen *et al.,* 1964). In brief summary of hippocampal function, the anatomic and physiologic contributions of the hippocampus as a component of the limbic system, and more specifically of the Papez circuit, indicate it is involved in such complex phenomena as emotion, personality, visceral and endocrine changes, behavioral patterns, and recent memory.

Perhaps the veterinary student and veterinarian have more interest in the hippocampus than in other components of the limbic system. This is because gross sections are taken of the hippocampus from most animals in which rabies is suspected. As stated earlier, proper procedures of histologic technique and staining make it possible to demonstrate *Negri bodies,* intracytoplasmic inclusion bodies which are pathognomonic for rabies.

It should be remembered that the hippocampus is not the only brain area that will show Negri bodies, even in dogs, but, for some reason these inclusion bodies have a natural predilection for the hippocampus. The best sites for demonstration of Negri bodies vary with the species; for example, in some animals the semilunar ganglion and in others the cerebellum is preferred for sectioning.

Fornix
(See Fig. 2-15 [p. 40]; Plate 25)

In a hemisected brain the septum pellucidum can be seen stretching between the fornix ventrally and the corpus callosum dorsally, to separate the rostral portions of the two lateral ventricles. The fornix was referred to previously as being a direct continuation of the efferent fibers of the hippocampus. As the bundle of fibers passes rostrally toward the area of the interventricular foramen and the anterior commissure, it splits to pass on the rostral and caudal sides of the commissure. Those fibers that descend from dorsal to ventral and pass rostral to the commissure constitute the *precommissural fornix,* whereas those passing behind the commissure form the *postcommissural fornix.*

Some fibers of the precommissural fornix pass to the septal nuclei, lateral hypothalamic area, and the preoptic region, whereas others join the postcommissural fibers. The postcommissural fornix passes within the wall of the third ventricle to the mammillary body where synapses occur with neurons extending to the anterior nucleus of the thalamus as discussed previously with the Papez circuit. In the cat some fibers of the postcommissural fornix have been traced as far caudad as the periaqueductal gray matter of the midbrain (Nauta, 1958).

Other anatomic structures of the limbic system are discussed in the appropriate chapter covering the division of the brain to which each belongs.

Cerebral Cortex

The cerebral cortex has been discussed in varying detail in preceding chapters. Phylogenetically the archicortex (primarily hippocampus and the dentate gyrus), paleocortex (the primitive olfactory lobe), and the neocortex (neopallium) have been discussed in their sequential order of importance from lower to higher vertebrates. The general pattern of cortical dominance is that the neocortex becomes superior and forces the older cortices into lesser ranks of importance anatomically and physiologically.

Embryologically the neocortical dominance is exemplified by the very early formation of the rhinal fissure (sulcus) which separates the olfactory from the non-olfactory cortex. In the dog brain the ventrolateral position of the rhinal fissure indicates the tremendously large surface area of the neocortex in comparison with the olfactory cortex. The area of the canine neocortex is 84.2 per cent of the entire hemispheric area, whereas the combined paleocortex, archicortex, and intermediate cortex (cortex intermedius) equals 15.8 per cent (Adrianov and Mering, 1964). These authors consider the intermediate cortex to be those regions that lie between cortical areas of dissimilar origin (*i.e.,* between neocortex and archicortex, or between neocortex and paleocortex).

General Significance of Neocortical Dominance

Because of his neocortical supremacy, man leads the animal kingdom in such qualities as the ability to think, communicate, remember, associate, and analyze input to the central nervous system. These faculties are obvious in man, but it should be realized that dogs and cats have a sufficiently well-developed neocortex to possess these qualities, although admittedly to a much lesser extent than man. To carry the point one step further: the dog is endowed with a relatively larger and better developed convoluted neocortex than the rat, with its smooth, or lissencephalic, neocortex.

The fundamental advantages of the better-developed neocortex in general are referable to the cortex as the 'brain power' which permits more flexible and sophisticated responses to environmental changes.

The lower the animal on the phylogenetic scale, the more stereotyped its reactions to stimuli; that is, the lower animals have a very limited and largely inherited reflexive behavior method of response or adaptation to their surroundings. In contradistinction, the higher animals with a better cortex are able to vary responses on the basis of learned and remembered past experiences of a similar nature.

It is interesting that dog was used in the well-known, classic conditioned reflex experiment by Pavlov (1927), which involves such complex phenomena as learning, communication by signal, interpretation, and association. Briefly stated, he first caused pricking of the dog's ears by ringing a bell. Some of the impulses in this neuronal circuit are transmitted by collaterals to other circuits, but are not effective because of synaptic resistance. Secondly, he gave food at the same time the bell was rung. These combined stimuli caused salivation. The impulses transmitted by this circuit causing salivation lower the resistance and, if they are repeated frequently, the synaptic resistance will become lower than that of the 'bell-ear circuit.' Therefore a pathway is established, after repeated usage, between hearing of the bell and salivation. Lastly, even after the food was removed, the ringing of the bell caused salivation.

The concept of basic importance in reference to such behavior is that there are many neuronal arcs or circuits in the cortex which may come in contact with each other and with appropriate usage (training) may become joined so that a stimulus of one arc may cause a response which originally depended on impulses transmitted by a different circuit. The revolutionary impact of this is that only by means of a cerebral cortex can an animal vary the inherent stereotyped responses to stimuli and learn by experience to respond in a manner best suited for him under a given set of circumstances. Obviously this principle is extended far beyond a simple 'bell-salivation'

response and applies to various theories and ramifications of learning, memory, and communication which reach their extreme development in man.

Structure of the Cerebral Cortex

Gross Anatomy

The cerebral cortex is the gray cellular layer that covers the surface of the cerebral hemispheres. It has been mentioned that the surfaces of the cerebral hemispheres of lower animals, including lower mammals, are smooth, whereas those of higher mammals are *convoluted,* that is, contain folded cortices with rounded elevations called *gyri* and the intervening indentations called *sulci.* In this context, the higher mammalian orders with convoluted cerebri include Ungulata, Carnivora, and Primates. The cerebral patterns among species vary, and, unlike the morphology of the brain stem, the gyri and sulci cannot be easily compared among animals (*e.g.,* dog and man). The morphologic differences parallel functional differences, although there are some similar overlapping relationships in dog and man, as discussed below.

Cerebral Pattern of the Dog

An outstanding example of the difficulties encountered in attempting to compare dog and human cortices is that the same four cerebral lobes are present in each, but in dogs there are no specific sulci or notches that separate the lobes comparable to those which are well known in the human brain. The lobes have the same names as the major bones of the calvaria, namely, frontal, parietal, temporal, and occipital. Since there are no specific anatomic demarcations, there is variance among authors regarding the lobar boundaries.

The general gross relationships of the lobes remain rather uniform in dog and human brains, but the relative sizes of the lobes differ. In a dorsal view (Fig. 16-8) the relative size of the surface area of the frontal lobe in the dog is not strikingly different from that in a human brain, in that it extends almost half the distance from the frontal (rostral) pole to the occipital (caudal) pole. The occipital lobe, however, extends much farther rostrally in the dog than in a human brain. This results in the parietal lobe being reduced to a relatively very small lobe, as illustrated in the figure.

From the lateral view, the temporal lobe extends farther dorsally in the dog brain when compared with that of man. This factor, too, results in a much smaller parietal lobe. The relative smallness of the parietal lobe in the dog is in general agreement with Adrianov and Mering (1964) and McGrath (1960). Adrianov and Mering, however, illustrate a caudolateral extension of the parietal lobe and a central occipito-parietal area on each side of the longitudinal fissure caudal to the true parietal lobe, which is not illustrated in the above figures.

The following description of the cerebral patterns of convolutions is focused on the major sulci. It seems unnecessary to describe individual gyri, which most frequently bear names identical with those of the adjacent sulci. The figures referred to in the discussion should aid the reader in locating the major sulci and gyri. It should be realized, however, that there are many variations among individual brains and therefore not every dog brain is going to correspond with the illustrations in all respects.

Lateral Surface (Fig. 16-9 A)

The orientation of gyri and sulci on the lateral surface of the dog and cat brain is described with reference to the *Sylvian (lateral cerebral) fissure (sulcus)* as illustrated in the figure. Nomina Anatomica Veterinaria (1968) lists the term *pseudosylvian fissure* for Carnivora, where it is primitive and not formed from a process of opercularization

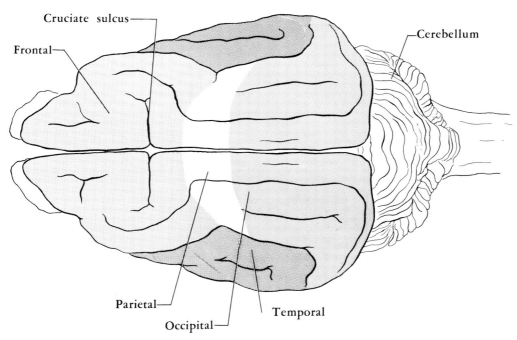

Fig. 16-8. Diagrammatic dorsal view of the dog brain, illustrating the relative positions of the four cerebral lobes. (Based on McGrath, 1960.)

as the Sylvian fissure is in Ungulata and Primates. On the lateral surface this fissure is near the midpoint of the ventrolateral base at the junction of the *rostral (anterior)* and *caudal (posterior) rhinal sulci.*

The constant Sylvian (pseudosylvian) fissure has four arciform sulci which curve around it, at intervals peripheral from it, as relatively constant semicircular grooves, namely, the ectosylvian, suprasylvian, ectomarginal, and marginal sulci (Fig. 16-9). Note that the sulci and gyri tend to be divided into three segments—rostral, middle, and caudal, and that adjacent gyri and sulci have the same names.

Variations in the pattern of convolutions exist among brains; for example, it is not uncommon to find the caudal ectomarginal sulcus extending much farther rostrally than is shown in Figure 16-9 A. It may extend into the middle suprasylvian gyrus and extend to join the marginal sulcus close to the ansate sulcus. The longitudinal division of the middle suprasylvian gyrus would permit identification of definite middle ectomarginal gyri and sulci, that is, the ectomarginal gyrus medial (dorsal) to the ectomarginal sulcus, and the middle suprasylvian gyrus lateral (below) the middle ectomarginal sulcus.

Dorsal Surface (Fig. 16-9 B)

The most conspicuous sulcus on the dorsal surface of the dog brain is the *cruciate sulcus.* This deep sulcus extends transversely in an almost perpendicular course from the dorsal longitudinal fissure. The sulcus extends to the medial surface of each hemisphere, where usually it may be seen to fuse with the cingulate sulcus (see Fig. 16-12). In many brains there may be two small inconsistent, poorly defined sulci which extend parallel to the cruciate sulcus on the dorsal surface. The sulcus on the rostral side is the *precruciate sulcus;* the one caudal to the cruciate sulcus is the *postcruciate sulcus.* The precruciate sulcus is barely visible and was

completely absent in 35 per cent of one series of dog brains studied (Adrianov and Mering, 1964). The same authors state that the postcruciate sulcus is much more constant, being absent in only 4 per cent of the dog brains studied.

CORTICAL MOTOR AND SENSORY AREAS IN THE DOG (FIG. 16-10)

The *coronal sulcus,* which has been defined as the rostral segment of the marginal sulcus, is lateral to and curves slightly away from the cruciate sulcus. The connection between the coronal and the marginal sulci is not clear in all dog brains. Functionally the coronal sulcus is important, because according to some cortical maps it separates the sensory and motor areas (Fig. 16-10). One should be very cautious in considering the cruciate sulcus of the dog as truly analogous to the central sulcus of primates as stated is possible in Nomina Anatomica Veterinaria (1968).

It is well established that in man the precentral gyrus (Brodmann area 4) is the chief voluntary motor area, and the postcentral gyrus (Brodmann area 3–1–2) is well known as the chief somesthetic cortical area. There are differences of opinion about functionally comparable areas in dogs and cats. Woolsey (1960) states in reference to dog and cat brains that there is no indication of a central sulcus between the 'precentral' motor and the 'postcentral' sensory areas. He maintains that the cruciate sulcus lies entirely within the precentral agranular field and that the ansate sulcus is at the caudal border of the 'postcentral' area.

Ariëns Kappers *et al.* (1967) state that in carnivores the anterior and posterior sigmoid gyri, in front of and behind the cruciate sulcus, contain motor cortex and belong to the area gigantopyramidalis. They also state that in carnivores the caudal border of the gigantopyramidal (motor) region lies somewhat dorsal to the ansate and coronal sulci. In general terms this agrees with

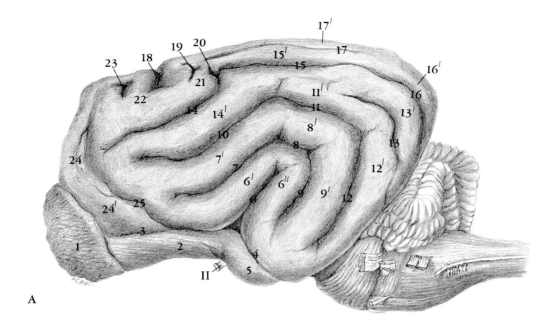

A

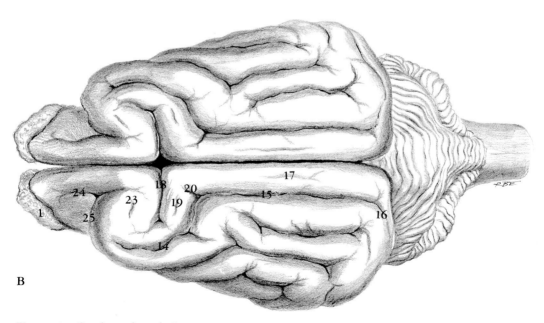

B

Fig. 16-9. See legend on facing page.

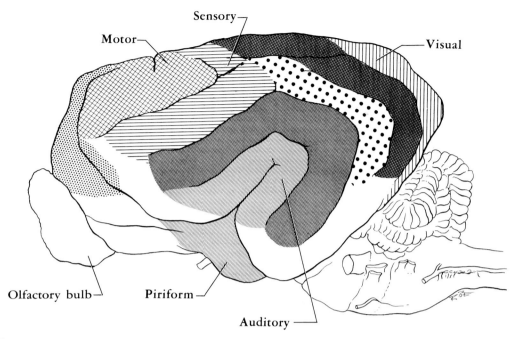

Fig. 16-10. Cerebral cortical motor and sensory areas in the dog. (After Campbell, 1905.)

Figure 16-10, which is based on the cortical map of Campbell (1905).

A more recent study in dogs using electrical stimulation localized the motor area for the forelimb musculature primarily within the posterior sigmoid (postcruciate) gyrus (Breazile and Thompson, 1967). These authors present other somatotopic motor localizations for the dog. Adrianov and Mering (1964) refer to a "motor analysor"

area in both the precruciate and postcruciate gyri. They agree that their motor cortical mapping corresponds more with Campbell (1905) than with any other reports.

CORTICAL MOTOR AND SENSORY AREAS IN THE CAT (FIG. 16-11)

The primary motor and sensory cortical areas in the cat are rostral to those in the

Fig. 16-9. Gyri and sulci of the cerebral cortex of the dog. A. Lateral view. B. Dorsal view of the brain, labeled to identify gyri and sulci on the dorsal surface that may be difficult to identify from the lateral view as shown in A. The same numbers are used for identification of the sulci in both views. 1, olfactory bulb; 2, olfactory tract; 3, rostral rhinal sulcus; 4, caudal rhinal sulcus; 5, piriform area; 6, (pseudo)sylvian fissure; 6', rostral sylvian gyrus; 6'', caudal sylvian gyrus; 7, rostral ectosylvian sulcus; 7', rostral ectosylvian gyrus; 8, middle ectosylvian sulcus; 8', middle ectosylvian gyrus; 9, caudal ectosylvian sulcus; 9', caudal ectosylvian gyrus; 10, rostral suprasylvian sulcus; 11, middle suprasylvian sulcus; 11', middle suprasylvian gyrus; 12, caudal suprasylvian sulcus; 12', caudal suprasylvian gyrus; 13, ectomarginal (ectosagittal) sulcus; 13', caudal ectomarginal (ectosagittal) gyrus; 14, coronal sulcus; 14', coronal gyrus; 15, marginal (sagittal) sulcus; 15', marginal (sagittal) gyrus; 16, caudal marginal (sagittal) sulcus; 16', caudal marginal (sagittal) gyrus; 17, endomarginal (endosagittal) sulcus; 17', endomarginal (endosagittal) gyrus; 18, cruciate sulcus; 19, postcruciate sulcus; 20, ansate sulcus; 21, postcruciate gyrus; 22, precruciate gyrus; 23, precruciate sulcus; 24, prorean sulcus; 24', prorean gyrus; 25, presylvian sulcus.

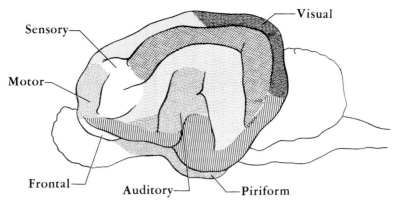

Fig. 16-11. Cerebral cortical areas in the cat. Note that the primary motor and sensory areas are rostral to those illustrated in Figure 16-10 for the dog. (After Campbell, 1905.)

dog (Campbell, 1905). As illustrated in Figure 16-11, these rostral positions result in a reduced frontal area in the cat. There is need for more precise experimental methods for cortical mapping in animals. Breazile and Thompson (1967) state that factors contributing to the controversies are the anesthetic used and the depth of anesthesia, both of which are known to appreciably alter the excitability of the cerebral cortex.

Medial Surface (Fig. 16-12)

The most consistent sulcus on the medial surface is the *sulcus of the corpus callosum* (*callosal sulcus*), which borders the dorsal surface of the rostral, middle, and caudal segments of the corpus callosum. At its caudal end the callosal sulcus tends to fuse with the *hippocampal sulcus*, which extends ventrolaterorostrally to the ventral surface of the temporal lobe. Approximately halfway between the callosal sulcus and the dorsal edge of the hemisphere is the *callosomarginal sulcus*, which tends to parallel particularly the caudal half of the corpus callosum. The rostral portion of the callosomarginal sulcus, which frequently may be disconnected or sinuous, is the *genual sulcus*. In general this sulcus parallels the contour of the genu of

the corpus callosum. The remaining middle and caudal segments of the callosomarginal sulcus are collectively referred to as the *splenial sulcus*. More in keeping with the comparable location of the cingulate sulcus in the human brain, the middle portion of the callosomarginal sulcus is frequently named the *cingulate sulcus*, and the caudal portion, closer to the splenium of the corpus callosum, is then called the splenial sulcus.

Fiber Connections of Cerebral Cortex

The types of fibers, based on their course, have been mentioned in previous chapters, especially in Chapter 2 as components of the medullary substance of the cerebral hemispheres. Based on the above-mentioned criterion, the fibers are divided into the three groups: (1) projection, (2) association, and (3) commissural fibers.

Projection Fibers

Projection fibers are those that either originate or terminate in the cerebral cortex. Those that originate there are referred to as *corticifugal, or efferent, projection fibers*. The classic example of this type is the corticospinal fibers, most of which arise in the motor cortical area. Other important corticif-

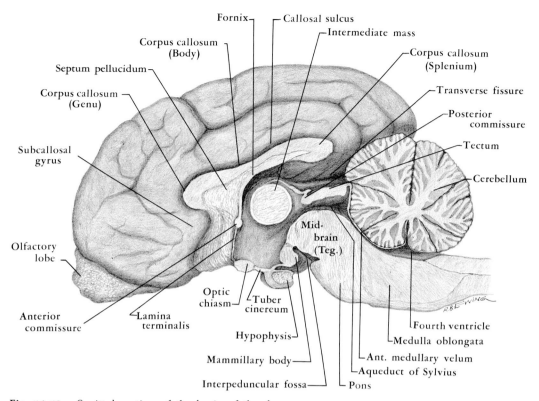

Labels in figure:
Fornix — Callosal sulcus — Intermediate mass
Corpus callosum (Body) — Corpus callosum (Splenium)
Septum pellucidum — Transverse fissure
Corpus callosum (Genu) — Posterior commissure
Subcallosal gyrus — Tectum — Cerebellum
Olfactory lobe — Mid-brain (Teg.)
Optic chiasm — Tuber cinereum
Anterior commissure — Lamina terminalis — Fourth ventricle — Medulla oblongata
Hypophysis — Ant. medullary velum
Mammillary body — Aqueduct of Sylvius
Interpeduncular fossa — Pons

RBLEWING

Fig. 16-12. Sagittal section of the brain of the dog.

ugal fibers are those of the corticopontine tracts, which, like the corticospinals, are exceptionally well developed in man. The corticobulbar fibers descending from the cortex mainly to the motor nuclei of the cranial nerves have been mentioned previously and are discussed in Chapter 17. In addition there are projection fibers from wide cortical areas to the red nucleus, caudate nucleus and putamen, reticular formation, and the thalamus.

The second group of projection fibers, those that terminate within the cortex, are the ascending, or sensory, fibers, often referred to as the *corticipetal fibers*. Most, if not all, of these sensory fibers are thalamocortical projections, in that they arise from the thalamus and are projected as the *thalamic radiations* to the cortex. The projection fibers, both ascending to and descending

from wide areas of the neocortex, form a fan-shaped radiation, the *corona radiata*, as they converge toward the brain stem to form the *internal capsule.*

This structure, the internal capsule, is composed of both ascending (sensory) and descending (motor) projection fibers, which anatomically can be grouped into three capsular areas. The *anterior limb (crus)* passes through the corpus striatum, between the head of the caudate nucleus and the putamen. In sections where the putamen is not present or is not extensive, the internal (and external) capsules separate the head of the caudate nucleus and the claustrum (Plates 26 and 27). The middle area, the *genu* (L.—knee), is a short bend where the two limbs meet. The caudal portion of the capsule is the *posterior limb (crus)*, which extends along the lateral side of the thalamus. The fibers of the internal capsule are intersected by the fibers of the corpus callosum at the dorsolateral edge of the lateral ventricle.

Lesions in the internal capsule are functionally significant because of the number of fibers participating in the corona radiata, and therefore a very small lesion in this area may damage a large number of fibers. In man this is the site of the lesion in a typical patient with a 'stroke,' which usually is a result of hemorrhage of one or more of the lenticulostriate arteries, branches of the middle cerebral artery. The available data are meager for such cerebrovascular accidents in dogs. McGrath (1960) comments that both hypertension and cerebral arteriosclerosis are rare in the dog. Since these are common predisposing factors for stroke in man, it seems logical to expect very few clinical syndromes similar to those in the human stroke patients. McGrath proposes that two favorable anatomic characteristics in dog, in comparison with man, may account for the rarity of cerebrovascular syndromes: (1) the rich potential anastomoses of the canine cerebral circulation, and (2) the similar plane of the head and heart in quadrupeds.

Association Fibers

Association fibers are those which connect one area of the brain with another on the same side. There are long association fibers connecting different lobes, and short ones connecting adjacent gyri. The human central nervous system has the greatest number of association fibers in correlation with the most highly integrated and complex connections. Recall that this type of fiber is not restricted to the brain, as exemplified by the ipsilateral internuncial neurons of the spinal reflex arc, which were also called association fibers in Chapter 7.

Commissural Fibers

Commissural fibers connect the two cerebral hemispheres across the midline. The greatest commissure is the corpus callosum. Examples of other commissural structures are: the rostral (anterior) commissure, the caudal (posterior) commissure, and the commissure of the hippocampus. In the spinal cord region there also are commissural fibers which are again exemplified by a kind of internuncial neuron of the spinal reflex arc. In this case the internuncial neuron crosses the midline to define a contralateral, or crossed, reflex arc (see Chapter 7).

Histology of the Cerebral Cortex

The intricate neuronal interaction of the cortical neurons, which is imperfectly understood, is largely dependent on the interneuronal relationships. The cellular arrangements are not identical in all areas of the cortex. Cytoarchitectonics is a very specialized study by which the cerebral cortex can be mapped into areas based on the cellular structures and arrangements of the various cell layers. As one might expect, the cytoarchitectonic information for the human cerebral cortex is very abundant. However, cerebral cortices have been mapped for other mammals; for example, Adrianov and Mering (1964) include the cytoarchitectonics of the dog.

The basic pattern of cortical histologic structure is the lamination of the neurons. In histologic sections of cortex stained with a Nissl stain, one can identify the cell bodies which are arranged in horizontal laminae parallel to the cortical surface. Each layer is identified by the types, density, and arrangement of its cells. In certain cortical areas an entire layer may be totally absent or very sparse. In some areas the cardinal identifying feature is the difference in size of the cell bodies when compared with those of other areas.

Fundamentally, in most cortical areas there are six layers, as illustrated in Figure 16-13. Beginning at the cortical surface and proceeding inwardly, these layers are: (I)

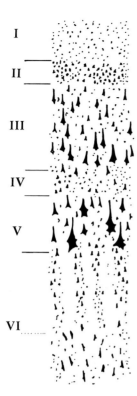

Fig. 16-13. Diagram showing the basic cyto-architecture of the cerebral cortex when stained with a Golgi stain. Fundamentally there are six cytologic layers, shown here numbered from the cortical surface inwardly: I, molecular, or plexiform; II, external granular; III, external pyramidal; IV, internal granular; V, internal pyramidal, or ganglionic; VI, multiform, or fusiform, layer. The patterns vary with cortical areas to the extent that certain layers may be entirely absent in specific cortical loci.

molecular, or plexiform, (II) external granular, (III) external pyramidal, (IV) internal granular, (V) internal pyramidal, or ganglionic, (VI) multiform, or fusiform.

Without going into detailed description of the individual layers, a few points should be emphasized: (1) The molecular layer contains relatively few cells. It contains mainly the dendritic ramifications of the deeper neurons which pass tangentially to form a plexus-like formation, giving it the

name plexiform. (2) The neurons are mostly pyramidal or spindle-shaped (fusiform), designating the name of respective layers having a preponderance of either cell shape. (3) The external pyramidal layer contains small cells, whereas the internal pyramidal cells are larger. In the motor area the large pyramidal cells (layer V) are extremely large and are known as *Betz cells;* they are the origin of the great projection fibers contributing to the long descending (motor) tracts, that is, the corticospinal tracts.

Functions of the Cerebral Cortex

The cerebral cortex reaches its peak of development and structure in man. The interneuronal connections, both intracortical and subcortical, are much more complex and intricate than in subhuman animals. Based on the general principle that function is dependent on structure, it is not surprising that human cerebral cortical function is far more complex and intricate than that in subhuman mammals. Many cortically dependent functions of man which, on casual consideration, seem to be restricted solely to him are often, in fact, only a matter of a more sophisticated degree of expression than lower mammals, such as the dog, are capable of. Present-day knowledge of human cerebral cortical structure and functions is far from complete, but in many respects our understanding of cortical function in subhuman mammals is even less adequate.

The rationale for beginning a consideration of cerebral functions of subhuman mammals with such a comparison is to avoid having a student think that there is absolutely no relationship between human and subhuman cortices. Truly, data from animal research, particularly in this area, must be evaluated in proper perspective, but much of our knowledge concerning human cortical function has been derived from animal experimentation.

In consideration of the functions of any

specific area of the central nervous system, one should realize that there is no single area of the central nervous system that functions independently. Therefore, one should be very cautious about referring to any isolated locus, either a brain stem nucleus or a cortical area, as a 'center' for a specific function as implied in a classic man-made functional cortical map as illustrated in Figure 16-10. The importance of such a cortical map is that it does indicate that certain cortical areas can be designated according to a particular function based on origins (motor) or terminations (sensory) of a preponderance of fibers transmitting impulses of a functional type common to those fibers.

A more sophisticated cortical map than that illustrated in Figure 16-10 would include *association, or secondary, areas* of the specific modalities labeled. The association areas show a phylogenetic progression which reaches its peak in the human brain. These areas are secondary in the sense that they contain neuronal connections associated with the primary areas, but do not contain the axons of the modality itself; for example, lesions in the association visual areas do not cause blindness, but will cause ocular disturbances. Secondary acoustic areas have been located in cat brains (Ades, 1943), and secondary visual areas have been described for rabbit (Thompson *et al.,* 1950).

A common method of cortical mapping is based on Brodmann's method of designating areas by numbers. The numbers, ranging from 1 to approximately 50, are consistently used for specific areas. It seems unnecessary for purposes of this text to give specific Brodmann numbers for corresponding cortical areas, but it should be realized that this system has been used for the rat, cat, and monkey, as illustrated by Krieg (1955). Casual observation of the illustrations in that work reveals that cortical areas differ among animals, those in the monkey corresponding very closely to the Brodmann areas of the human cortex as may

be found in most standard textbooks of human neuroanatomy.

Fundamental Functional Significance of Lobes of Dog Brain

The following is a brief summary of the functional aspects of each lobe of the cerebral cortex, with examples of their basic clinical correlations. Much of this discussion is based on the work of McGrath (1960).

Frontal Lobe (Fig. 16-8)

An outstanding area of the frontal lobe is the motor cortex, which contains the cell bodies of the descending projection fibers of the pyramidal system. In addition to the corticospinal (pyramidal) tract fibers *per se,* there are corticobulbar fibers which descend to the brain stem and terminate by synapses primarily in the motor nuclei of the cranial nerves. The relatively lesser importance of the dog's motor cortex and the pyramidal tracts for voluntary movement, in comparison with those of primates, has been discussed earlier in this Chapter. Experimental removal of the entire neocortex in dog does not result in paralysis, and the animal is still capable of performing many complex actions (*e.g.,* walking, running, righting itself, and eating if food is placed in its mouth). There are psychic changes, such as not recognizing food as such, and failure of the animal to recognize its owner or attendant. In the motor decorticate dog also there is a permanent loss of tactile sensation and optic placing reactions, plus a deficiency in the hopping reaction. In unilateral lesions the effects are shown on the opposite side of the body.

The rostral portion of the frontal lobe is referred to as the *prefrontal, or frontal, association area.* McGrath cites references to early literature describing animals with clinical and experimental lesions in this area. The outstanding common feature of effects of such lesions appeared to be psychomotor disturbances. Such changes range from apathy, stupidity, and inactivity to aggressiveness, ill-temperament, and hyperactivity. The hyperactivity varied in different animals from aimless pacing to aggressive propulsive movements. One should therefore realize that lesions in this general area of the brain may manifest themselves in a variety of ways among different animals even of the same species.

Parietal Lobe (Fig. 16-8)

The parietal lobe is an ill-defined, relatively small lobe in the dog brain which functions in a sensory capacity and as an association area. It is extremely difficult to evaluate parietal lobe functions or dysfunctions in animals.

Temporal Lobe (Fig. 16-8)

The cortical representation of the auditory system is in the temporal lobe. It should be remembered that unilateral lesions of the auditory cortex produce bilateral diminished hearing rather than unilateral deafness, because of the inconsistent crossing in the auditory pathway. In addition, portions of the rhinencephalon are included as temporal structures with the piriform area used as a general cortical landmark. The cerebral complexities are realized in the difficulty experienced in attempting to localize cerebral lesions by observing behavioral changes in an animal. Similar to the psychomotor changes described above which result from lesions in the prefrontal area, it is believed that convulsive seizures constituting a syndrome of 'psychomotor epilepsy' originate in the temporal lobe. Experimental lesions in this lobe in cats and dogs have resulted in a form of this syndrome. The 'psycho' portion of the term refers to the change in behavior with an expression of fear or panic accom-

panied by either attempts to run wildly or panic-like immobility. The 'motor' portion of the syndrome includes masticatory movements, salivation, pupil dilatation similar to the 'chewing gum' convulsions sometimes seen in a dog with distemper and followed by grand mal seizures. The Klüver-Bucy syndrome described earlier in this Chapter results from bilateral lesions of the temporal lobe. It should be remembered that in this syndrome, too, there are changes in the rage threshold, sexual habits, eating habits, and oral behavior.

Occipital Lobe (Fig. 16-8)

The occipital lobe contains the visual area of the cerebral cortex. Unilateral lesions in the primary visual area produce blindness in the contralateral eye. Bilateral removal of the dorsal portion of the occipital lobes results in 'psychic blindness,' that is, the animal is not blind but can no longer recognize what it sees. Experimental electric stimulation in specific areas of this lobe results in conjugate eye movement to the opposite side. It has been reported that bilateral ablation of the rostral portion of this lobe results in loss of the optic placing reflex, that is, the animal fails to place his forelegs on a table when it is brought close to the edge. The animal appears to have normal vision, but walks into objects rather than avoiding them.

Bibliography

Ades, H. W.: A secondary acoustic area in the cerebral cortex of the cat. J. Neurophysiol. 6: 59–63, 1943.

Adrianov, O. S., and Mering, T. A.: *Atlas of the Canine Brain.* Moscow, Govt. Publication of Medical Literature, 1959 (translated by E. Ignatieff, edited by E. F. Domino). Ann Arbor, Edwards Brothers, Inc., 1964.

Allen, W. F.: Effect of ablating the frontal lobes,

hippocampi, and occipito-parieto-temporal (excepting pyriform areas) lobes on positive and negative olfactory conditioned reflexes. Am. J. Physiol. *128:* 754–771, 1940.

Andersen, P., Blackstad, T. W., and Lömo, T.: Location and identification of excitatory synapses on hippocampal pyramidal cells. Exp. Brain Res. *1:* 236–248, 1966.

Andersen, P., Eccles, C., and Loyning, Y.: Pathway of postsynaptic inhibition in the hippocampus. J. Neurophysiol. *27:* 608–619, 1964.

Ariëns Kappers, C. U., Huber, G. C., and Crosby, E. C.: *The Comparative Anatomy of the Nervous System of Vertebrates, Including Man.* New York, Hafner Publishing Co., 1967. Vol. 3.

Breazile, J. E., and Thompson, W. D.: Motor cortex of the dog. Am. J. Vet. Res. *28:* 1483–1486, 1967.

Broca, P.: Anatomie comparee circonvolutions cerebrales. Le grand lobe limbique et la scissure limbique dans la serie des mammiferes. Rev. Anthropol. ser. 2, *1:* 384–498, 1878.

Brodal, A.: *Neurological Anatomy in Relation to Clinical Medicine,* 2nd ed. New York, Oxford University Press, 1969.

Campbell, A. W.: *Histological Studies on the Location of Cerebral Function.* New York, Cambridge University Press, 1905.

Clark, C. H.: Basic Concepts in Neurologic Diagnostics. Chapter 2 in *Canine Neurology— Diagnosis and Treatment* (Hoerlein, B. F.). Philadelphia, W. B. Saunders Co., 1965. Pp. 7–24.

Everett, N. B.: *Functional Neuroanatomy,* 6th ed. Philadelphia, Lea & Febiger, 1971.

Fox, C. A., Hillman, D. E., Siegesmund, K. A., and Sether, L. A.: The Primate Globus Pallidus and Its Feline and Avian Homologues. In *Evolution of the Forebrain. Phylogenesis and Ontogenesis of the Forebrain* (Hassler, R., and Stephans, H., Eds.). Stuttgart, Georg Thieme Verlag, 1966.

House, E. L., and Pansky, B.: *A Functional Approach to Neuroanatomy,* 2nd ed. New York, McGraw-Hill Book Co., Blakiston Division, 1967.

Kaada, B. R., and Ursin, H.: Further localization of behavioral responses elicited from the amygdala in unanesthetized cats. Acta Physiol. Scand. *42* (Suppl. 145): 80–81, 1957 (abstract).

Klüver, H., and Bucy, P.: Preliminary analysis of functions of the temporal lobes in monkeys. Arch. Neurol. Psychiat. *42:* 979–1000, 1939.

Krieg, W. J. S.: *Brain Mechanisms in Diachrome.* Evanston, Ill., Brain Books, 1955.

Liles, S. L., and Davis, G. D.: Athetoid and choreiform hyperkinesias produced by caudate lesions in the cat. Science *164:* 195–197, 1969.

McGrath, J. T.: *Neurologic Examination of the Dog,* 2nd ed. Philadelphia, Lea & Febiger, 1960.

Nauta, W. J. H.: Hippocampal projections and related neural pathways to the mid-brain in the cat. Brain *81:* 319–340, 1958.

Noback, C. R.: *The Human Nervous System. Basic Elements of Structure and Function.* New York, McGraw-Hill Book Co., Blakiston Division, 1967.

Nomina Anatomica Veterinaria, Adopted by the General Assembly of the World Association of Veterinary Anatomists, Paris, 1967. Vienna, Adolf Holzhausen's Successors, 1968.

Palmer, A. C.: *Introduction to Animal Neurology.* Philadelphia, F. A. Davis Co., 1965.

Papez, J. W.: *Comparative Neurology.* New York, Thomas Y. Crowell Co., 1929.

Papez, J. W.: A proposed mechanism of emotion. Arch. Neurol. Psychiat. *38:* 725–743, 1937.

Pavlov, I. P.: *Conditioned Reflexes* (translated by G. U. Anrep). New York, Oxford University Press, 1927.

Peele, T. L.: *The Neuroanatomic Basis for Clinical Neurology,* 2nd ed. New York, McGraw-Hill Book Co., Blakiston Division, 1961.

Ruch, T. C., Patton, H. D., Woodbury, J. W., and Towe, A. L.: *Neurophysiology,* 2nd ed. Philadelphia, W. B. Saunders Co., 1965.

Shealy, C. N., and Peele, T. L.: Studies on amygdaloid nucleus of cat. J. Neurophysiol. *20:* 125–139, 1957.

Terzian, H., and Ore, G. D.: Syndrome of Klüver and Bucy reproduced in man by bilateral removal of the temporal lobes. Neurology *5:* 373–380, 1955.

Thompson, J. M., Woolsey, C. N., and Talbot, S. A.: Visual areas I and II of the cerebral cortex of the rabbit. J. Neurophysiol. *13:* 277–288, 1950.

Trautmann, A., and Fiebiger, J.: *Fundamentals of the Histology of Domestic Animals* (translated and revised from 8th and 9th German edition by R. E. Habel and E. L. Biberstein). Ithaca, N.Y., Comstock Publishing Associates, Division of Cornell University Press, 1957.

Webster, K. E.: Cortico-striate interrelations in the albino rat. J. Anat. *95:* 532–544, 1961.

Webster, K. E.: The cortico-striatal projection in the cat. J. Anat. *99:* 329–337, 1965.

Woolsey, C. H.; Brain Fissuration in Relation to Cortical Localization of Function. In *Structure and Function of Cerebral Cortex* (Tower, D. B., and Schade, J. P., Eds.). New York, Elsevier Publishing Co., 1960.

chapter 17

Cranial Nerves

Introduction

The nerves that are attached to the brain are referred to as the cranial nerves. There are twelve pairs, all of which are attached to the brain stem, except for the first pair.

The nerves may be referred to by number beginning with I at the rostral attachment to the brain and passing caudally to XII at the caudal portion of the medulla oblongata (Fig. 17-1). Some of the names are descriptive of certain characteristics, for example, function—olfactory, abducens; distribution—facial, hypoglossal, optic; appearance—trigeminal. For the beginning student who may find it difficult to remember the correlation of cranial nerve numbers and names, Table 17-1 lists the names of the nerves in numerical order, together with a mnemonic which may be useful.

The so-called olfactory and optic nerves are not typical cranial nerves, as will be explained later in this Chapter. All of the nerves, except olfactory (I) and trochlear (IV), may be identified at their attachment to the brain stem from the ventral view of the brain (Fig. 17-1). The olfactory 'nerves' are within the nasal cavity and pass through the foramina of the cribriform plate of the ethmoid bone to terminate and synapse in the olfactory bulb. In routine removal of the brain from the calvaria the olfactory nerves are severed at their connection with the olfactory bulb. The trochlear nerve (IV) is

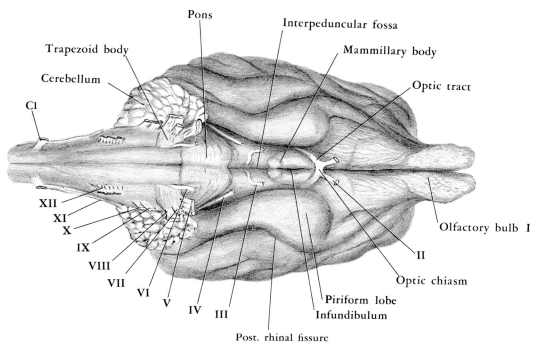

Fig. 17-1. Ventral view of the dog brain, illustrating the gross attachments of the cranial nerves.

the only nerve which grossly emerges from the dorsal surface of the brain stem.

Comparison of Cranial and Spinal Nerves

It should be remembered that the spinal nerves, in general, have a common basic anatomic organization in their attachment to the spinal cord. Each spinal nerve, with the possible exception of the first, grossly has a dorsal root with a sensory ganglion, and a ventral motor root. Both roots unite near the intervertebral foramen to form the spinal nerve proper (see Fig. 6-3, p. 93), which functionally is mixed, in that it contains both sensory and motor fibers. It is not surprising that as a result of cephalization during phylogeny the cranial nerves became much more complex functionally and anatomically and consequently deviated from the typical spinal nerve structure.

Some of the differences in cranial nerves, when compared with spinal nerves, are the following:

1. The attachments of cranial nerves to the brain are not equally spaced linearly as are those of the spinal nerves to the spinal cord.

2. There is no separate attachment of dorsal and ventral roots as in spinal nerves. In mixed cranial nerves the sensory and motor axons attach to the brain at the same point, or very close to each other.

3. Not all cranial nerves have sensory ganglia—only nerves V, VII, VIII, IX, and X.

4. Some cranial nerves are considered purely sensory: I, II, and VIII; although sensory, the first two cranial nerves have no sensory ganglia.

5. Some nerves are considered functionally as purely motor; these are III, IV, VI, XI, and XII, although proprioceptive (GSA) fibers are assumed to be present.

TABLE 17-1. Numbers and Names of Cranial Nerves, Including a Mnemonic. (After Miller, 1962.)

Number of Nerve	Name of Nerve	Mnemonic
I	Olfactory	Only
II	Optic	Orderly
III	Oculomotor	Organized
IV	Trochlear	Thoughts
V	Trigeminal	That
VI	Abducens	Are
VII	Facial	Frequent
VIII	Acoustic (Vestibulocochlear)	Accomplish
IX	Glossopharyngeal	Great
X	Vagus	Valor
XI	Accessory	And
XII	Hypoglossal	Honor

6. Cranial nerves contain 'special' functional components, in addition to four 'general' components of spinal nerves. This gives a total of seven functional components distributed among the cranial nerves.

7. Correlated with the above, the four functional columns of the spinal nerves in the spinal cord gray matter are linearly fragmented into smaller segments as true nuclei in the brain stem. Within the brain stem the 'special' component nuclei are present in addition to the 'general' four.

8. The internal ramifications of cranial nerve fibers are much more complicated than are those of spinal nerves.

Gross Anatomy of Cranial Nerves

This discussion will be restricted to the basic gross anatomy as a guide for the student as he studies a mammalian brain, skull, and cadaver, for example, of a dog or cat. It is not intended that this discussion be complete as a dissecting guide; therefore, only selected branches of nerves are considered and dissection directions are not given. The basic fundamentals of general distribution with the target organs innervated are illustrated in Figure 17-2. Functional and dysfunctional aspects are considered in the following portion of this

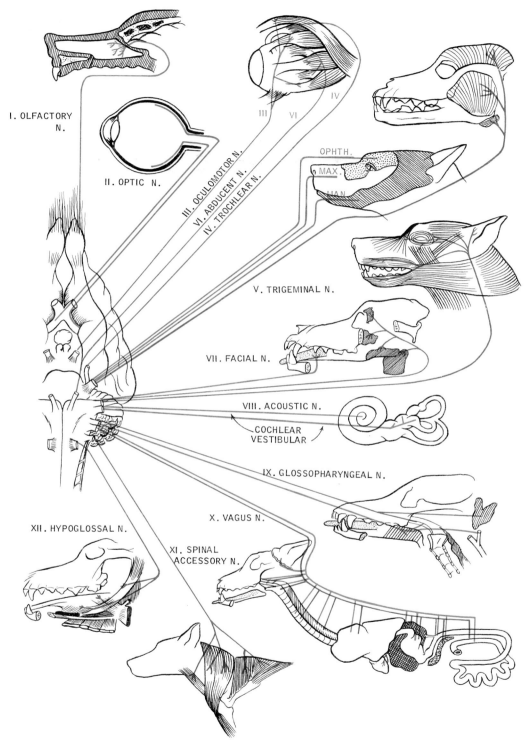

I. OLFACTORY N.

II. OPTIC N.

III. OCULOMOTOR N.

VI. ABDUCENT N.

IV. TROCHLEAR N.

III

VI

IV

OPHTH.

MAX.

MAN.

V. TRIGEMINAL N.

VII. FACIAL N.

VIII. ACOUSTIC N.

COCHLEAR

VESTIBULAR

IX. GLOSSOPHARYNGEAL N.

X. VAGUS N.

XII. HYPOGLOSSAL N.

XI. SPINAL
ACCESSORY N.

Fig. 17-2. The origin and major distribution of the cranial nerves in the dog. (From Hoerlein, 1965, by permission of W. B. Saunders Company.)

Chapter. In keeping with our plan of study of the brain stem from caudal to rostral, the cranial nerves are considered in the same direction.

Hypoglossal Nerve (XII)

The hypoglossal nerve (XII) arises as a longitudinal series of fine rootlets from the ventrolateral surface of the caudal medulla lateral to the pyramids (Fig. 17-1). The filaments coalesce to form a relatively thick nerve which leaves the cranial cavity via its own foramen, the *hypoglossal foramen* (Fig. 17-3). Peripherally the nerve passes ventrad, medial to the digastric muscle, and crosses lateral to the external carotid artery near the bifurcation of the common carotid artery. The main portion enters the tongue as its sole motor innervation. In the neck the descending branch leaves the parent nerve to pass caudally and anastomoses with the first cervical nerve on the lateral surface of the larynx to form the *ansa hypoglossi (ansa*

cervicalis). Rami from this loop are distributed to some of the infrahyoid muscles (*e.g.,* sternohyoid and sternothyroid).

Accessory Nerve (XI)

The 'accessory' name of this nerve is based on its being so close anatomically and in fact partly joined with the vagus nerve. Consideration of the accessory nerve varies among authors. It is customary to describe two divisions of the nerve: the cranial, or bulbar, portion from the medulla, which forms the internal branch, and the spinal portion, which forms the external branch.

Generally the bulbar portion, or *internal branch,* is reported to leave the accessory nerve to join the vagus at the level of the jugular foramen. There is evidence that the accessory fibers are within the recurrent laryngeal nerve of the vagus and convey motor impulses in the caudal laryngeal nerve to the intrinsic muscles of the larynx. This has been reported for the cat (DuBois

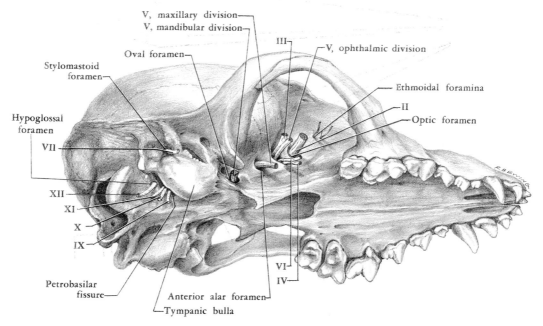

Fig. 17-3. A ventrolateral view of the dog skull, showing the foramina through which cranial nerves II to VII and IX to XII pass.

and Foley, 1936) as well as for man (Truex and Carpenter, 1969).

The spinal portion, or *external branch*, of the accessory nerve arises from the gray matter in the ventral horn of the upper six or seven cervical segments of the spinal cord in the dog (McClure, 1964). Rootlets pass dorsolaterally to form a common bundle (external branch), which ascends alongside the cord dorsal to the denticulate ligament (Fig. 17-1). This spinal portion enters the cranial cavity via the foramen magnum to join the bulbar portion of the accessory nerve on the caudal side of the vagus nerve. The glossopharyngeal (IX) and the vagus (X), including fibers of the bulbar portion of the accessory (XI), and the external branch of the accessory nerve enter the jugular foramen and exit from the cranium via the petrobasilar fissure. Peripherally the external branch of the accessory nerve passes lateral to the hypoglossal nerve and enters the combined cleidomastoideus and sternomastoideus muscles. After innervation of these muscles it extends to anastomose with the second, third, and fourth cervical nerves before reaching the trapezius muscle, which it innervates.

Vagus Nerve (X)

The vagus nerve (X) is the longest cranial nerve, with the greatest area of distribution. It is the only cranial nerve to extend to body regions caudal to the head and neck, passing through the neck and thorax, and into the abdomen. Evidence of its complexity is that it contains fibers of the greatest number of functional components identified in any single nerve, that is, five (six, counting two types of GSA).

The vagus nerve attaches to the medulla by fine rootlets which coalesce to form a bundle that enters the jugular foramen and leaves the cranium via the petrobasilar (petro-occipital) fissure. Within the jugular foramen the nerve has a localized enlargement, the *superior (jugular) ganglion.* Immediately distal to the petrobasilar fissure ventromedial to the tympanic bulla there is a prominent distal swelling, the *inferior (nodose) ganglion.* A short distance caudal to the inferior ganglion the vagus nerve is joined by the sympathetic trunk. The two fuse to form the *vagosympathetic trunk,* which is enveloped by the carotid sheath, along with the common carotid artery, and extends caudally to the thoracic inlet.

Vagal Branches in the Cervical Region

At the level of the jugular ganglion, the *auricular nerve* passes through the caudal end of the petrous temporal bone to meet the facial nerve within the facial canal. Vagal fibers extend to the deep area of the external acoustic meatus for innervation of the skin close to the tympanic membrane. In young animals impulses of the general somatic afferent component are generated in the vagus nerve by inflammation caused by mites, resulting in vomiting. Apparently the young animal is unable to differentiate the input of the functional components of the vagus nerve to the brain, which interprets the impulses as coming from the stomach, also innervated by the vagus nerve, and vomiting occurs (Miller, 1962).

Vagal rami join those of the glossopharyngeal nerve and the combined ramifications form the *pharyngeal plexus* medial to the digastric muscle on the dorsolateral side of the pharynx in the vicinity of the nodose and cranial (superior) cervical ganglia.

At the level of the nodose ganglion the *cranial laryngeal nerve* leaves the vagus nerve and divides into external and internal branches. The *external branch* passes caudad to innervate the cricothyroideus muscle and continues to the area of the thyroid gland. The *internal branch* passes deep to enter the larynx at the cranial border of the thyroid cartilage. Prior to its penetration into the deep laryngeal wall, the internal branch sends a ramus caudally to anastomose with the caudal laryngeal nerve.

The left and right recurrent laryngeal nerves leave the respective vagus nerves at different levels. The *left recurrent laryngeal nerve* arises from the left vagus nerve within the thorax at the level of the aortic arch and loops around the ligamentum arteriosum. From here the recurrent laryngeal nerve passes cranially between the trachea and esophagus to terminate as the *caudal laryngeal nerve*. The *right recurrent laryngeal nerve* arises from the right vagus nerve at the level of the thoracic inlet and loops around the subclavian artery. The course and termination of this recurrent nerve are similar to those of the left. The caudal laryngeal nerves are the motor innervation for all of the intrinsic laryngeal muscles except the cricothyroid, which is innervated by the external branch of the cranial laryngeal nerve, as described previously.

Caudal to the thoracic inlet the vagus nerve contributes parasympathetic fibers to the thoracic and much of the abdominal viscera, as discussed in Chapter 9.

Glossopharyngeal Nerve (IX)

The glossopharyngeal nerve (IX) is attached to the ventrolateral surface of the brain stem immediately rostral to the rootlets of the vagus (Fig. 17-1). The close relationship between the glossopharyngeal and vagus nerves is apparent internally (as discussed in the following section of this Chapter, under Functional Components), at the surface of the brain stem, and peripherally in the pharyngeal region where rami from both nerves contribute to the pharyngeal plexus. In the dog there is only one discernible sensory ganglion, the *petrosal*, which is at the jugular foramen–petrobasilar fissure. The cell bodies of the somatic afferent neurons which are located in the superior ganglion in other species, including man, may be present within the nerve itself rather than being aggregated in sufficient quantity to form a true ganglion.

Peripherally, this nerve sends branches to the middle ear, where they contribute to the tympanic plexus. From this plexus the minor petrosal nerve extends along the dorsolateral side of the eustachian tube to transport preganglionic parasympathetic fibers to the *otic ganglion.* Postganglionic fibers from here innervate the parotid and zygomatic salivary glands.

Of special interest is the *carotid sinus nerve,* which leaves the glossopharyngeal nerve to follow the internal carotid artery in the neck to the carotid sinus near the bifurcation of the common carotid artery. This is the afferent (GVA) nerve of the carotid sinus reflex.

Vestibulocochlear Nerve (VIII)

The vestibulocochlear nerve (VIII) is the shortest cranial nerve and the only one that does not leave the skull. The two divisions, vestibular and cochlear, arise from the respective parts of the inner ear within the petrous portion of the temporal bone and enter the brain as a grossly united nerve at the lateral end of the trapezoid body (Fig. 17-1). The only portions of this nerve that can be observed grossly are at the sites of its attachment to the brain and at its entrance through the internal acoustic meatus.

Facial Nerve (VII)

The facial nerve (VII) is attached to the brain stem very close to the medial side of the vestibulocochlear nerve at the rostrolateral side of the trapezoid body (Fig. 17-1). It remains in close proximity with the vestibulocochlear nerve as both nerves enter the internal acoustic meatus. Within the petrous portion of the temporal bone the facial nerve leaves the vestibulocochlear nerve to enter the facial canal. The facial nerve within the canal makes a sharp turn caudally, to form the *external genu of the facial nerve.*

At the genu of the facial nerve lies the *geniculate ganglion,* the only sensory ganglion of the facial nerve. The nerve emerges from the skull via the stylomastoid foramen (Fig. 17-3). From this location deep under the parotid gland the facial nerve divides into a number of peripheral branches, the names of which designate the general area of distribution; for example, the cervical and retroauricular nerves ramify in the ventral and dorsal cervical regions, respectively. Shortly distal to its emergence the facial nerve divides into its terminal branches of the face: the auriculopalpebral and the dorsal and ventral buccal nerves (Fig. 17-4). This figure illustrates the superficial distribution of the peripheral branches. It should be remembered that the muscles innervated by the facial nerve are derived from the second embryonic branchial arch and are the so-called muscles of facial expression.

Abducens Nerve (VI)

The abducens nerve (VI) leaves the ventral surface of the brain stem at the level of the trapezoid body at the lateral border of the pyramids medial to the seventh and eighth cranial nerves (Fig. 17-1). The abducens nerve has a long rostral course as it passes through the cavernous sinus to enter the orbit via the orbital fissure. This nerve serves as the motor innervation for the lateral rectus and retractor bulbi muscles of the eye.

Trigeminal Nerve (V)

The trigeminal nerve (V) is the largest of the cranial nerves based on the diameter of the attachment at the pons immediately rostral to the trapezoid body and nerves VII and VIII (Fig. 17-1). As the name indicates, the nerve has three divisions, named for the

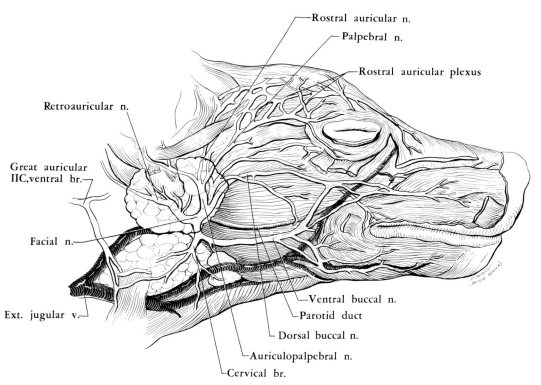

Fig. 17-4. Branches of the facial nerve on the lateral surface of the head and face. The parotid gland has been dissected. (Modified from McClure, 1964.)

respective areas of distribution: ophthalmic, maxillary, and mandibular. The nerve is mixed but is primarily sensory, in that all three divisions are sensory but only the mandibular division contains motor fibers. These innervate the muscles of mastication, which were derived embryonically from the first branchial arch. Intracranially at the rostral apex of the petrous portion of the temporal bone lateral to the cavernous sinus is located the large sensory *semilunar, or Gasserian, ganglion,* which serves all three divisions. Each of the three divisions leaves the cranium via a separate foramen: the ophthalmic division through the orbital fissure in company with cranial nerves III, IV, and VI; the maxillary division via the anterior alar foramen; and the mandibular division via the oval foramen (Fig. 17-3).

Ophthalmic Division

The ophthalmic division is the smallest of the three divisions of the trigeminal nerve. Intracranially this division may be traced rostrally within the lateral wall of the cavernous sinus and beyond as it enters the orbit via the orbital fissure. The names of the primary sensory branches indicate the areas of innervation: the supraorbital (frontal) nerve from the skin area above the orbit (Fig. 17-5), the lacrimal nerve from the lacrimal gland, and the nasociliary from the nasal turbinate and eyeball areas.

Maxillary Division

The terminal branches of the maxillary division are sensory for the general maxillary and adjacent areas. This division serves the cheek, side of the nose and muzzle, maxillary sinus, nasal mucosa, and soft and hard palate, plus the upper teeth and gingiva. Intracranially this division can be followed from the semilunar ganglion rostrally as it also enters the lateral wall of the cavernous sinus and extends to the round foramen, through which it passes to enter

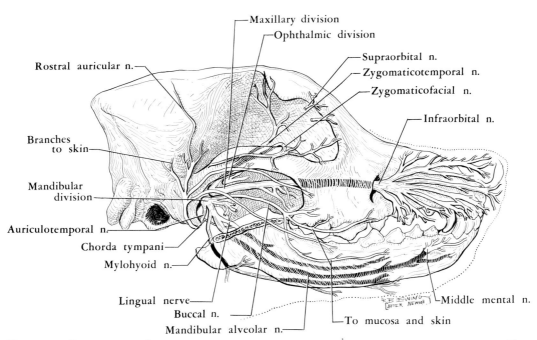

Fig. 17-5. Diagrammatic lateral view showing peripheral distribution of the trigeminal nerve. The mandible has been cut as illustrated. (After McClure, 1964.)

the alar canal. Within the canal it turns rostrally to pass through the anterior alar foramen to traverse the pterygopalatine fossa (Fig. 17-3), where it branches to supply the areas stated above. The large peripheral extension of the maxillary division is the *infraorbital nerve,* which may easily be identified as it leaves the infraorbital canal. From here the nerve ramifies over the skin area of the lateral side of the upper lip and muzzle. Nerves may be followed to individual sinuses surrounding the tactile hairs (vibrissae).

Mandibular division

The mandibular division of the trigeminal nerve is the only division which is mixed, in that it is the only one containing motor (SVE) fibers to all of the muscles of mastication, plus a few other branchiomeric muscles. This division emerges from the cranium through the oval foramen (Fig. 17-3), and immediately distal to the foramen divides into many branches. With careful dissection the nerves transmitting the motor impulses to the muscles of mastication may be identified. In general the nerve innervating the muscle has the same name as the muscle, for example, the pterygoid, deep temporal, and the masseteric. A relatively large mandibular alveolar nerve is easily identified as it leaves the mandibular division to enter the mandibular canal of the mandible. The nerve may be traced through the mandibular canal as it gives off sensory rami to the alveoli of the teeth of the lower jaw. Distally this nerve emerges from the canal via the mental foramina as the *mental nerves,* which are sensory rami to the skin at the rostrolateral area of the mandible (Fig. 17-5). The large lingual nerve can be traced into the musculature of the tongue where it ramifies and anastomoses with the hypoglossal nerve. It should be remembered, however, that the hypoglossal nerve is the sole motor innervation for the tongue, whereas the lingual is general sensory in function.

Trochlear Nerve (IV)

The trochlear nerve (IV) is the thinnest of all cranial nerves, and is the only one to emerge from the dorsal surface of the brain stem. The right and left trochlear nerves emerge from the extreme caudal end of the midbrain to decussate immediately within the rostral tip of the rostral (anterior) medullary velum. The nerve passes lateroventrally within the transverse fissure to appear on the ventral surface in the cerebellopontine angle (Fig. 17-1). From here it passes rostrally along the basal surface of the brain with the abducens, oculomotor, and the ophthalmic division of the trigeminal nerve, with which it passes through the orbital fissure (Fig. 17-3). Inside the orbit the trochlear nerve passes to innervate only one of the extrinsic ocular muscles, the dorsal oblique.

Oculomotor Nerve (III)

The oculomotor nerve (III) emerges from the ventromedial area of the mesencephalon within the interpeduncular fossa (Fig. 17-1). Its fibers bend rostrally to pass within the cavernous sinus wall and extend through the orbital fissure (Fig. 17-3). Within the orbit it divides into a dorsal and a ventral ramus which are distributed for innervation of the dorsal, medial, and ventral recti muscles, the ventral oblique muscle, and the levator palpebrae muscle. Preganglionic parasympathetic fibers pass to the ciliary ganglion for synapse and postganglionic fibers extend to penetrate the sclera of the eye for autonomic innervation.

Optic Nerve (II)

The optic nerve (II) is atypical in at least three respects: (1) It is composed of axons which have their cell bodies in the retina;

(2) the fibers have no neurilemma; and (3) it is entirely enveloped by all three meninges, which are continuous with the intracranial meninges. Within the orbit the nerve is the central axis of a cone formed by the extrinsic ocular muscles. Adjacent to the optic nerve within the cone are the ciliary ganglion, nerves, and blood vessels which are embedded in the orbital fat. The nerve with its meninges passes through the optic foramen to enter the cranial cavity and shortly joins its mate to form the optic chiasm at the rostral portion of the hypothalamus (Fig. 17-1). Caudal to the chiasm each optic tract transmits impulses from both optic nerves to the lateral geniculate body, rostral colliculi, and the pretectal area (see Chapter 18).

Olfactory Nerves (I)

Many small bundles of nerve fibers arise in the olfactory mucosa and pass caudally through the numerous foramina of the cribriform plate of the ethmoid bone to terminate in the olfactory bulb (Fig. 17-6). These are collectively referred to as the olfactory, or first cranial, nerve. As they pass through the cribriform plate, the bundles are enveloped by all three meninges. Because of these peculiar anatomic features, the nerve is considered atypical.

The *vomeronasal nerves* (Fig. 17-6) serve to convey a special type of olfactory input to the brain. They are axons which arise from the neuroepithelium in the mucosa of the vomeronasal organ and pass caudally through the cribriform plate. Intracranially they extend along the ventromedial edge of the longitudinal fissure to terminate in the accessory olfactory bulb, located on the medial surface of the olfactory tract and caudal edge of the olfactory bulb (McClure, 1964). The vomeronasal organ is absent in man and higher primates, bats, and various aquatic animals, but persists as a functional organ in almost all other mammals (Romer, 1962). In the dog, which is typical of most

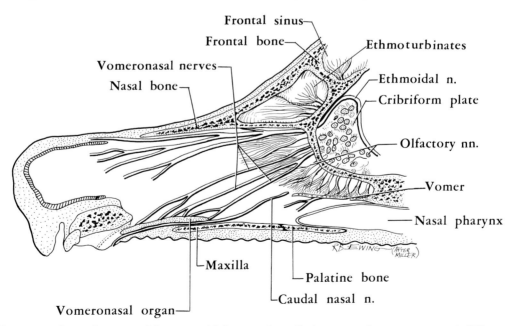

Frontal sinus
Frontal bone
Vomeronasal nerves
Nasal bone
Ethmoturbinates
Ethmoidal n.
Cribriform plate
Olfactory nn.
Vomer
Nasal pharynx
Maxilla
Palatine bone
Caudal nasal n.
Vomeronasal organ

Fig. 17-6. Sagittal section of the nose with bony and cartilaginous nasal septum removed. Olfactory nerve fascicles are shown in relation to the cribriform plate of the ethmoid bone. Other major nerves and structures are shown for orientation. (After McClure, 1964.)

mammals, the vomeronasal organ is a spindle-shaped structure near the midline in the floor of the nasal cavity caudal to the incisive bone. The organ connects with the mouth cavity via the nasopalatine duct, which pierces the palate as illustrated in Figure 17-6.

Functional Components

The comparisons between the spinal cord and the cranial region of the neural tube were discussed in the first portion of Chapter 2. It should be recalled that the same general characteristics of differentiation of gray and white matter appear in both regions, but the cranial region demonstrates a more complicated organization. In the spinal cord the general cylindrical shape of the neural tube is retained, with the ependymal cells remaining to line the central canal, the mantle layer forming the gray matter, and the marginal layer differentiating into the peripheral white area. The sulcus limitans indicates the separation of the dorsal alar (sensory) and the ventral (motor) plates as the sites of columns (nuclei) of termination and origin, respectively. The representations of the four functional components of a typical spinal nerve are evolved as four uninterrupted columns throughout most of the spinal cord (Fig. 17-7).

In the brain stem region of the neural tube the increased complexity is indicated by changes that occur in the original columns representing the four functional components. (1) The original columns are fragmented longitudinally to form isolated nuclei. Although the columns are interrupted, there still tends to be a stratum for each component, and nuclei of the same functional component tend to be in the same relative location in transverse sections of the brain stem. (2) In the brain stem three of the original four columns, namely general somatic afferent (GSA), general visceral

afferent (GVA), and general visceral efferent (GVE), each have an additional component with a 'special' designation which accompanies them, as follows: special somatic afferent (SSA), special visceral afferent (SVA), and special visceral efferent (SVE). We will consider the fourth column, the general somatic efferent (GSE), as not having a 'special' component counterpart, and, as stated elsewhere, will drop the 'general' and refer to it simply as the somatic efferent (SE) component. This results in the seven functional components, SSA, GSA, SVA, GVA, GVE, SVE, and SE (Fig. 17-8), which are distributed among the twelve pairs of cranial nerves.

As shown in Table 17-2, no single cranial nerve contains all seven functional components. The greatest number for a single nerve is five (six, if one considers the two types of GSA) as observed in each of cranial nerves VII, IX, and X. Obviously some of the same functional components and internal nuclei may be shared by many cranial nerves, but the peripheral distribution of a functional component varies among the nerves. An attempt will be made to correlate the target organs with the nuclei and functional components based on a criterion, *e.g.,* the embryologic origin of the target organ, which is common to a functional component regardless of the individual nerves transmitting that component.

Correlation of Functional Components and Target Organs

Because a few general statements with reference to the correlation of the seven functional components and target organs may help in understanding the detailed information which follows, a brief explanation of the functional components is given here, based on the common property of the target organs innervated by a single component.

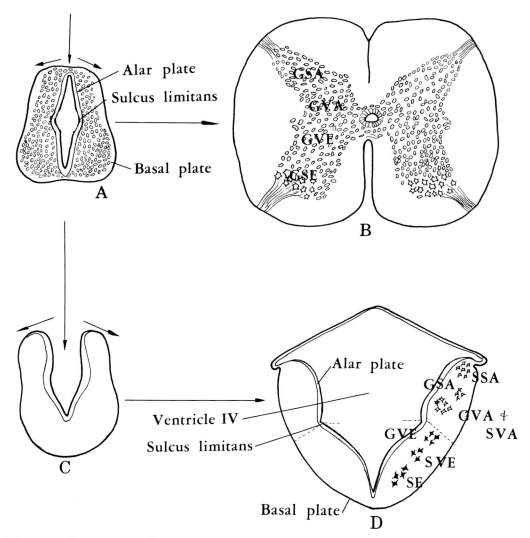

Fig. 17-7. Diagrammatic illustration of the differentiation of the alar and basal plates of the mantle layer of the neural tube (A). The four functional components of a typical spinal nerve are specifically represented in the spinal cord gray matter (B). Development from the neural tube to medulla (D) is more complicated, and passes through an intermediate stage (C). The mid-dorsal area of the neural tube splits and the dorsal walls migrate laterally so that the somatic afferent components (which were located dorsally as in B) become located dorsolaterally in the medulla. The original neural canal (in A) becomes the fourth ventricle in the rhombencephalon and the sulcus limitans remains visible in the adult and continues to separate sensory (alar plate) from motor (basal plate) nuclei (D).

Sensory (Afferent)

The nerves containing sensory fibers which supply target organs embryonically derived from ectoderm, somatopleure, or somites and located in the body wall and limbs are designated as *somatic*. As pointed out previously, there are two types of somatic afferent fibers: general and special. The general somatic afferent fibers are

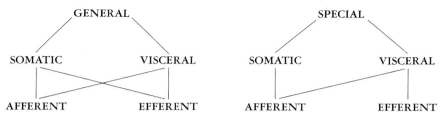

Fig. 17-8. Diagram illustrating the seven functional components of cranial nerves. The designation of 'general' or 'special' for somatic efferent may be omitted, as discussed in the text.

defined the same as the afferent fibers for spinal nerves. The neurons transmitting impulses which arise in the skin are labeled *GSA exteroceptives*, and those from muscles, tendons, and joints are classified as *GSA proprioceptives*. The sensory fibers transmitting impulses which originate in the retina and inner ear are classified as *special somatic afferent* (*SSA*). An exception to the above criteria occurs in consideration of the olfactory nerve (I). Although the nasal pits are ectodermal in derivation, the olfactory nerve is classified as *visceral* because of its functional relationship with feeding, a dominantly visceral function. Olfaction and taste, which are both concerned with feeding, are classified as *special visceral afferent* (*SVA*). The *general visceral afferent* (*GVA*) is similar in function to the same component in spinal nerves, that is, in general they accompany the GVE (autonomic) fibers for visceral sensibility. This component is more complex in cranial nerves, as shown later in discussion of GVA fibers involved in important cardiovascular reflexes.

Motor (Efferent)

The efferent nerve fibers innervating skeletal muscle derived from somites are classified as somatic efferent (SE). As stated earlier, this component is the only one considered in this text as having no 'general' or 'special' designation. Those efferent fibers classified as *visceral* are of two types: (*a*) *General visceral efferents* (*GVE*) are defined the same as for spinal nerves, namely,

autonomic. For cranial nerves this is parasympathetic for innervation of smooth and cardiac muscle, plus glands. (*b*) *Special visceral efferents* (*SVE*) are indeed special, because this is a visceral component which innervates skeletal muscle. The embryologic explanation is that the skeletal muscles involved are those which are derived from the branchial (visceral) arches and which are therefore referred to as branchial or branchiomeric muscles. The arches are anatomically associated with the pharynx, larynx, mouth, and face—all serving a dominantly visceral function. The brachiomeric muscles are special in another respect, in that the proprioceptive fibers coming from these muscles are classified the same as all proprioceptive fibers throughout the body, that is, general somatic afferent (GSA). This results in a situation where the motor innervation to the muscle is visceral (SVE), but the proprioceptive afferents from the muscles are general somatic afferent (GSA).

Nuclei

Anatomic and Physiologic Organization

The nuclei of the brain stem classified as 'general' are rostral segmented continuations of the sensory and motor columns of the spinal cord gray matter. In addition, there are three 'special' functional components (SSA, SVA, SVE) which are added to account for the seven functional components discussed above and illustrated in

TABLE 17-2. Summary of Cranial Nerves, Components, Distribution, Nuclei, Ganglia, and Symptoms Resulting from Loss of Functional Components

Nerves	Components	Distribution	Nuclei	Ganglia	Typical Symptoms of Dysfunction (Loss)
I Olfactory	SVA	Olfactory mucosa	—	—	Loss of sense of smell (anosmia)
II Optic	SSA	Retina	—	—	Blindness (Anopia)
III Oculomotor	SE	Med., Dorsal, Vent. recti, Vent. oblique, Lev. palpebrae muscles	Oculomotor	—	Lat. strabismus, diplopia, ptosis
	GSA proprio.		Mesencephalic V (?)	—	
	GVE	Intrinsic eye muscles	Edinger-Westphal (parasympathetic III)	Ciliary	Mydriasis (dilation of pupil)
IV Trochlear	SE	Dorsal oblique eye muscle	Trochlear	—	
	GSA proprio.		Mesencephalic V (?)	—	
V Trigeminal	SVE	Muscles of mastication	Motor V (masticatory)	—	Asymmetrical chewing
	GSA proprio.		Mesencephalic V	—	
	GSA extero.	Skin of face and mucous membranes of head	Main sensory V, Spinal V (NDRV)	Semilunar (Gasserian)	Anesthesia of face
VI Abducens	SE	Lat. rectus, Retractor bulbi muscles	Abducens	—	Internal (medial) strabismus
	GSA proprio.		Mesencephalic V (?)	—	
VII Facial	SVE	Muscles of facial expression	Facial	—	Facial paralysis
	GVE	Mandibular, Sublingual, Lacrimal glands	Rostral (Sup.) Salivatory (parasympathetic VII)	Mandibular Pterygo-palatine	Partial dry mouth, dry eye

SVA	Ant. ⅔ tongue	Solitarius	Geniculate	Loss of taste ant. ⅔ tongue
GVA	Same as GVE	Solitarius	Geniculate	Difficult to test in animals.
GSA proprio.	Same as SVE	(?)	Geniculate	
GSA extero.	Cutaneous from ear	Spinal V (NDRV)	Geniculate	Anesthesia—small area of ear
VIII Vestibulo-cochlear SSA proprio.	Semicircular canals, utricle, sacculus	Vestibular (4)	Vestibular	Disequilibrium
SSA extero.	Spiral organ of Corti	Cochlear (2)	Spiral	Deafness
IX Glosso-pharyngeal SVE	Stylopharyngeus m.	Ambiguus	—	Paralysis-stylo-pharyngeus muscle (minor)
GVE	Parotid and Zygomatic (orbital) salivary glands	Caudal (inf.) salivatory (Parasym. IX)	Otic	Partial dry mouth
SVA	post. ⅓ tongue	Solitarius	Inferior IX (Petrosal)	Loss of taste post. ⅓ tongue
GVA	Carotid sinus, pharynx, post. ⅓ tongue	Solitarius	Inferior IX (Petrosal)	Tachycardia, incr. bl. press., anesthesia upper pharynx, loss of gag reflex
GSA proprio.	Stylopharyngeus m.	Mesencephalic V (?)	—	-minor-
GSA extero.	cutaneous from ear	Spinal V (NDRV)	Superior IX (Not dog)	Anesthesia—small area of ear (ext. acoustic meatus)
X Vagus SVE	Skeletal muscles of pharynx, larynx (Branchiomeric)	Ambiguus	—	Hoarseness, dysphonia, dysphagia

TABLE 17-2 (Continued)

Nerves	Components	Distribution	Nuclei	Ganglia	Typical Symptoms of Dysfunction (Loss)
X Vagus (cont.)	GVE	Viscera of cervical, thoracic, abdominal regions	Dorsal efferent X (Parasym. X)	Terminal— (Intramural)	Tachycardia, decreased peristalsis, indefinite visceral disturbances
	SVA	Taste buds epiglottis, root of tongue	Solitarius	Inferior X (Nodose)	Loss of taste— epiglottis, root of tongue (minor)
	GVA	General sensory from larynx, pharynx & viscera of GVE innerv.	Solitarius	Inferior X (Nodose)	Loss of cough reflex, vomiting reflex
	GSA proprio.	Muscles of pharynx, larynx	(?)	—	
	GSA extero.	Cutaneous from ear	Spinal V (NDRV)	Superior X (Jugular)	Anesthesia—external acoustic meatus, tympanic membrane
XI Accessory	SVE	Trapezius, sterno-cleidomastoid (Brachiocephalic)	Ambiguus	—	Atrophy of neck muscles, inability to lift shoulder
	GSA proprio.		(?)	Cell bodies in XI (?) & spinal nerves joining XI (?)	
XII Hypoglossal	SE	Tongue muscles	Hypoglossal	—	LMNL (lower motor neuron lesion) ipsilateral. Tongue deviates to side of lesion on protrusion
	GSA proprio.	Tongue muscles	(?)	Cell bodies in nerve XII (?)	

Key to abbreviations:
NDRV—Nucleus of the descending root of the trigeminal nerve.
Mesencephalic V—Mesencephalic nucleus of the trigeminal nerve.

Figure 17-7. The same figure illustrates schematically that in passing from the spinal cord to the medulla one should imagine a wedge-shaped cleavage from the dorsal surface through the dorsal median sulcus to the central canal of the spinal cord. This spreads the dorsal horn (sensory) lateral to the ventral horn (motor) and the relative positions of the nuclei in the adult brain stem are established as illustrated in Figure 17-7. The opened central canal in the rhombencephalon is the fourth ventricle. The floor of the fourth ventricle in the adult still retains the sulcus limitans, which separates the sensory nuclei laterally from the motor nuclei medially.

In the discussion which follows, the cranial nerves are organized according to the seven functional components and their respective nuclei. Due to the fact that a great portion of the internal structure of the brain stem is concerned with these nuclei, much of this material has been discussed in varying detail throughout Chapters 11, 12, and 13. Here the purpose is to correlate the functional aspects of the nuclei with specific cranial nerves and the peripheral organs innervated.

Somatic Efferent Nuclei

Cranial Nerves XII, VI, IV, III

As stated earlier, somatic efferent nuclei are not designated either 'general' or 'special,' the simple designation 'SE' being used in this text. All of the SE nuclei are close to the midline and near the dorsal surface of the brain stem.

The nucleus of the hypoglossal nerve (XII) (Plates 10 and 11) is a long nucleus in the floor of the fourth ventricle, extending from a site caudal to the obex rostrally into the open portion of the ventricle. The fibers of the hypoglossal nerve extend ventrally to cross the lateral border of the inferior olive and emerge at the ventrolateral surface of the caudal medulla. Microscopic study of

a brain section stained properly reveals many fibers in and around the nucleus. The hypoglossal nucleus receives fibers from many sources which mainly serve in reflexes of the tongue. The SE fibers of the hypoglossal nerve furnish the sole motor innervation for the muscles of the tongue.

The nucleus of the abducens nerve (VI) is not as conspicuous as that of the hypoglossal nerve. The abducens nucleus has the same relative position as the hypoglossal, but is rostral to it at the level of the internal genu of the facial nerve (Plate 14). The abducens fibers pass ventrally along the medial border of the superior olive and emerge at the lateral surface of the pyramids at the level of the trapezoid body (Fig. 17-1). It should not seem surprising that the abducens nucleus is smaller and less distinct than the hypoglossal. The mass of muscle of the tongue, innervated by the hypoglossal nerve, is much greater than that of the extrinsic ocular muscles innervated by the abducens nerve, that is, the lateral rectus and the retractor bulbi muscles.

The nucleus of the trochlear nerve (IV) (Plate 19) is in the midbrain at the level of the caudal colliculus ventral to the central (periaqueductal) gray matter and immediately lateral to the medial longitudinal fasciculus (MLF). As stated earlier, this thin nerve is the only nerve to emerge from the dorsal surface of the brain stem, and it decussates within the rostral medullary velum before passing to the ventral surface of the brain. This nerve innervates only one extrinsic ocular muscle, the dorsal oblique.

The nucleus of the oculomotor nerve (III) (Plate 20) is at the level of the rostral colliculi of the midbrain immediately rostral to the trochlear nucleus. The oculomotor nucleus borders the ventromedial portion of the periaqueductal gray matter and is dorsal to the medial longitudinal fasciculus (MLF). Actually, this is a nuclear complex composed of many smaller nuclei. It should be remembered that the oculomotor nerve contains parasympathetic (GVE) fibers to the intrinsic muscles of the eye. The cell bodies of the preganglionic parasympathetic fibers are in a separate nucleus, the Edinger-Westphal, located within the oculomotor complex. The SE fibers of the third cranial nerve pass ventrally to emerge on the medial side of the cerebral peduncles at the interpeduncular fossa (Fig. 17-1). They innervate all of the extrinsic ocular muscles except the lateral rectus, the dorsal oblique, and the retractor bulbi in the dog.

FUNCTIONS OF EXTRINSIC OCULAR MUSCLES

The extrinsic muscles of both eyes innervated by cranial nerves III, IV, and VI must work together as a unit so that both eyes move in the same direction at the same time. Such action is referred to as *conjugate ocular movement.* As an example, as the lateral rectus of the right eye contracts the medial rectus of the left eye contracts. In addition, the medial rectus of the right eye and the lateral rectus of the left eye must relax in order to permit conjugate eye deviation to the right. The actions of the medial and lateral recti muscles constitute the simplest example of such ocular movements because they act in direct opposition to each other in a horizontal plane. The actions of the other muscles are more complicated because of their insertions on the eyeball and the fact that all of the rest of the extrinsic ocular muscles except the dorsal oblique and the retractor bulbi are innervated by the oculomotor nerve. Figure 17-9 illustrates the normal action of individual extrinsic ocular muscles.

General Visceral Efferent Nuclei

Cranial Nerves X, IX, VII, III

As stated earlier, this functional component (GVE) is synonymous with autonomic nervous system, which is composed of sympathetic and parasympathetic divisions

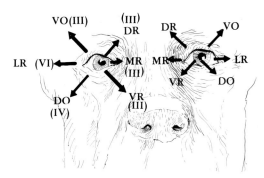

Fig. 17-9. Direction of pull on the eyeball by individual extrinsic ocular muscles (indicated by arrows). *DO,* dorsal oblique, *LR,* lateral rectus, *VO,* ventral oblique, *DR,* dorsal rectus, *MR,* medial rectus, and *VR,* ventral rectus (muscles); *III,* oculomotor, *IV,* trochlear, and *VI,* abducens (nerves). Due to conjugate movements, the resultant tensions of the various muscles, and innervation of most of the muscles indicated above by the oculomotor nerve, the above diagram should be used with caution for clinical testing. A unilateral lesion of the abducens nerve causes an ipsilateral internal strabismus ('crossed eye'). A total lesion of the oculomotor nerve causes the ipsilateral eye to look ventrolaterally.

(see Chapter 9). It will be remembered that the vagus, glossopharyngeal, facial, and oculomotor are the only cranial nerves that give rise to presynaptic (preganglionic) parasympathetic fibers. This means that the cell bodies of the preganglionic fibers are located in the nuclei of origin discussed below and the fibers leave the brain stem as intrinsic portions of the respective nerves. One outstanding characteristic of GVE innervation is that there are two neurons between the neuraxis and the target organ. Therefore, the preganglionic axons synapse with postganglionic cell bodies within autonomic ganglia located between the neuraxis and the effector.

The dorsal efferent nucleus of the vagus (parasympathetic nucleus of X) (Plate 11) is located immediately dorsolateral to the hypoglossal nucleus in the caudal area of the medulla oblongata. Some authors maintain that this nucleus also contains cell bodies of GVE fibers of the bulbar portion of the accessory nerve (XI) (Crosby *et al.,* 1962). We will agree with what seems to be the majority and not include cranial nerve XI with the parasympathetic component. Therefore, all of the fibers from this nucleus pass peripherally via the vagus nerve to viscera of the cervical, thoracic, and most of the abdominal area. The vagal parasympathetic ganglia are intramural or terminal within the walls of the viscera and the postsynaptic neurons are microscopic in size.

The caudal (inferior) salivatory nucleus (parasympathetic nucleus of IX) is very difficult to identify as a separate entity. It is considered in this discussion to be the rostral extension of the dorsal efferent nucleus of X. The caudal salivatory nucleus contains cell bodies of preganglionic fibers within the ninth cranial nerve which terminate in the otic ganglion by synapses with postganglionic neurons which innervate the parotid and zygomatic (orbital) salivary glands.

The rostral (superior) salivatory nucleus (parasympathetic nucleus of VII) also is difficult to identify as a separate nucleus. This is considered to be the rostral extension of the caudal salivatory nucleus. Preganglionic neurons from this nucleus pass peripherally within branches of the facial nerve to two autonomic ganglia as illustrated in Figure 17-10. (1) Preganglionic fibers pass via the major petrosal nerve to the *pterygopalatine ganglion,* from which postganglionic fibers pass to the lacrimal gland plus the nasal and palatine glands in the dorsal portion of the roof of the mouth. (2) Other preganglionic fibers pass peripherally to enter the chorda tympani which joins the lingual nerve, a branch of the mandibular division of the trigeminal nerve. Preganglionic fibers from VII (now with the lingual nerve of V) pass to the *mandibular ganglion,* where they synapse with postganglionic fibers which pass to the mandibular and sublingual glands.

The Edinger-Westphal nucleus (parasym-pathetic nucleus of III) is in the tegmentum of the midbrain at the rostral portion of the oculomotor nucleus. GVE parasympathetic fibers pass within the oculomotor nerve to the orbit and terminate within the *ciliary ganglion* (Fig. 17-10), from which postganglionic fibers extend rostrally to enter the eyeball for innervation of the intrinsic muscles of the eye. The dominant effect of parasympathetic innervation on the muscles of the iris is pupillary constriction (miosis).

Special Visceral Efferent Nuclei

Cranial Nerves XI, X, IX, VII, V

This functional component (SVE) is confined to those cranial nerves that innervate the skeletal muscles which are derived embryonically from the branchial arches and are therefore referred to as 'branchiomeric muscles.'

The general position of the special visceral efferent nuclei is lateral to the somatic efferent and general visceral efferent, but also deeper within the reticular formation in the general zone indicated schematically in Figure 17-7. The SVE fibers tend to pass dorsally and arch laterally before they emerge from the ventrolateral area of the brain stem. The facial nerve (VII) illustrates the greatest dorsal arching, as described below.

The nucleus ambiguus is the nucleus of origin for the SVE fibers of cranial nerves XI, X, and IX. The nucleus is well named because it is an ill-defined gray cylindrical mass within the mid-lateral area of the reticular formation throughout most of the medulla (Plate 10). The muscles innervated by the SVE component of nerves IX, X, and XI are those derived from the last two or three branchial arches. These are the striated muscles of the larynx and pharynx innervated by the glossopharyngeal and vagus nerves. The sternocleidomastoid muscle group plus the trapezius are innervated by the accessory nerve. There are

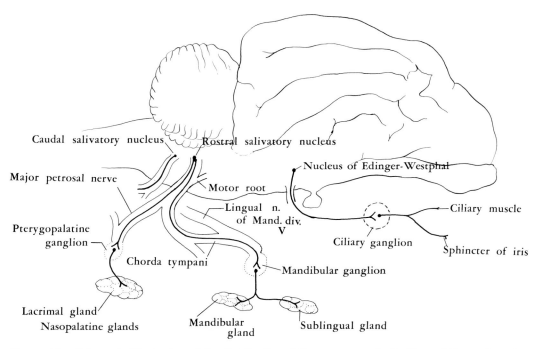

Caudal salivatory nucleus

Rostral salivatory nucleus

Nucleus of Edinger-Westphal

Major petrosal nerve

Motor root

Lingual n. of Mand. div. V

Ciliary muscle

Pterygopalatine ganglion

Ciliary ganglion

Sphincter of iris

Chorda tympani

Mandibular ganglion

Lacrimal gland

Nasopalatine glands

Mandibular gland

Sublingual gland

Fig. 17-10. Schematic illustration of the distribution of the parasympathetic fibers of the oculomotor and facial nerves. Note that the oculomotor nerve has only one parasympathetic ganglion (ciliary), whereas the facial nerve has two (pterygopalatine and mandibular). The chorda tympani of VII joins the lingual nerve of the mandibular division of cranial nerve V prior to the preganglionic termination in the mandibular ganglion. The caudal salivatory nucleus (parasympathetic nucleus of IX) is shown. Preganglionic fibers from this nucleus synapse in the otic ganglion, from which postganglionic fibers are distributed to the parotid and zygomatic (orbital) salivary glands. (Modified from Stromberg, 1964.)

differences of opinion as to whether or not these muscles innervated by the eleventh cranial nerve are actually of branchiomeric origin.

The motor nucleus of the facial nerve (VII) (Plate 14) is the SVE nucleus of origin for the fibers innervating the muscles derived from the second branchial arch. These are the muscles of facial expression, which are located very superficially in the head and neck regions. This nucleus in dog brain is located immediately caudal to the superior olive and almost at the ventrolateral surface about midway between the pyramid and the spinal tract of the trigeminal nerve (Plate 14). From the ventrolateral position of the nucleus the SVE axons pass dorsorostro-medially toward the floor of the fourth ven-

tricle. At this location near the midline the fibers sweep laterally around the nucleus of the abducens nerve (VI) to form the *internal genu of the facial nerve* (Plate 14). The fibers continue ventrolaterally to exit from the brain stem at the lateral side of the superior olive at the level of the trapezoid body (Plate 15). The fibers of the abducens nerve (SE) pass ventrally and slightly laterally, on the medial side of the superior olive, and emerge at the lateral edge of the pyramids at the level of the trapezoid body (Plate 15).

The facial nucleus receives afferent fibers from many sources. The termination of fibers from the corticobulbar tract puts the facial muscles under voluntary control. Many afferents are involved in reflexes; for example, fibers from the trapezoid nuclei (in the auditory pathway) pass to the facial nucleus for reflexes which involve changes of facial expression as a result of loud noises.

The motor nucleus of the trigeminal nerve (*masticatory nucleus*) is located at the level of the pons medial to the principal sensory nucleus and the motor fibers of the trigeminal nerve (V) (Plate 15). SVE fibers of the trigeminal nerve may be identified within the middle cerebellar peduncle (brachium pontis) as they cross its base to pass into the peripheral nerve (Plate 16). These fibers, as stated earlier, are within only the mandibular division which leaves the cranium via the oval foramen. Peripherally they innervate all of the muscles of mastication, plus the tensor tympani, tensor palati, and the mylohyoid muscles. Similar to the other SVE nuclei, this nucleus receives corticobulbar fibers for voluntary control of mastication. Other afferents to this nucleus are from nuclei of other cranial nerves, in addition to a major input from the sensory nuclei of the trigeminal nerve for reflexes.

The Visceral Afferent Nucleus

Cranial Nerves X, IX, and VII

All of the visceral afferent fibers in cranial nerves, of both general and special types, except those for olfaction are contained in cranial nerves VII, IX, and X. All three of these nerves share one nucleus of termination, the *nucleus of the tractus solitarius* (Plates 11 and 12). This nucleus and its tract are long cylindrical structures which extend longitudinally throughout most of the medulla oblongata. Similar to the nucleus ambiguus, which is shared by three cranial nerves, the nucleus solitarius has a rostrocaudal localization of cell bodies, with those of nerve VII located rostrally, those of IX centrally, and those of X caudally. It should be kept in mind that since this is a nucleus of termination (sensory), the cell bodies in the nucleus are not those of the neuron

processes in the peripheral nerves. The cell bodies of the peripheral GVA and SVA fibers of nerves IX and X are located in the inferior ganglia of the respective nerves, that is, in the petrosal ganglion of the glossopharyngeal and the nodose ganglion of the vagus nerve. The facial nerve (VII) has only one ganglion, the *geniculate*, which contains the cell bodies of the GVA, SVA, and GSA fibers in the facial nerve.

The SVA component is taste from the anterior two-thirds of the tongue via VII, from the posterior one-third via IX, and from the epiglottic region via X. The GVA component is represented in nerves IX and X by the general sensibility from the posterior portion of the oral cavity, including the tongue, larynx, and pharynx. GVA fibers in X convey impulses from the thoracic and abdominal viscera. The carotid sinus nerve (GVA) is a branch of IX which innervates the carotid sinus and functions as the afferent portion of the carotid sinus reflex arc. The GVA fibers have been reported as conveying deep sensibility from the face (Ranson and Clark, 1959).

General Somatic Afferent Nuclei

Exteroceptive—Cranial Nerves X, IX, VII, V

Proprioceptive—Cranial Nerves XII, XI, X, IX, VII, VI, V, IV, III

The general somatic afferent nuclei are joined in a rostrocaudal sequence to form a continuous column extending from the mesencephalon caudally to the cervical spinal cord, where it becomes continuous with the substantia gelatinosa. The trigeminal nerve (V) has the greatest GSA representation, and the three brain stem nuclei for this component are named trigeminal nuclei regardless of the contributions from the other nerves.

The two *exteroceptive nuclei of the trigeminal nerve* are within the lateral portions of the reticular formation of the pons and medulla and extend caudally into the upper cervical segments of the spinal cord. These nuclei receive fibers chiefly from all three divisions of the trigeminal nerve which convey impulses from the skin and ectodermal mucous membranes of the head. The cell bodies of the peripheral neurons are within the *semilunar, or Gasserian, ganglion.* The fibers enter the brain stem via the trigeminal nerve at the level of the middle cerebellar peduncle (Plate 16).

Internally, at the level of entrance of the nerve and extending slightly rostrally is the *pontine (chief, principal, or main) sensory nucleus of the trigeminal nerve.* This is the middle of the three GSA nuclei, which is continuous caudally with the *nucleus of the spinal (descending) tract of the trigeminal nerve.* Caudally, at the level of the second cervical segment the descending tract of the trigeminal nerve is continuous with the *dorsolateral fasciculus of Lissauer* of the spinal cord. The nucleus of the descending tract of the trigeminal nerve is continuous caudally with the *substantia gelatinosa* of the spinal cord.

The spinal tract of the trigeminal nerve contains fibers which convey impulses of pain and temperature sense plus probably touch from the entire area of the face on the same side. These are descending fibers of neurons of the first order which enter its nucleus caudal to the level of entrance. A destructive lesion of this tract will therefore result in loss of pain and temperature sense, and possibly touch, from the same side of the face. Fibers of the second order arise within the spinal nucleus of the trigeminal nerve and chiefly cross to the opposite side and ascend to the posterior ventral area of the thalamus. As in other sensory tracts, the final neuron in the pathway is a thalamo-cortical neuron which reaches the cortex via the internal capsule.

From cell bodies in the pontine and spinal nuclei there are longitudinal fibers which enter the reticular formation, from which collaterals are dispersed to motor nuclei of cranial nerves.

The skin of the ear and the area of the external auditory meatus is innervated by GSA fibers of cranial nerves VII, IX, and X. These fibers also have their cell bodies in ganglia of the respective nerves and the axons enter the spinal tract of the trigeminal nerve to descend and terminate within its nucleus. Therefore, it should be noted that the spinal, or descending, tract of the trigeminal nerve also contains fibers from the seventh, ninth, and tenth nerves.

The proprioceptive nucleus is the *mesencephalic nucleus of the trigeminal nerve* (Plate 16) and it probably contains the GSA cell bodies of fibers from the muscles of mastication and the extrinsic muscles of the eye. This nucleus is unique in that it is an exception to the rule that somatic afferent neurons have their cell bodies within ganglia outside the neuraxis. Another unique feature is that the cell bodies of the mesencephalic nucleus are unipolar. Although some authors agree that nerves III, IV, V, and VI have proprioceptive cell bodies in the mesencephalic nucleus of the trigeminal nerve, the location of proprioceptive cell bodies for nerves VII, IX, X, XI, and XII is highly controversial. Some authors believe that the proprioceptive cell bodies for some of these nerves are embedded in the peripheral nerves (see Crosby *et al.,* 1962).

Special Somatic Afferent Nuclei

The functional component special somatic afferent is associated with the optic (II) and vestibulocochlear (VIII) nerves. These two nerves are complicated and are considered in separate chapters: the optic system in Chapter 18, and the vestibulocochlear nerve with its internal connections in Chapter 19. It should be noted that the vestibular division of the eighth cranial nerve serves as the principal proprioceptive nerve of the body and is classified as special somatic afferent (see Table 17-2).

Lesions of the Cranial Nerves

It should be apparent from the discussions above that some cranial nerves are much more complicated than others anatomically and physiologically, and not all cranial nerves are equally important clinically. Some of the isolated functional components of a nerve are virtually of no clinical significance, whereas other components of the same nerve may be extremely important. Table 17-2 summarizes the symptoms caused by loss of some of the functional components as a result of destructive lesions.

Many cranial nerves are anatomically so close together intracranially that such a lesion as a skull fracture or a tumor will affect more than one nerve. Examples of nerves which share the same foramen are numerous; for example, cranial nerves IX, X, and XI all pass through the petrobasilar fissure; nerves VII and VIII enter the internal acoustic meatus; nerves III, IV, the ophthalmic division of V, and VI enter the orbit via the orbital fissure.

Following is a brief summary of clinical symptoms resulting from cranial nerve lesions. This discussion is very basic, because additional considerations are presented in various locations: in this Chapter under the individual cranial nerve, and in Chapters 18 and 19 for lesions of the optic and vestibulo-cochlear nerves.

Hypoglossal Nerve (XII)

Since the hypoglossal nerve is the sole motor innervation of the tongue, a destructive lesion of the nucleus or nerve produces a lower motor neuron paralysis with atrophy of the muscles on the ipsilateral half of the tongue. There is a deviation of the tongue toward the paralyzed side (Fig. 17-11). In contradistinction, if a brain lesion destroys the corticobulbar fibers to the hypoglossal nucleus, the tongue deviates to the side opposite the lesion. This is because a majority of the corticobulbar fibers to the hypoglossal nucleus are contralateral and the nucleus of the side opposite the lesion is deprived of stimulation. Therefore, the contralateral muscles are paralyzed and the

Fig. 17-11. Injury to the right hypoglossal nerve in a dog hit by a car about four years previously. Notice the atrophy of the right side of the tongue which, when protruded, is drawn toward the same side as the injury. The dog submerges his muzzle to lap water. (From Hoerlein, 1965, by permission of W. B. Saunders Company.)

tongue deviates to the paralyzed side, that is, the side opposite the lesion. Note that in both types of lesions the tongue deviates to the paralyzed side.

Accessory Nerve (XI)

Loss of the portion of the accessory nerve that serves as the motor innervation for the brachiocephalic and trapezius muscles results in a typical lower motor neuron lesion of the muscles.

Vagus Nerve (X)

As stated previously, the vagus nerve has the greatest distribution of all cranial nerves. It joins the glossopharyngeal nerve (IX) to form the glossopharyngeal-vagus complex for innervation of the soft palate, larynx, and pharynx. Lesions of this portion of the vagus cause: loss of cough reflex, loss or impairment of voice (dysphonia), and difficulty in swallowing (dysphagia). Destructive lesions of the parasympathetic component to the heart result in tachycardia (increased heart rate), whereas irritative lesions produce bradycardia (decrease in heart rate). Parasympathetic loss for the gastrointestinal tract results in decreased peristalsis and decreased secretion by gastrointestinal glands.

Glossopharyngeal Nerve (IX)

General visceral afferent loss from the area of the soft palate and pharyngeal mucosa results in loss of the gag reflex. A destructive lesion of the carotid sinus nerve produces uncontrolled tachycardia and increased blood pressure. Loss of GVE innervation to the parotid gland may produce detectable dryness of the mouth.

Vestibulocochlear Nerve (VIII)

See Chapter 19.

Facial Nerve (VII)

Loss of the somatic visceral efferent component to the facial muscles results in facial (Bell's) palsy, or paralysis. The facial muscles on the same side as the affected nerve sag and saliva may drip from the corner of the mouth. The condition may be idiopathic, that is, occur spontaneously with an unknown etiology. Cocker Spaniels and Boxers between the ages of 6 and 9 years are the dogs most commonly affected (McGrath, 1960). McGrath reports that facial paralysis is the most common cranial nerve syndrome seen in the clinic of the Veterinary Hospital of the University of Pennsylvania. The SVE component of this nerve also serves as the efferent portion of the corneal reflex. Therefore, in such a lesion the eyelid remains open and the cornea becomes dry.

The chorda tympani, through which fibers pass from the facial nerve to the tri-geminal, is a classic example of how a functional component may be lost as the result of a lesion to a nerve which does not contain that functional component internally. The lingual nerve is a branch of the mandibular division of the trigeminal, which does not have intrinsic nuclei for either GVE or SVA (see Table 17-2). Cutting the lingual nerve distal to its junction with the chorda tympani (see Fig. 17-10) will cause loss of taste (SVA) to the anterior two-thirds of the tongue, plus a loss of secretion from sublingual and submaxillary glands because of the SVA and GVE fibers of the chorda tympani, which connects the facial nerve (which has SVA and GVE) and the lingual nerve. Transection of the lingual nerve proximal to its junction with the chorda tympani (see Fig. 17-10) will, however, result in no loss of taste or of secretory activity of the salivary glands.

Abducens Nerve (VI)

A destructive lesion of the abducens nerve produces an internal strabismus of the eye of the same side due to paralysis of the lateral rectus muscle.

Trigeminal Nerve (V)

Unilateral SVE deficiency produces ineffective chewing with slight deviation of the jaw toward the paralyzed side. It should be remembered that the GSA component of the ophthalmic division of the trigeminal nerve serves as the afferent portion of the corneal reflex. Therefore lack of a corneal reflex could result from a trigeminal deficiency.

Trochlear Nerve (IV)

The trochlear nerve, which innervates the dorsal oblique muscle of the eye, is very seldom, if ever, injured by itself. Therefore it will not be discussed separately.

Oculomotor Nerve (III)

Loss of the somatic efferent component to the extrinsic ocular muscles results in drooping of the eyelid (ptosis) due to paralysis of the levator palpebrae, and deviation of the ipsilateral eyeball ventrolaterally because of paralysis of the remaining muscles innervated by this nerve. The GVE innervation of the intrinsic ocular muscles is significant in reflexes, for example, the light, or pupillary, reflex (see Chapter 18). The dominance of parasympathetic stimulation produces miosis (constriction of the pupil). Therefore, a total lesion of the oculomotor nerve produces mydriasis (pupillary dilatation). It should be remembered that a lesion does not have to be accidental, or result from a physical blow or wound. Mydriasis may be associated with acute glaucoma and encephalitis (Bunce, 1958). Presumably this is because of the increased pressure on the long ciliary nerves.

Optic Nerve (II)

A lesion of the optic nerve results in blindness, but certainly this is not the only site where a lesion will cause blindness (see Chapter 18).

Olfactory Nerve (I)

Destruction of the olfactory nerve results in anosmia (loss of olfactory sense). Due to the continuation of the meninges and subarachnoid space through the cribriform foramina, rhinitis may easily extend directly to the brain and result in encephalitis.

Bibliography

Bunce, D. F. M.: The neurologic examination of the dog. Mod. Vet. Pract. *39:* 10, 1958.

Crosby, E. C., Humphrey, T., and Lauer, E. W.: *Correlative Anatomy of the Nervous System.* New York, Macmillan Co., 1962.

DuBois, F. S., and Foley, J. O.: Experimental studies on the vagus and spinal accessory nerves in the cat. Anat. Rec. *64:* 285–307, 1936.

Hoerlein, B. F.: *Canine Neurology—Diagnosis and Treatment.* Philadelphia, W. B. Saunders Co., 1965.

McClure, R. C.: The Cranial Nerves. Chapter 10 in *Anatomy of the Dog* (Miller, M. E., Christensen, G. C., and Evans, H. E.). Philadelphia, W. B. Saunders Co., 1964. Pp. 544–571.

McGrath, J. T.: *Neurologic Examination of the Dog,* 2nd ed. Philadelphia, Lea & Febiger, 1960.

Miller, M. E.: *Guide to the Dissection of the Dog.* Ann Arbor, Mich., Edwards Brothers, Inc., 1962.

Ranson, S. W., and Clark, S. L.: *The Anatomy of the Nervous System—Its Development and Function,* 10th ed. Philadelphia, W. B. Saunders Co., 1959.

Romer, A. S.: *The Vertebrate Body,* 3rd ed. Philadelphia, W. B. Saunders Co., 1962.

Truex, R. C., and Carpenter, M. B.: *Human Neuroanatomy,* 6th ed. Baltimore, Williams & Wilkins Co., 1969.

chapter 18

Eye and the Visual System

It is well known that great differences in the anatomic and physiologic properties of the ocular system exist among mammals. Animal behaviorists and comparative anatomists study these differences and correlate them with the degree of dependency of a species on its visual system. This in turn is greatly influenced by the animal's mode of living, as exemplified in the following statements: (1) Comparative anatomists maintain that the higher the development of an animal, the farther forward its eyes, which favors greater intelligence and concentration plus a preference for daylight activity; (2) eye position in the skull is correlated with the feeding habits of different mammals. The herbivores and hunted animals (*e.g.*, rabbit) have retained the lateral eye position, whereas predators (*e.g.*, carnivores) have developed a forward eye position which is advantageous for hunting and permits great accuracy in striking prey because of the binocular visual fields.

Correlated with the normal anatomic differences of the ocular system among various breeds of dogs, there are predispositions among them to pathologic conditions of the eye. Magrane (1965) presents a summary of such predispositions.

In evaluation of the visual system of the domestic dog, as compared with some other mammals, including man, the dog actually has a poor visual apparatus. Roberts (1955) maintains that this is one undesirable effect of so much selection in breeding dogs for other qualities which man has considered desirable. When compared with man the domestic dog is nearsighted (myopic), has an uneven cornea (astigmatic), is color blind, and accommodates poorly (Magrane, 1965). Magrane does list some favorable anatomic and physiologic factors of the dog's eye which compensate for the deficiencies. When compared with man, the dog has a larger visual field, better peripheral vision, and a larger pupil. The dog can see better in semidarkness or darkness because of abundant rods in the retina and the presence of the tapetum lucidum. In addition, it also has a fair degree of binocular vision, permitting depth perception.

Embryology of the Eye

In a very young embryo the primary optic vesicle can be identified as an evagination of the diencephalon, to which it remains attached. The bulbous vesicle soon invaginates to form the optic cup (see Fig. 2-3, p. 15). The ectodermal lining of the cup, which is a continuation of the forebrain wall, becomes the retina, composed of photosensitive elements. The attachment to the brain, which in adult anatomy is the optic nerve, is not via a typical nerve but rather a tract.

The lining of the cup differentiates into two layers: the outer becomes invaded by melanin and forms the pigmented retinal layer; the inner becomes thickened and forms the sensory layer of the retina. The lens is formed from an ingrowth of the superficial ectoderm. The extrinsic ocular muscles form from the primitive head somites (see Chapter 17). Finally the eyelids are formed from folds in the skin after the eye has differentiated beneath the surface.

Dogs are born with their eyelids fused, and the palpebral fissure does not open

until approximately two weeks after birth (Fox, 1963). At birth the eyes have not developed into a light-sensitive organ, as indicated by the histologically underdeveloped cornea and retina (Andersen, 1970).

The embryonic relationships of the eye and brain are retained in the adult. The adult optic nerve is not considered a true nerve, based on the following:

1. The optic nerve has no neurilemmal sheath of Schwann, as is present in a typical craniospinal nerve. Therefore, regeneration of this nerve is impossible (Everett, 1965).

2. The optic nerve is similar to other brain tracts in that it contains various types of neuroglia present in tissue of the central nervous system.

3. The optic nerve fibers are all myelinated, although they are of various diameters.

4. The meninges extend peripherally along the entire optic nerve and the dura mater fuses with the sclera of the posterior portion of the eyeball. Accompanying the meninges to the eye, the cerebrospinal fluid thereby furnishes a possible 'portal of entry' for infectious bacteria and viruses, facilitating development of meningitis and encephalitis as a sequel to eye infections.

Anatomy of the Eye

Gross Anatomy

The general pattern of mammalian relationships is present in the eye of the dog (Fig. 18-1), but in comparison with the human eye there are three special features of the dog's eye that should be discussed briefly.

1. *The nictitating membrane (palpebra tertia)*, or third eyelid, protrudes from the ventromedial border of the palpebral fissure to extend about three-fourths of the distance toward the lateral canthus above the edge of the lower eyelid. Embedded within this membrane is a plate of hyaline cartilage

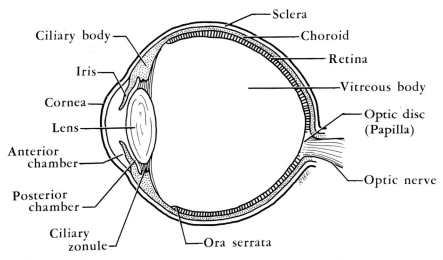

Fig. 18-1. Schematic illustration of a sagittal section of the eye, to indicate its anatomic structure.

which is surrounded by glandular tissue, the *nictitans gland.* In some species, in addition to this superficial gland there is a deeper gland, referred to as *Harder's gland,* but this is not present in the dog. To further complicate this gross anatomic feature of the dog's eye, on the bulbar surface of the nictitating membrane is an elevated area of lymphoid tissue which may become infected.

2. The choroid (posterior two-thirds of the uvea, or middle coat) of the dog differs from that of man in that in the dog, dorsal to the site of exit of the optic nerve (optic disc, or papilla), there is an irregularly shaped area of iridescent luster, the *tapetum lucidum* (Fig. 18-2). This varies in size, color, and brightness among different species and even among different breeds of dogs. Wyman and Donovan (1965) present tables giving the average size of the tapetum lucidum, and hair color of various breeds in relation to tapetal coloring. The optic disc usually is at the borderline of the tapetum lucidum dorsally and the tapetum nigrum ventrally (Fig. 18-2). Magrane (1965) refers to the *tapetum nigrum* as the *non-tapetal fundus,* in contrast to the larger *tapetal fundus,* re-

ferred to above as the tapetum lucidum.

As illustrated in Figure 18-2, the tapetum nigrum extends dorsally as a narrow peripheral border around the tapetum lucidum. The pigment of the tapetum nigrum is usually melanin derived from the underlying choroid or the retina (Smythe, 1958). The color of the tapetum nigrum, particularly in the ventral fundus, is dark brown to black. The cat has a brighter and more colorful tapetum lucidum than the dog. Horses and ruminants have a fibrous tapetum, whereas dogs and cats have a cellular form. It is this structure that is responsible for the luster of the eyes of some species, especially at night. The uvea of human and swine eyes lack a tapetum (Trautmann and Fiebiger, 1957).

3. The drainage of the aqueous humor from the anterior and posterior chambers is more extensive in dog than in man. Rather than by a single *canal of Schlemm,* as found in man, the aqueous drainage is by way of many vessels, known as the *scleral venous plexus.* In the dog there are commonly four vessels up to 60 micra in diameter, or three larger ones up to 180 micra in diameter (Prince and Ruskell, 1960).

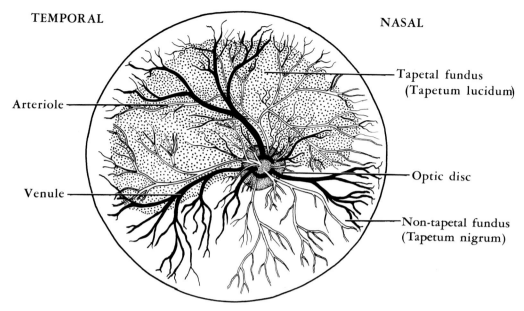

TEMPORAL   NASAL

Arteriole

Venule

Tapetal fundus
(Tapetum lucidum)

Optic disc

Non-tapetal fundus
(Tapetum nigrum)

Fig. 18-2. Diagram of the fundus of the right eye of the dog, showing the optic disc (papilla), blood vessels, tapetum lucidum, and tapetum nigrum. (After Magrane, 1965.)

Histology

Since the histologic details of the canine eye are fundamentally similar to those of the typical mammal, including man, they may be found in almost any histology text. The general structure of the eyeball is illustrated in Figure 18-1. The wall of the eye consists of three tunics, or coats, in direct contact with each other, as follows:

1. The external *fibrous tunic* is divisible into a larger posterior portion, the *sclera*, and a smaller anterior portion, the *cornea*. The sclera is opaque fibrous connective tissue which protects the delicate inner layers and gives support and form to the eyeball. The cornea is devoid of blood vessels and lymphatics (Finerty and Cowdry, 1960). This property of the cornea plus its transparency permits unobstructed vision.

2. The middle tunic is the uvea, or *vascular tunic*. This coat furnishes the means for proper nutrition of the ocular tissues, permits alteration in size of the pupil by action of muscles of the iris and ciliary muscles, and controls accommodation (which is poor in the dog). The posterior portion of the uvea, in direct apposition to the optic retina, is the *choroid* (Fig. 18-1). The anterior uveal segment directly outside the nonoptic retina is composed of the *ciliary body* and the *iris*. As discussed previously, the tapetum lucidum and the tapetum nigrum are special features of the canine uvea which are not present in man.

3. The internal (third) tunic of the eyeball is the *retina*, which is composed of two portions: (*a*) the optic (neural) retina, and (*b*) the nonoptic (distal) retina. The optic retina contains the neural elements and extends from the entrance of the optic nerve (optic disc) forward to the *ora serrata*, where it terminates at the caudal end of the ciliary body. This portion of the retina covers the choroid, as illustrated in Figure 18-1. The retina is loosely attached to the choroid except at the optic disc and the ora serrata. The pressure of the vitreous humor is suffi-

cient to hold the retina in contact with the choroid under normal conditions, but they may be traumatically or pathologically separated, except perhaps at the two points stated above (optic disc and ora serrata). Such separation is referred to clinically as 'detached retina.'

The nonoptic, or distal, portion continues from the ora serrata over the inside surface of the ciliary body and the iris. This portion does not contain neurons.

Histology of the optic retina

The classical ten layers of the optic retina may be identified in the dog (Prince *et al.,* 1960). The complexity of the retina may be condensed for better understanding, without loss of effectiveness, by considering that it is composed of three-neuron relays as given by Elliott (1969) and included in the illustration of the dog's retina by S. Ramon y Cajal, which was cited by Polyak (1957).

Figure 18-3 is a schematic representation of the three relay layers. It should be noted that the light must pass through the cornea, anterior and posterior chambers (aqueous humor), lens, and vitreous humor (body) before reaching the retina. The optic nerve fibers are on the inside retinal surface as axons of the ganglionic cells, but the normal stimulation for the ganglionic cells is from the *rod-and-cone layer* to *bipolar-cell layer* then to the *ganglionic layer.* Therefore the energy of the light must pass through each layer twice.

The energy on the first trip is not actually the same type as that on the return, because the rods (and cones) convert the stimulus of the so-called light energy into nerve impulses. The *rods* are very sensitive photoreceptors and function during low illumination for 'night vision.' They contain a chemical substance, *rhodopsin,* or *visual purple,* which rapidly is reduced or disappears when illuminated. In order for rhodopsin to regenerate it is necessary for the rods to maintain contact with the pigment layer of

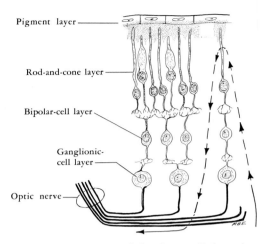

Pigment layer

Rod-and-cone layer

Bipolar-cell layer

Ganglionic-cell layer

Optic nerve

Fig. 18-3. Diagram of the three cellular relay centers of the retina. The arrows indicate the direction of light as it passes through the retina. Note that the energy must pass through each layer twice. After the first penetration the energy of light is converted to nerve impulses by the rod-and-cone layer for transmission through the other two layers. The optic nerve is composed of axons of ganglionic cells which are normally stimulated only by impulses received from the bipolar cells. The deeper synaptic layer between the rod-and-cone layer and the bipolar cells is the *external plexiform layer,* and the layer between the bipolar and ganglionic cells is the *internal plexiform layer.* (Modified from Elliott, 1969.)

the retina. The process of rhodopsin reduction is thought to be responsible for generation of the nerve impulse.

The *cones* have a higher threshold of excitability than the rods and are activated by high illumination, therefore functioning primarily for 'day vision.' Although the cones do not contain rhodopsin, they apparently contain a somewhat similar photosensitive chemical.

In the canine retina there are almost 20 times as many rods as cones, and there is no retinal area (other than the optic disc, described below) that is devoid of rods (Magrane, 1965). However, Magrane recognizes a small area of greatest sensitivity and sharpest vision approximately 3 mm lateral to the site of exit of the optic nerve. He prefers the label *area centralis* for this area, which designates position and not function. Prince *et al.* (1960) call the area 'macula,' because of its similarity to that structure in the human eye. It should be remembered that the macula lutea of the human eye contains all cones of a specialized type; it is the area of greatest visual acuity and color distinction in man (Elliott, 1969).

The Visual Pathway

The neuroanatomic details of the visual pathway in dog are similar to those in man. As mentioned, the optic nerve is composed of axons from the ganglionic cells in the retina. The nonmyelinated axons from the inside surface of the optic retina and the great multitude of fibers converge at the *optic disc,* or *papilla,* to form the optic nerve (Fig. 18-3). The optic disc is referred to as the blind spot of the retina, because it is composed of only nerve fibers and no photoreceptors and is therefore insensitive to light.

From the disc the fibers pass through the *lamina cribrosa,* a sievelike perforated area of the sclera. Here they form the optic nerve and become myelinated and enveloped by meninges and cerebrospinal fluid. The optic nerve passes through the optic canal of the skull to reach the base of the diencephalon, where it meets its counterpart from the opposite side to form the *optic chiasm,* which is considered a portion of the hypothalamus. Approximately 75 per cent of the fibers cross in the dog's optic chiasm, but variations exist among different breeds of dogs. The dolichocephalic breeds have less binocular vision because of their long nose, and it is thought that they have fewer uncrossed chiasmatic fibers than have the brachycephalic breeds (de Lahunta and Cummings, 1967). The bundles of fibers passing caudally from the chiasm and laterally around the cerebral peduncles are the *optic tracts,* which terminate in the lateral geniculate bodies.

As illustrated in Figure 18-4, only the fibers arising from the nasal (medial) retina cross within the chiasm, whereas the fibers arising from the temporal (lateral) retina remain uncrossed on the lateral side of the chiasm to pass within the ipsilateral optic tract to the lateral geniculate body on the same side. The final neurons in the optic pathway arise in the lateral geniculate body. Their axons enter the white matter of the hemisphere and pass caudally as the *optic radiation* to terminate within the occipital lobe of the cerebral hemisphere (visual cortex), as illustrated in Figure 18-4.

Summary of Pathway for Visual Impulses

The pathway traversed by impulses giving rise to visual sensations may be summarized as follows: retina → optic nerve → optic chiasm (medial, crossed; lateral, uncrossed) → optic tract → lateral geniculate body → optic radiation → visual (occipital) cortex.

McGrath (1960) gives evidence that perhaps the higher visual areas are less important in dog than in man. He reports that a dog retains the ability to see following a brain-stem transection above the level of the rostral colliculus.

Interruption of the Visual Pathway
Visual Defects

As indicated previously, lesions in different loci throughout the visual pathway will result in different degrees of blindness and involvement of different areas of the visual fields. If the visual-field segments of blindness (*scotomas*) can be determined by means of a clinical ophthalmologic examination, it would be possible to intelligently predict the site of the lesion. Unfortunately, the veterinarian has great difficulty in deter-

VISUAL FIELDS

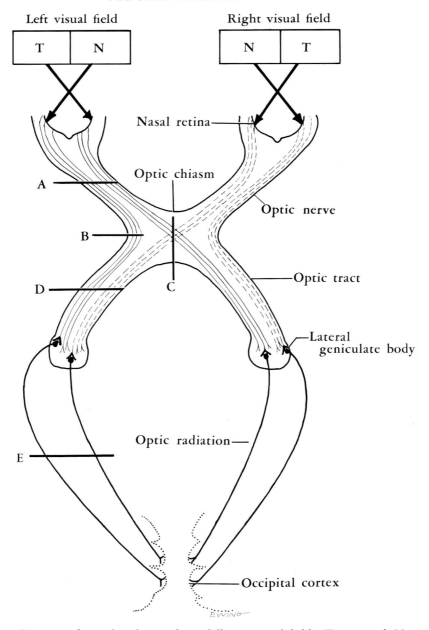

Left visual field

T	N

Right visual field

N	T

Nasal retina

Optic chiasm

A

Optic nerve

B

Optic tract

D

C

Lateral geniculate body

Optic radiation

E

Occipital cortex

EWING

Fig. 18-4. Diagram of visual pathways from different visual fields (T, temporal; N, nasal). Note that the nasal retina is stimulated from the temporal visual field and the temporal retina from the nasal visual field. Lines A, B, C, D, and E indicate sites of lesions producing visual field defects shown and labeled in Figure 18-5.

mining the visual field deficiencies because the animal cannot tell the examiner whether or not it sees an object at a certain location while the eyes are fixed straight ahead.

Electroretinography as a clinical aid in veterinary ophthalmology shows promise in determining nonfunctional areas of the retina and thereby may help in localization of lesions of the central visual pathway (Keller, 1969). It should be kept in mind, however, that visual defects as used in human ophthalmology refer to the *visual field* segment that is blind. As illustrated in Figure 18-4, the nasal area of the retina receives the light stimulus from the temporal visual field for the respective eye, that is, the right nasal

retina receives its light stimulus from the right temporal visual field.

The loss of vision in half of each visual field is referred to as *hemianopia* (hemianopsia). If the same side, right or left half, of each visual field is blind, it is known as *right* or *left homonymous hemianopia* (Figs. 18-4 and 18-5). If the left half of the visual field of one eye and the right half of the visual field of the other are involved, it is known as either a *bitemporal* or a *binasal heteronymous hemianopia,* depending on which visual field is lost in each eye.

By studying Figures 18-4 and 18-5, it should be readily apparent that a total unilateral lesion of the visual pathway any-

Visual Field Defects

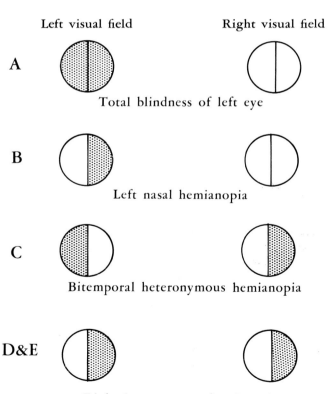

Left visual field   Right visual field

A   Total blindness of left eye

B   Left nasal hemianopia

C   Bitemporal heteronymous hemianopia

D&E   Right homonymous hemianopia

Fig. 18-5. Diagram illustrating specific visual field defects produced by lesions in the visual pathways at the levels bearing corresponding letters in Figure 18-4.

where caudal to the optic chiasm will result in homonymous hemianopia, that is, if either the left optic tract, left geniculate body, left optic radiation, or left visual cortex is totally destroyed it will produce a *right homonymous hemianopia,* and a total lesion in the pathway caudal to the chiasm on the right side will result in a *left homonymous hemianopia.*

A lesion encroaching on the middle portion of the optic chiasm, frequently a tumor of the hypophysis (pituitary gland), would first destroy the decussating fibers and hence result in a *bitemporal heteronymous hemianopia* (Figs. 18-4 and 18-5).

The human optic radiation has been defined in such detail that small lesions destroying only portions of the radiation are positively correlated with quadrantic homonymous defects (*i.e.,* blindness in certain quadrants of the visual field). Such sophistication has not been achieved in veterinary neuro-ophthalmology.

In study of the visual pathway as described above and illustrated in Figure 18-4, it should be realized that the pathway extends from a rostral point beyond the brain itself (retina) caudally through the entire length of the brain to the occipital cortex. Based on this extensive anatomic plan, the visual system is vulnerable to injuries which may be localized anywhere from the eye to the external occipital protuberance of the skull. It should be remembered that frequently vision is affected by lesions located outside the optic system, for example, the bitemporal heteronymous hemianopia mentioned previously, resulting from tumor of the hypophysis.

Optic Reflexes

Most of the fibers in the optic tract terminate in the lateral geniculate body, but some bypass without synapsing and form the *brachium of the rostral (superior) colliculus.* These fibers synapse either in the rostral colliculus or in the pretectal area, which is at the midbrain–diencephalon junction. These areas are functionally concerned with optic reflexes, as explained below.

Pupillary (Light) Reflex
(Cranial Nerves II and III)

Stimulus: Illumination of one eye. (A pen-light is commonly used.)

Effect: Constriction of the pupil in both eyes.

Afferent neurons: The axons pass from the retina, within the optic nerve, optic tract, through the brachium of the rostral colliculus to the pretectal area. Here they synapse with neurons whose axons pass bilaterally to the *Edinger-Westphal nuclei* located within the tegmentum of the upper or rostral mesencephalon. These nuclei are parasympathetic nuclei of the oculomotor nerve.

Efferent neurons: (1) Parasympathetic (GVE) neurons arise within the Edinger-Westphal nuclei to enter the oculomotor nerves as presynaptic (preganglionic) neurons to the *ciliary ganglia* where they synapse with (2) postsynaptic GVE parasympathetic neurons whose nonmyelinated axons pass into the eyeballs to stimulate the constrictor pupillae muscles (of the irises), which results in constriction of the pupils. Note that both pupils constrict. The reflex resulting in constriction of the pupil of the stimulated eye is referred to as *direct light reflex,* whereas the constriction of the pupil of the contralateral (nonstimulated) eye is known as *consensual light reaction,* or *reflex.*

If either the optic or the oculomotor nerve is destroyed, or both, the reflex will not be elicited. A definitive test for distinguishing which nerve is involved when the pupillary reaction is absent in only one eye is based on the following principles. A lesion of the oculomotor nerve or the ciliary ganglion results in absence of the light reflex in the eye of the same side because the impulses are prevented from reaching that pupillary sphincter. In contradistinction, illumination of the eye on the side of the affected oculomotor nerve will produce only a consensual reaction (*i.e.*, constriction of the pupil of the contralateral eye). These facts are due to the afferent impulses conveyed by the optic nerve being relayed to the Edinger-Westphal nuclei on both sides of the mesencephalon. In a complete unilateral optic nerve lesion, illumination of the contralateral (intact) eye causes both direct and consensual reaction to light (*i.e.*, constriction of the pupil in both eyes). If, however, the eye on the side of the optic nerve lesion is illuminated, there will be no pupillary change in either eye.

Another clinical interpretation of the light reflex relates to localization of the lesion in correlation with blindness or hemianopia. Loss of the pupillary light reflex without loss of vision often indicates a lesion in the pretectal area, third nerve, or ciliary ganglion, as considered previously; a positive pupillary light reflex (tested in both eyes), but loss of conscious perception of vision may result from a lesion in the lateral geniculate body, optic radiation, or the visual (occipital) cortex.

In evaluation of the consensual portion of the light reflex, the veterinarian must be extremely cautious, because frequently, in normal dogs, the reflexive pupillary constriction occurs only in the eye actually stimulated by light (Smythe, 1958).

Corneal Reflex

(Cranial Nerves V and VII)

Stimulus: Touching the cornea with a foreign object. (A small piece of cotton is commonly used.)

Effect: Prompt closing of both eyes.

Afferent neurons: The cornea is innervated by rami of the ophthalmic division of the trigeminal nerve. Sensory (GSA) impulses are conveyed by this division to the rostral portion of the spinal or descending nucleus of the trigeminal root. The unipolar cell bodies of these peripheral neurons are in the semilunar ganglion. From the nucleus of the spinal root of the trigeminal nerve in the upper medulla, neurons of the second order pass to the ipsilateral and contralateral facial nuclei.

Efferent neurons: Cell bodies in each facial nucleus (SVE) give rise to axons which pass in the facial nerve to the orbicularis oculi muscle which closes the respective eye (both sides are stimulated).

Similar to the pupillary reflex, the reaction in the stimulated eye is referred to as the *direct corneal reflex,* and that of the contralateral eye as the *consensual corneal reflex.* A totally destructive lesion which interrupts the trigeminal innervation from the cornea touched abolishes the reflex in both eyes. If the ipsilateral facial nerve is destroyed, there will be a consensual corneal reflex, but no direct reflex. If the contralateral rami to the orbicularis oculi muscle are destroyed there will be a direct corneal reflex, but no consensual reflex. Therefore it should be obvious why both eyes should be tested for this reflex.

Blink Reflex

(Cranial Nerves II and VII)

This reflex is also known as the *menace, opticofacial,* or *visual-orbicularis* reflex.

Stimulus: A sudden visual stimulus, for example, a sudden thrust of the hand toward the face. A glass or plexiglass shield should be in front of the face, so that the compressed air waves do not hit the cornea and elicit a corneal reflex.

Effect: Reflexive closure of the eyes.

Afferent path: Retinae to optic nerves, to optic tracts (bypass the lateral geniculate bodies) directly to the rostral colliculi in the midbrain tectum.

Efferent path: Cell bodies in the rostral colliculi give rise to descending axons which form the *tectobulbar tracts* to the motor nuclei of cranial nerves, that is, in this reflex to the motor nuclei of the facial nerves. The SVE fibers in the facial nerve pass to the orbicularis oculi for closure of the eyes.

The blink reflex is a protective reaction to a relatively weak stimulus. A very strong stimulus, such as a violent thrust of the hand toward the face, causes a 'startle' response with turning of the head and perhaps reaction of the whole body. The pathway for this reflex is a reinforcement of that given for the basic blink reflex.

Afferent path: Same as for blink reflex.

Efferent path: In addition to the tectobulbar fibers mentioned for the blink reflex, there are additional tectobulbar fibers that pass to the accessory (XI) nuclei. Lower motor neurons from these nuclei pass to the brachiocephalic muscles for turning of the head. Also from the rostral colliculi are descending fibers which form the *tectospinal tract.* These descend to the spinal cord and synapse with the lower motor neurons, which innervate body musculature.

In addition to being a powerful reflex center, the rostral colliculus in the dog is also an integrative center subserving visual perception. A dog (or cat or monkey) with a unilateral extirpation of the rostral colliculus does not acknowledge stimuli from the visual field of the opposite side, even though it is not blind to these stimuli. It continually turns its head to the side of the lesion, and seems to be overresponsive to stimuli from that side (Noback, 1967).

Clinical Comments

Horner's Syndrome

Many of the symptoms of Horner's syndrome involve the eye and therefore it should be mentioned here. It should be remembered that this syndrome results from a destructive lesion in the sympathetic system, commonly in the cervical sympathetic ganglia. Three cardinal symptoms are: unilateral miosis (constriction of the pupil), ptosis (drooping of the eyelid), and enoph-

thalmos (sinking of the eyeball into the orbit). In addition to the peripheral lesion mentioned, interruption of descending tracts of the central nervous system (lateral reticulospinal and tectotegmentospinal) to the intermediolateral horn cells of the upper thoracic levels (sympathetic GVE) or a lesion of the spinal cord at the upper thoracic levels may result in this syndrome.

Glaucoma in Dogs

The following brief, admittedly inadequate, consideration of glaucoma is based on the work of Magrane (1965) and Lovekin (1964), and reference is made to their reports for further important clinical aspects, including the pathology, treatment, such as surgical procedures, and management.

The term 'glaucoma' refers to an increased intraocular pressure resulting from one or a variety of abnormal conditions. In itself it is not a disease. Although this condition occurs with a relatively low incidence among different breeds, Magrane believes it to be the most common cause of blindness in the middle-aged dog. Irreversible blindness may result within a few hours to several days; therefore it is obviously important that the veterinarian not only be able to diagnose the condition, but know the treatment and care necessary to prevent loss of vision.

Etiology

In many cases it is necessary to report the condition as 'idiopathic,' which is an admission of ignorance of its cause. Magrane proposed two theories: (1) *Neurovascular,* which is based on abnormal ocular blood circulation. This need not be restricted to local circulatory disorders; for example, malfunction of the hypothalamus, the autonomic nervous system, or the endocrine system may be a possible contributor. (2) *Mechanical,* which is based on a physical blockage of drainage of the ocular fluid. As is true so often in medicine, one mechanism does not function (or fail to function) independently. Therefore, many ophthalmologists believe that a combination of the two theories may explain the etiology.

General Considerations

Magrane reported that primary glaucoma is found almost exclusively in one breed, Cocker Spaniel. Both Magrane and Lovekin agree on the probable inheritance of the condition in that breed.

Neither author offered an explanation for the seasonal variation in incidence. In dogs most attacks occur during the cold months of the year in the northern cities.

Examination in the Neonatal Dog

Postnatal ontogeny of the canine eye and visual system should be considered in ophthalmologic examination of the neonatal dog. Many conditions observed in the adult dog as the result of neurologic conditions appear as normal characteristics of the visual system of young pups. The eyes open at approximately two weeks of age and respond normally to reflex stimulation by about three weeks. Recognition of objects, however, is very poor until 25 days. This normal latency resembles psychic blindness as detected in the adult dog with various types of encephalopathies or neoplasms in the optic tracts or visual cortex (Fox, 1963).

Bibliography

Andersen, A. C.: Eye/Gross and Subgross. Chapter 16A in *The Beagle As an Experimental Dog* (Andersen, A. C., Ed.). Ames, Ia., Iowa State University Press, 1970. Pp. 374–385.

de Lahunta, A., and Cummings, J. F.: Neuro-ophthalmologic lesions as a cause of visual deficit in dogs and horses. J.A.V.M.A. *150:* 994–1011, 1967.

Elliott, H. C.: *Textbook of Neuroanatomy,* 2nd ed. Philadelphia, J. B. Lippincott Co., 1969.

Everett, N. B.: *Functional Neuroanatomy,* 5th ed. Philadelphia, Lea & Febiger, 1965.

Finerty, J. C., and Cowdry, E. V.: *A Textbook of Histology,* 5th ed. Philadelphia, Lea & Febiger, 1960.

Fox, M. W.: Postnatal ontogeny of the canine eye. J.A.V.M.A. *143:* 968–974, 1963.

Keller, W. F.: Personal communication, 1969.

Lovekin, L. G.: Primary glaucoma in dogs. J.A.V.M.A. *145:* 1081–1091, 1964.

McGrath, J. T.: *Neurologic Examination of the Dog,* 2nd ed. Philadelphia, Lea & Febiger, 1960.

Magrane, W. G.: *Canine Ophthalmology.* Philadelphia, Lea & Febiger, 1965.

Noback, C. R.: *The Human Nervous System.* New York, McGraw-Hill, 1967.

Polyak, S.: *The Vertebrate Visual System.* Chicago, The University of Chicago Press, 1957.

Prince, J. H., Diesem, C. D., Eglitis, I., and Ruskell, G. L.: *Anatomy and Histology of the Eye and Orbit in Domestic Animals.* Springfield, Charles C Thomas, 1960.

Prince, J. H., and Ruskell, G. L.: The use of domestic animals for experimental ophthalmology. Am. J. Ophth. *49:* 1202–1207, 1960.

Roberts, S. R.: Animal vision. J.A.V.M.A. *127:* 236–239, 1955.

Smythe, R. H.: *Veterinary Ophthalmology,* 2nd ed. London, Baillière, Tindall and Cox, 1958.

Trautmann, A., and Fiebiger, J.: *Fundamentals of the Histology of Domestic Animals* (translated and revised from 8th and 9th German edition by R. E. Habel and E. L. Biberstein). Ithaca, N. Y., Comstock Publishing Associates, Division of Cornell University Press, 1957.

Wyman, M., and Donovan, E. F.: The ocular fundus of the normal dog. J.A.V.M.A. *147:* 17–26, 1965.

chapter 19

The Ear—Hearing and Equilibrium

Introduction

The ear of the dog, as that of all higher mammals, is anatomically composed of three connected divisions, so individualized that each is referred to as an 'ear,' namely, the outer (external) ear, middle ear, and inner (internal) ear (labyrinth). These anatomic divisions serve two important functions which to the inexperienced reader appear to be unrelated: *hearing*, as a special exteroceptive sense of the cochlear division, and *equilibrium*, as a special proprioceptive sense of the vestibular division.

Phylogenetically the vestibular portion of the inner ear is much older than the auditory; its function, equilibrium, is basic and has been relatively unchanged from fish to man (Romer, 1970). Embryonically, the ear develops from superficial ectoderm, the *auditory placode,* which thickens, invaginates, and becomes vesicular to form the *otocyst* (otic vesicle). Comparative anatomists consider these early developmental stages of the inner ear comparable to an internal form of the external lateral line system found in fishes.

In consideration of the close anatomic and physiologic relations between hearing and equilibrium, it should be realized that this is true only with reference to the inner ear, and that even here the two divisions, cochlea (hearing) and semicircular canals (equilibrium), are individually identifiable although joined anatomically. The eighth cranial nerve, which innervates both parts, is composed of two individual portions and appropriately is named the *vestibulocochlear nerve*. Although this is now the official name (Nomina Anatomica, 1966; Nomina Anatomica Veterinaria, 1968), the eighth cranial nerve is also known as the statoacoustic, auditory, or acoustic nerve. As indicated previously, the external and the middle ear are functionally concerned only with hearing.

Based on functional components, both the cochlear (hearing) and the vestibular (equilibrium) divisions of the eighth cranial nerve are *special somatic afferent* (SSA). This is readily understandable for the cochlear division, where the stimulus obviously arises from outside the body. It may appear confusing, however, that the vestibular division is SSA, since this division is stimulated to initiate proprioceptive impulses by changes within the body.

It should be remembered from Chapter 7 that proprioceptive fibers as considered in the spinal nerves associated with the body wall have been designated as general somatic afferent (GSA). Actually, the proprioceptive sense of posture and equilibrium in reference to the vestibular nerve is special rather than general primarily because: (1) the eighth cranial nerve is designated as a nerve of special sense, (2) the inner ear and the central nervous system are intimately associated in both the embryo and the adult, and (3) the posture and equilibrium of the body subserve and are dependent on the position of the head and neck, which is directed by the vestibulocochlear nerve.

Hearing

Outer Ear

The outer, or external, ear consists of the *pinna* (auricle) and the *external auditory meatus*

(ear canal). This portion of the ear has least to do directly with the neurophysiologic faculty of hearing and, as stated previously, it normally contributes nothing to the vestibular apparatus.

From the clinical point of view, there have been many controversies directed toward the comparative incidence of otitis in pendant-eared and erect-eared dogs. For details of anatomy of the external ear see Getty *et al.* (1956) and Getty (1964). A function of the external ear is generally said to be the funneling of air vibrations into the external auditory meatus to the tympanic membrane (ear drum), which is the partition between the external ear and middle ear.

Middle Ear (Fig. 19-1)

Internal to the tympanic membrane (ear drum) is the middle ear, or *tympanic cavity*. Most of the lateral wall of the middle ear, therefore, is formed by the *tympanic membrane*. This thin sheetlike structure is derived from all three germ layers: the superficial ectoderm invaginates as a lining of the external auditory meatus and external surface of the membrane; the endoderm from the pharynx lines the auditory (eustachian) tube, the tympanic cavity, and the internal surface of the membrane; and the middle layer of the tympanic membrane is derived from mesoderm.

The *auditory (eustachian) tube* connects the tympanic cavity (middle ear) with the nasopharynx. Functionally, this connection permits an equalization of air pressure on the two sides of the tympanic membrane. Although the middle ear and external ear are both in contact with air, the media for the transmission of sound differ in the two areas. The sound waves are carried to and through the external ear by air, whereas their energy is transmitted through the middle ear by the three small bones (auditory ossicles).

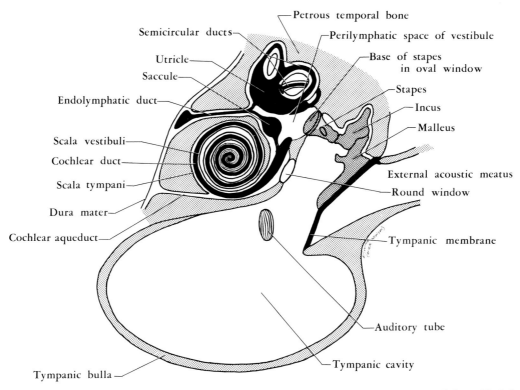

Fig. 19-1. Diagram of the middle ear (tympanic cavity) and inner ear (labyrinth) of dog. (Slightly modified from Getty *et al.*, 1956.)

Bones of the Middle Ear (Figs. 19-1, 19-2)

The bones of the middle ear (auditory ossicles) are three small bones that articulate in a linear fashion so that they are movable. From the tympanic membrane to the oval window in the bony labyrinth the bones are: the malleus (hammer), incus (anvil), and stapes (stirrup).

The largest of the three is the *malleus,* which has its manubrium (handle) attached to the tympanic membrane. At the base of the manubrium near the neck of the malleus is a muscular process to which the *tensor tympani muscle* attaches. Immediately ventral to this process the chorda tympani crosses the medial surface of the malleus on its way to join the lingual nerve (see Chapter 17 for

functional neuroanatomic characteristics of the chorda tympani).

Within the most dorsal portion of the tympanic cavity, the *epitympanic recess,* the head of the malleus articulates with the body of the middle ossicle, the *incus.* This bone is much smaller (approximately 4 mm by 3 mm) and shaped somewhat like a tooth, with two diverging roots. The longer crus is directed caudally but has a lenticular process which points rostrally and medially for articulation with the third ossicle, the *stapes.* This is the smallest bone in the body, measuring approximately 2 mm in length (Getty *et al.,* 1956). The stapes is shaped like a stirrup, a small head being connected to a larger base by two crura which border a very thin connective tissue membrane, the *obturator membrane.*

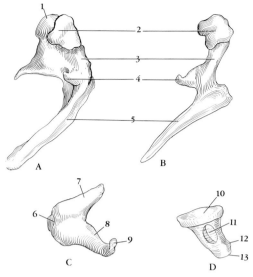

Fig. 19-2. The ossicles of the right ear (dog). *A,* malleus, medial view; *B,* malleus, caudal view; *C,* incus, mediorostral view; *D,* stapes, mediocaudal view. *1,* head; *2,* articular surface for incus; *3,* neck; *4,* muscular process; *5,* manubrium; *6,* articular surface for malleus; *7,* short crus; *8,* long crus; *9,* lenticular process; *10,* base; *11,* obturator membrane; *12,* muscular process; *13,* head. (After Getty *et al.,* 1956.)

The base of the stapes faces medially and lies within the oval window of the labyrinth to articulate by a syndesmosis with the thin cartilage covering the edge of the oval window. The caudal crus has a muscular process to which attaches the tendon of the *stapedius muscle,* the smallest skeletal muscle in the body.

The middle ear actually functions as a transformer, in that it converts the large-amplitude, low-force vibrations at the tympanic membrane to low-amplitude, large-force vibrations at the foot of the stapes within the oval window. The two small muscles mentioned above, the tensor tympani and stapedius, function as safety mechanisms by regulating the amount of tension exerted on the tympanic membrane and the oval window by the ossicles.

Inner Ear

The inner ear (labyrinth), as mentioned above, is the only one of the three divisions of the ear which normally participates directly in both functions—equilibrium and hearing.

The inner ear (Fig. 19-3) is embedded within the petrous portion of the temporal bone; anatomically it may be divided into three segments—the cochlea, vestibule, and semicircular canals.

1. The *cochlea* is coiled like a snail's shell, from which its name is derived. The number of turns varies among mammals, the number for the dog being three and one-fourth (Getty, 1956). Grossly the base of the cochlea is widest and throughout its length is a central bony core, the *modiolus,* which has many ganglia dispersed near the inside border of the coiled tubule. Each of these ganglia contains bipolar cell bodies characteristic of the special sensory ganglia. Since these form a continuous ganglionic mass as they accompany the spirals of the cochlear nerve within the modiolus, they are collectively referred to as the *cochlear,* or *spiral, ganglion* of the eighth cranial nerve. The apex (cupula) of the coiled tubular cochlea is directed ventrorostrally and slightly laterally in the petrous portion of the temporal bone. This is the rostral portion of the labyrinth and functions neurologically as the hearing portion of the ear.

2. The *vestibule* is the middle portion of the inner ear and contains two sensory areas deep within the membranous labyrinth: (*a*) the rostral *saccule* (sacculus), and (*b*) the caudal *utricle* (utriculus). The lateral wall of the vestibule contains the *oval window,* which is closed largely by the foot of the stapes. Ventral to the oval window is the *round window,* which contains a thin membrane, the *secondary tympanic membrane.* This separates the air in the cavity of the middle ear from the perilymph in the scala tympani of the labyrinth. Functionally this is a safety mechanism involved with the cochlear di-

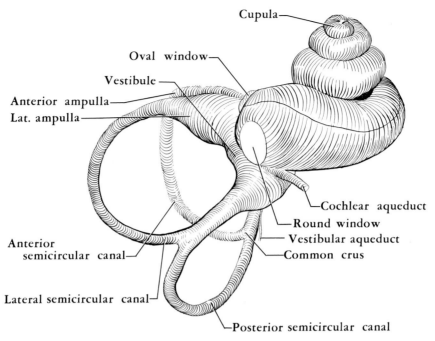

Cupula

Oval window

Vestibule

Anterior ampulla

Lat. ampulla

Cochlear aqueduct

Round window

Vestibular aqueduct

Common crus

Anterior
semicircular canal

Lateral semicircular canal

Posterior semicircular canal

Fig. 19-3. Ventroposteromedial view of the right osseous labyrinth. (After Getty *et al.,* 1956.)

vision of the labyrinth. The membrane is at the end of the scala tympani and receives the 'excess' vibratory waves of the perilymph which were originally generated by the foot of the stapes 'compressing' the perilymph within the scala vestibuli.

3. The three *semicircular canals* (Fig. 19-3), each in a plane approximately perpendicular to the other two, lie in the caudal portion of the labyrinth. Each canal forms an arc of approximately two-thirds of a circle and is connected to the vestibule at both ends, but not separately (*i.e.,* there are less than six connections with the vestibule).

At one end of each canal there is a localized swelling, the *ampulla.* The non-ampullated ends of the *posterior (caudal)* and *anterior (dorsal)* canals join to form a *common crus,* which attaches to the vestibule. The non-ampullated end of the *lateral (horizontal) canal* and the ampullated end of the posterior canal are united for a short distance caudal to the vestibule. Within the membranous labyrinthine portion of the ampullae there

are sensory end-organs, the *cristae ampullares,* which are discussed later in connection with the membranous labyrinth.

The above three portions of the labyrinth, which are observable within the petrous temporal bone and somewhat carved out of the bone substance, constitute the outside *osseous labyrinth.* Obviously this varies somewhat in size with the breed of dog, but the median length is approximately 15 mm (Getty, 1956). Within the osseous labyrinth is a fluid, the *perilymph,* which separates the osseous labyrinth from an inside model of the same shape, but much more delicate, the *membranous labyrinth.* This inner model also contains a fluid, the *endolymph,* which differs chemically from perilymph. It should be emphasized that the perilymph and endolymph are within all three labyrinthine segments.

The endolymph is in direct contact with the sensory end-organs inside the membranous labyrinth of both the vestibular (semicircular canals) and cochlear portions, as well as the vestibule (saccule and utricle). As shown in Figure 19-4, a cross section of the spiral cochlea reveals three compartments of which two, the *scala vestibuli* and the *scala tympani*, contain perilymph. These two compartments border a central triangular-shaped *cochlear duct*, which contains endolymph. The perilymphatic-filled channels communicate freely with one another at the apex (*helicotrema*) of the cochlea.

There is a connecting tube, the *cochlear aqueduct*, which connects the scala tympani with the cranial cavity. Within this tube is the *perilymphatic duct*, which permits exchange with the cerebrospinal fluid within the subarachnoid space and thereby permits the derivation of perilymph from cerebro-

spinal fluid. The *vestibular aqueduct* contains the *endolymphatic duct*, which is formed by the *utricular* and *saccular ducts* and extends to the extradural space, where it terminates in a dilated endolymphatic sac adjacent to the dura mater. The saccule is joined to the cochlea by the *ductus reuniens*.

Cochlear Duct

As illustrated in Figure 19-4, the triangular-shaped cochlear duct is bordered by the scala vestibuli above and the scala tympani below. An extremely thin membrane, the *vestibular*, or *Reissner's, membrane* extends obliquely from the upper end of the stria vascularis to the limbus spiralis. This mem-

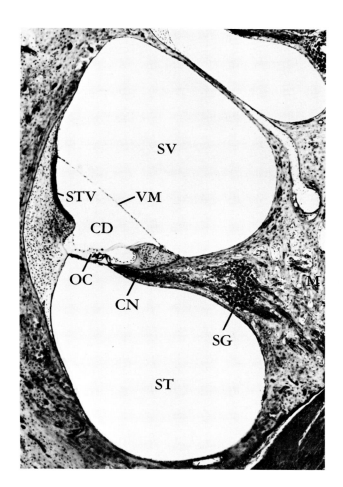

Fig. 19-4. High-power photomicrograph of a cross section through the cochlea in cat. *SV,* scala vestibuli; *VM,* vestibular (Reissner's) membrane; *STV,* stria vascularis; *CD,* cochlear duct; *OC,* organ of Corti (on basilar membrane); *CN,* cochlear nerve; *ST,* scala tympani; *SG,* spiral (cochlear) ganglion; *M,* modiolus.

brane forms the upper border of the cochlear duct and separates the perilymph of the scala vestibuli from the endolymph of the cochlear duct.

The lateral, or outside, wall (away from the modiolus) of the cochlear duct is formed by the *stria vascularis* of the spiral ligament. The endolymph of the cochlear duct is produced by this vascular wall. Many authorities believe that the organ of Corti receives its nutritive and oxygen supplies from the endolymph.

The base of the cochlear duct is formed largely by the sensory end-organ of the cochlear nerve, the organ of Corti, which rests on the basilar membrane. This partition separates the cochlear duct above from the scala tympani below. The base of the cochlea, although the widest portion, contains the shortest basilar membrane with its organ of Corti and responds to sounds of higher frequency. The partition increases in width progressively along the duct toward the apex (cupula), whereas the coils of the cochlea diminish in width, that is, the gross width of the cochlea progressively decreases, but the width of the floor of the inside cochlear duct increases from base to apex of the cochlear coils.

Organ of Corti

The organ of Corti is the complicated end-organ of the cochlear nerve which rests on the basilar membrane as it extends throughout the length of the cochlear duct. The cellular construction of the spiral organ of Corti in the dog is similar to that in man (Arey, 1968). As in typical neuroepithelia, there are sensory hair cells and sustentacular (supporting) cells. The hair cells are in direct contact with the terminal arborizations of the dendrites of the spiral ganglion cells. Directly above the hair cells there is a shelflike gelatinous *tectorial membrane* attached to the limbus at the inner side (toward the modiolus) with a free edge at approximately halfway across the cochlear

duct. Most authorities believe that the hair cells are actually embedded in the undersurface of the tectorial membrane. In most histologic preparations the tectorial membrane appears free within the endolymph above the hair cells, but this is probably a fixation artifact (Bloom and Fawcett, 1968).

As the sound wave energy passes from the foot of the stapes to the perilymph of the scala vestibuli, up to the helicotrema, and down the scala tympani, it causes the basilar membrane to vibrate. Since the organ of Corti rests on this membrane, it vibrates and causes the hairs of the hair cells to move within the tectorial membrane. In this manner, the hair cells act as mechanoreceptors and convert the mechanical energy of sound waves into nerve potentials which pass as nerve impulses along the many fibers of the cochlear nerve toward the spiral ganglion. From the bipolar cells within the spiral ganglion the nerve impulses pass through the internal auditory meatus via fibers of the cochlear nerve which terminates in the dorsal and ventral cochlear nuclei at the pons–medulla junction of the brain stem (Plate 14).

Auditory Pathway

Summary of Pathway of Impulses from External Ear to Brain Stem

Before discussing the true auditory pathway in the central nervous system, it seems appropriate to review how the impulses set up by sound vibrations reach the brain. Figure 19-5 is a summary of the functional anatomy of the ear relative to the transmission of 'sound' from the external ear to the brain stem.

Central Auditory Pathway

As one studies the central auditory pathway he should be impressed by the inconsistent courses, synapses, and groupings of

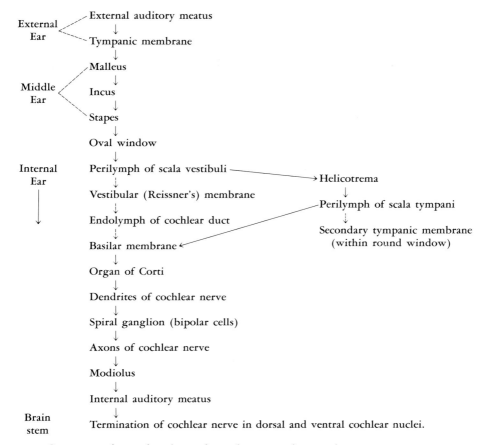

External
Ear
— External auditory meatus
↓
— Tympanic membrane
↓

Middle
Ear
— Malleus
↓
Incus
↓
— Stapes
↓
Oval window
↓

Internal
Ear
Perilymph of scala vestibuli ——————→ Helicotrema
↓   ↓
Vestibular (Reissner's) membrane   Perilymph of scala tympani
↓   ↓
Endolymph of cochlear duct   Secondary tympanic membrane
↓   (within round window)
Basilar membrane ←
↓
Organ of Corti
↓
Dendrites of cochlear nerve
↓
Spiral ganglion (bipolar cells)
↓
Axons of cochlear nerve
↓
Modiolus
↓
Internal auditory meatus
↓

Brain
stem
Termination of cochlear nerve in dorsal and ventral cochlear nuclei.

Fig. 19-5. Summary of sound pathway from the external ear to brain stem.

fibers from specific nuclei to higher specific loci. It seems apparent that the impulses reaching the cortex would not be in uniform waves, owing to the variable subcortical distributions of specific neurons. In contrast to most major central pathways, it is impossible to state that there is a definite number of neurons in the auditory pathway. Likewise, this pathway is neither strictly contralateral nor ipsilateral, because of the variability of neurons crossing or not crossing at various levels. These features account for the fact that an isolated central lesion in this pathway results in a diminution of hearing in both ears rather than total deafness in either one or both of them.

Secondary Auditory Neurons (*Fig. 19-6*)

The pathway of 'sound' stimuli and the nerve impulses generated by them to the dorsal and ventral cochlear nuclei as the entrance into the central nervous system via the neurons of the first order (cochlear nerve) was outlined in Figure 19-5. Secondary neurons arise within both cochlear nuclei and pass to any of the following: (1) the superior olivary nucleus (Plate 15), (2) the trapezoid body, which crosses to either the superior olivary nucleus or the lateral lemniscus of the opposite side (Plate 15), and (3) the lateral lemniscus of the same side. Note these variable pathways in Figure

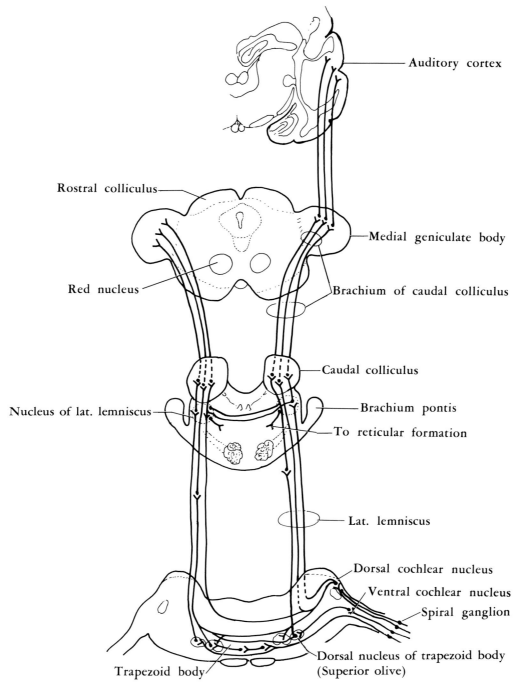

Auditory cortex

Rostral colliculus

Medial geniculate body

Red nucleus

Brachium of caudal colliculus

Caudal colliculus

Nucleus of lat. lemniscus

Brachium pontis

To reticular formation

Lat. lemniscus

Dorsal cochlear nucleus

Ventral cochlear nucleus

Spiral ganglion

Dorsal nucleus of trapezoid body
(Superior olive)

Trapezoid body

Fig. 19-6. Diagram of the auditory pathway. An important characteristic of this pathway is its variable number of neurons which may cross to the opposite side at various levels or remain on the same side (see text for details).

19-6. The lateral lemniscus ascends through the medulla and pons (Plate 16) to reach the caudal (inferior) colliculus of the mesencephalon (Plates 17, 18). Embedded within the substance of the lateral lemniscus is the nucleus of the lateral lemniscus. There may be more than one nucleus, and fibers may or may not synapse within the nucleus on their way to the caudal colliculus. Immediately proximal to the caudal colliculus some fibers emerging from the nucleus of the lateral lemniscus are thought to cross over and enter the lateral lemniscus and caudal colliculus of the opposite side.

Acoustic fibers migrate to the dorsolateral surface of the caudal colliculus and form the brachium of the caudal colliculus (Plate 19) as they pass directly to the medial geniculate body.

From the medial geniculate body on each side fibers radiate as the auditory radiations to reach the Sylvian gyrus of the cerebral cortex (see Fig. 16-10, p. 255). According to Papez (1929), this gyrus represents the primary auditory cortical area in carnivores and probably the ectosylvian and suprasylvian gyri function as auditory association areas. If so, note the large cortical area of the dog brain which represents the auditory function. Recent degeneration studies indicate that most auditory fibers terminate in the rostral portion of the ectosylvian gyrus in the dog (Tunturi, 1970); he found no degenerating fibers in the Sylvian gyrus.

In contrast, Rose and Woolsey (1949, 1950) identified fibers in this gyrus in the cat. Indications of better auditory armamentaria in these animals, in comparison with man, are: (1) A relatively larger auditory area in the cerebral cortex, as discussed above. (2) Three and one-fourth turns of the cochlea compared with approximately two and one-half turns in man (Everett, 1965). These two factors contribute supporting evidence to the place theory of hearing, which in general terms proposes one representation of the cochlea in the cerebral cortex (Ruch *et al.,* 1965). (3) The dog has a frequency range of hearing with upper limits of approximately 35,000 cycles per second (Hertz, Hz), compared with an upper limit of 16,000 Hz for useful hearing in man (Prosser and Brown, 1961). The 'silent' dog whistles have a frequency well above the upper limit of man's sensitivity.

Equilibrium

The vestibular portion of the eighth cranial nerve is essential for relaying information from the labyrinth to the brain stem. Peripherally, the labyrinthine sensory endorgans are the three *cristae ampullares* of the membranous semicircular canals and the *maculae* of the utricle and saccule within the membranous vestibule.

The stimulus for the hair cells of the cristae ampullares is a turning of the head, which sets the fluid in motion and deflects the hairs within the cupula of the appropriate crista (Fig. 19-7 A). Any shift in position of the head serves as a stimulus for the maculae. The hair (sensory) cells of the maculae are actually stimulated by the effect of gravity on the *otoliths* (otoconia), the calcareous bodies that are embedded in the surface layer of the gelatinous masses (Fig. 19-7 B). The cristae ampullares are reported to function in dynamic equilibrium, whereas the maculae of the utricle and saccule are involved with static equilibrium.

Within the vestibular apparatus of the labyrinth there is a branch of the vestibular nerve arising from the hair cells in each of the five neuroepithelia. On stimulation, impulses pass from the labyrinth to the *vestibular (Scarpa's) ganglion,* located at the internal auditory meatus. Like the cochlear ganglion, the vestibular ganglion in the adult has retained the bipolar type of cell bodies.

The axons of vestibular ganglion neurons pass through the internal auditory meatus with the cochlear nerve and enter the pons–

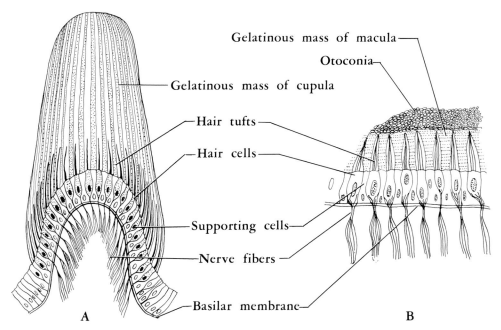

Gelatinous mass of macula

Otoconia

Gelatinous mass of cupula

Hair tufts

Hair cells

Supporting cells

Nerve fibers

Basilar membrane

A   B

Fig. 19-7. Diagrams of high-magnifications of crista ampullaris (A) and macula (B). Note the basic similarities in structures except for the otoconia only in the macula. (See text for differences in location and stimuli for the two receptors.)

medulla junction of the brain stem, where most axons terminate in the vestibular nuclei. Some vestibular fibers pass through the nuclei without synapse to proceed directly to the cerebellum by way of the caudal cerebellar peduncle (restiform body) (see Chapter 12).

Vestibular Nuclei (Fig. 19-8, Plates 12, 13, 14)

Within the area of the rhomboid fossa (floor of the fourth ventricle) there are four vestibular nuclei on each side of the brain stem. (1) The *caudal, spinal,* or *inferior vestibular nucleus* (Plate 12) is located in the medulla on the medial side of the caudal cerebellar peduncle and dorsolateral to the tractus solitarius. (2) The *lateral vestibular (Deiters') nucleus* is immediately rostral to the spinal nucleus and is located medial to the caudal cerebellar peduncle and dorsomedial to the

spinal tract and nucleus of the trigeminal nerve (Plate 13). (3) The *medial vestibular (Schwalbe's) nucleus* is medial to the lateral nucleus and dorsolateral to the medial longitudinal fasciculus (MLF) (Plate 12). (4) The *rostral (superior) vestibular nucleus (of Bechterew)* (Plate 14) is the most rostral of the four nuclei and is found in the pons at the level of the internal genu of the facial nerve. This nucleus borders the ventrolateral corner of the fourth ventricle, which at this more rostral level is smaller as it approaches the aqueduct.

Efferents of Vestibular Nuclei

The lateral vestibular nucleus projects ipsilateral descending fibers as the *lateral vestibulospinal tract* (Fig. 19-8; see Fig. 10-7, p. 157), which extends in the ventral funiculus of the spinal cord to the lumbosacral levels in cat, monkey, and man (Brodal,

disappeared or have been compensated for by the time of clinical examination. (2) Many of the isolated symptoms are present as a result of lesions other than those in the labyrinth. As an example, cerebellar lesions result in motor deficiencies somewhat similar to those described for vestibular lesions.

Nystagmus certainly is not pathognomonic for any syndrome, but may result from cerebellar or certain brain-stem lesions as well as from labyrinthine disorders. Frequently an external otitis or perhaps mites on the tympanic membrane and in the external auditory meatus, especially in young animals, may result in head tilt and disequilibrium, with the animal favoring the affected side. Govons (1944) referred to the importance of the labyrinth in experimental head injury in dogs.

Nystagmus

Nystagmus is defined as involuntary rhythmic oscillations of the eyes. The 'to-and-fro' movements may be in a vertical plane or in a rotary arc, but is probably most commonly in a horizontal plane. This is something of a typewriter-like movement in the sense that there is a slow movement in one direction and a rapid return. The direction of nystagmoid movement is said to be that of the quick phase, even though this is the compensatory direction.

As stated above, nystagmus may be a symptom resulting from lesions in various loci, but it may also be easily demonstrated in a healthy animal (or human) by rotation of the body, or by irrigation of the external auditory meatus with either hot or cold water (Bárány's caloric test). It is thought that hot water causes the endolymph to move upward and cold water causes the opposite movement. Therefore the direction of the resulting nystagmus may be reversed by changing the temperature of the water used for irrigation.

Summary of Vestibular Nystagmus

Stimulation of one vestibular nerve or MLF results in deviation of the eye to the opposite side, followed by movement of the head, neck, and trunk. This, of course, is enhancement of the normal functioning of the stimulated vestibular apparatus. As discussed previously, a total destructive lesion results in deviation of the eyes, head, and trunk to the same side, that is, the animal (or human) has a tendency to fall to the side of the lesion.

When the head is turned to the right, the eyes tend to remain fixed, and as the head continues to the right the eyes are pushed to the left. The cupulae and cristae ampullares are activated by the head turning, but when the eyes are forced as far as possible to the opposite side proprioceptive impulses are initiated by ocular muscle stretch and the eyes snap in the direction of rotation to be fixed again in the midline. If head rotation continues the ocular oscillation is repeated and results in nystagmus due to the constant stimulation of the cristae.

The cristae are stimulated by activity and the maculae are important for posture (static). It is true that stimulation of the maculae (by change in head position) results in changed eye position, but since the stimulus is static there is no resulting nystagmus.

Vestibular Reflex Mechanisms

There is a closed neural circuit connecting the spinal cord, the vestibular nuclei, the cerebellum, and the extra-vestibular nuclear brain stem which influences the tonicity particularly of the extensor muscles. The classic example of exaggerated extensor muscle tonus is the 'decerebrate rigidity' resulting from transection of the brain stem immediately rostral to the vestibular nuclei. Dogs and cats do not demonstrate decerebrate rigidity when the midbrain is intact

(Ruch *et al.*, 1965). The animal demonstrates an exaggerated extension primarily of the limbs, tail, back, and neck. A decerebrate animal can stand when placed on its feet, but with overextension of the aforementioned body areas; it cannot walk or change position. Such a posture is the result of stimuli arising mainly from the muscle spindles of the neck and limb muscles, in addition to the effect on the vestibular portion of the labyrinth.

If the vestibular stimuli are generated chiefly by head movements or positions, then obviously the proprioceptive impulses from the neck muscles are important contributors to the mechanism. The effect of the neck proprioceptors may be eliminated by cutting the dorsal roots of the cervical nerves, or by use of a cast to immobilize the neck. If this is done to a decerebrate animal, the preparation is satisfactory for studying the 'isolated' tonic vestibular reflexes. When the animal is placed on its back the extensor tonus of all the limbs is greatest, due to the inverted position of the utricle and the great pull of the otoconia on the end-organs of the utricle. Various degrees of extensor tonus in the limbs may be demonstrated by turning the animal from the supine to the prone position.

Vestibular Righting Reflex

The vestibular righting reflex is fundamental in bringing the body into the upright position and also depends heavily on the neck muscles, which permit the head to retain its normal orientation in space. Once the neck proprioceptives are stimulated, the body follows the movement of the head. Visual stimuli are important compensatory factors, and therefore blindfolding of dogs, cats, and higher mammals is advisable. Rabbits and guinea pigs need not be blindfolded, according to Papez (1929). This reflex is demonstrated by lifting the animal at the pelvis vertically from the floor with the head down. The head tends to retain the normal relation to the body as the pelvis is twisted. In testing the righting reflex in larger dogs, the animal is held down on its side, making sure the head and neck are down; the hands are then released and the animal is observed as it rights itself. McGrath (1960) illustrates the proper procedures for testing various reflexes in the dog.

The vestibular righting reflex is commonly demonstrated by holding a cat in the supine position in mid-air; if it is held high enough from the landing surface, as it is dropped it quickly turns to land on its feet.

In conclusion, the higher the animal is on the phylogenetic scale, the less it depends on pure vestibular reflexes. 'Higher' mammals depend more on visual contributions for orientation in space, and man has 'learned' to suppress purely reflex actions and utilize cerebral cortical influence. Vestibular cortical centers have not, however, been located as precisely as have other sensory modalities (Crosby *et al.*, 1962).

Bibliography

Arey, L. B.: *Human Histology,* 3rd ed. Philadelphia, W. B. Saunders Co., 1968.

Bloom, W., and Fawcett, D. W.: *A Textbook of Histology,* 9th ed. Philadelphia, W. B. Saunders Co., 1968.

Brodal, A.: *Neurological Anatomy in Relation to Clinical Medicine,* 2nd ed. New York, Oxford University Press, 1969.

Crosby, E. C., Humphrey, T., and Lauer, E. W.: *Correlative Anatomy of the Nervous System.* New York, Macmillan Co., 1962.

Everett, N. B.: *Functional Neuroanatomy,* 5th ed. Philadelphia, Lea & Febiger, 1965.

Getty, R.: The Ear. Part of Chapter 17 in *Anatomy of the Dog* (Miller, M. E., Christensen, G. C., and Evans, H. E.). Philadelphia, W. B. Saunders Co., 1964. Pp. 847–863.

Getty, R., Foust, H. L., Presley, E. T., and Miller, M. E.: Macroscopic anatomy of the ear of the dog. Am. J. Vet. Res. *17:* 364–375, 1956.

Govons, S. R.: Experimental head injury produced by blasting caps. An experimental study. Surgery *15:* 606–621, 1944.

Jenkins, T. W., and Burns, R. D.: Specific subtotal labyrinthectomy in the kangaroo rat. I. Right lateral semicircular canal: procedure and effects. J. Comp. Neurol. *113:* 139–147, 1959.

McGrath, J. T.: *Neurologic Examination of the Dog.* 2nd ed. Philadelphia, Lea & Febiger, 1960.

Nomina Anatomica, Prepared by the International Anatomical Nomenclature Committee, 3rd ed. Amsterdam, Excerpta Medica Foundation, 1966.

Nomina Anatomica Veterinaria, Adopted by the General Assembly of the World Association of Veterinary Anatomists, Paris, 1967. Vienna, Adolf Holzhausen's Successors, 1968.

Palmer, A. C.: *Introduction to Animal Neurology.* Philadelphia, F. A. Davis Co., 1965.

Papez, J. W.: *Comparative Neurology.* New York, Thomas Y. Crowell Co., 1929.

Prosser, C. L., and Brown, F. A., Jr.: *Comparative Animal Physiology,* 2nd ed. Philadelphia, W. B. Saunders Co., 1961.

Romer, A. S.: *The Vertebrate Body,* 4th ed. Philadelphia, W. B. Saunders Co., 1970.

Rose, J. E., and Woolsey, C. N.: The relations of thalamic connections, cellular structure and evocable electrical activity in the auditory region of the cat. J. Comp. Neurol. *91:* 441–466, 1949.

Rose, J. E., and Woolsey, C. N.: Cortical Connections and Functional Organization of the Thalamic Auditory System of the Cat. In *Biological and Biochemical Basis of Behavior* (Harlow, H. F., and Woolsey, C. N., Eds.). Madison, University of Wisconsin Press, 1958. Pp. 127–150.

Ruch, T. C., Patton, H. D., Woodbury, J. W., and Towe, A. L.: *Neurophysiology,* 2nd ed. Philadelphia, W. B. Saunders Co., 1965.

Tunturi, A. R.: The pathway from the medial geniculate body to the ectosylvian auditory cortex in the dog. J. Comp. Neurol. *138:* 131–136, 1970.

Part II

Summary of Internal Structures of the Brain and Spinal Cord

Summary of Internal Structures of the Brain and Spinal Cord, Including Locations, Connections, and Functions

THE ABDUCENS NUCLEUS AND NERVE

1. Location of Nucleus and Course of the Nerve:
   a. At the level of the dorsal nucleus of the trapezoid body (superior olive) in the floor of the closed portion of the fourth ventricle immediately ventral to the internal genu of the facial nerve.
   b. The nerve fibers pass ventrally through the reticular formation and the trapezoid body to emerge between the dorsal nucleus of the trapezoid body laterally and the cerebrospinal tracts medially.
   c. Grossly from the ventral brain surface within the trapezoid body near the midline, the thin abducens nerve passes rostrally to enter the wall of the cavernous sinus and extends through the orbital fissure to enter the orbit for innervation of the retractor bulbi and the lateral rectus muscle.

2. Function:
   a. The abducens nucleus is classified as somatic efferent (SE). Note that, like the other somatic efferent nuclei (the hypoglossal, trochlear, and oculomotor), the abducens is not specified as either *general* or *special* somatic efferent. Some authorities (Crosby *et al.,* 1962) refer to these nuclei as *special* because the extrinsic ocular and the tongue muscles are derived from the embryonic head somites.
   b. The general somatic afferent (GSA) proprioceptive fibers from the target organs have not been identified with certainty in either dog or man. From a functional standpoint, the proprioceptive impulses must be present and the site of the cell bodies is logically the mesencephalic nucleus of the trigeminal nerve.
3. Lesion:
   A destructive lesion of the abducens nucleus or the nerve either within or outside the brain stem will result in a flaccid paralysis of the lateral rectus muscle of the same side. This is a typical lower motor neuron lesion (LMNL) which manifests itself as an ipsilateral internal strabismus (crossed eye).

NUCLEUS AMBIGUUS (VENTRAL MOTOR NUCLEUS OF X)
PLATE 10

1. Location: Medulla.
   a. At the level of attachment of nerves IX, X, the bulbar portion of XI, and XII to the brain stem.
   b. Within the reticular formation, dorsolateral to the caudal olivary nucleus, between it and the nucleus of the descending root of V.
   c. Dorsal to the lateral reticular nucleus.
2. Afferents:
   a. Receives afferent impulses from the sensory divisions of cranial nerves and ascending tracts of the spinal cord via secondary pathways of the trigeminal nerve.
   b. Corticobulbar fibers from the motor cortex (crossed and uncrossed) for voluntary swallowing and phonation.
   c. Secondary fibers from the nucleus and tractus solitarius as components of the following reflexes: Afferent impulses from the mucosa of the pharynx, larynx, and trachea traverse the glossopharyngeal and vagus nerves to terminate in the nucleus solitarius; its tract transmits impulses to the nucleus ambiguus which gives rise to peripheral fibers of cranial nerves IX and X for innervation of the branchiomeric muscles of the pharynx and larynx for the cough, gag, and a portion of the vomiting reflex.
3. Efferents and Function:
   a. SVE nucleus of origin for cranial nerves IX, X, and XI.
   b. Motor innervation to the muscles which arise from branchial arches 3 and 4 (muscles of pharynx and larynx).

BASAL GANGLIA (NUCLEI)
PLATES 23 TO 27

1. Location and Contents:
   a. This gray matter is composed of subcortical nuclei located deep within the basal portion of the cerebral hemispheres in close relationship to the diencephalon, but for the most part separated from it by the internal capsule.
   b. Authors differ in what they include as basal ganglia. The least number of structures that are generally accepted are: the caudate nucleus, the putamen, the globus pallidus, and the amygdala.
   c. In addition to the above, frequently the claustrum and sometimes the red nucleus, subthalamus, and substantia nigra are included in the basal ganglia. Sometimes the amygdala is not con-

sidered a component of the basal ganglia.

   d. The *corpus striatum* frequently is said to be a component of the basal ganglia. Corpus striatum is a collective term used to designate the gray and white striations formed by the rostral (anterior) limb of the internal capsule, partially separating the caudate nucleus and the lentiform nucleus. The latter nucleus is subdivided by the external medullary lamina, a thin sheet of white matter, into the lateral putamen and the medial globus pallidus. The globus pallidus is further separated from the entopeduncular nucleus by a medullated partition, the internal medullary lamina, which adds to the alternating gray and white configuration forming the striated area.

2. Fiber Connections:

   a. The fiber connections of the basal ganglia are not completely known. Because of such phenomena as feedback loops and reciprocating connections, the functional significance of known fiber connections is largely obscure.

   b. The afferents to the basal ganglia are from many sources, for example, (1) subthalamus, (2) thalamus, (3) cerebral cortex, (4) brain stem nuclei, and (5) substantia nigra.

   c. The efferents of the basal ganglia are dispersed, and projections include some of the structures cited as sources for the afferent fibers: (1) thalamus, (2) subthalamus, (3) hypothalamus, (4) brain stem nuclei, for example, red nucleus and caudal (inferior) olive, (5) substantia nigra, and (6) reticular formation.

Functions:

   a. There is evidence that generally the caudate nucleus and the putamen exert an inhibitory influence on somatic motor activity, whereas stimu-

lation of the globus pallidus produces an excitatory effect resulting in hypertonus; prolonged stimulation can produce tremor on the opposite side.

   b. Generally lesions of the basal ganglia result in abnormal gross bodily movements (*e.g.,* circling, rolling, or choreiform movements). Discrete unilateral lesions in the rostroventral region of the caudate nucleus in the cat produce permanent athetoid and choreiform hyperkinesias similar to those seen in man, although large generalized destruction of the caudate nucleus does not produce these symptoms.

   c. In the cat, bilateral ablation of the putamen results in a hyperactive animal that disregards its environment; bilateral ablation of the globus pallidus produces the opposite behavioral changes, that is, an animal that is hypoactive and somnolent, and exhibits hypotonus.

   d. It appears that pertinent literature offers no support for the statement by Clark (1965) that the basal ganglia "contain the voluntary motor system in the dog." Evidence based on investigations in animals indicates that "the basal ganglia can scarcely be considered as important motor centers" (Brodal, 1969). Brodal considers the cardinal concern of the basal ganglia to be the collaboration between the thalamus and the cerebral cortex via the closed circuit: basal ganglia to thalamus to cerebral cortex.

CEREBELLAR NUCLEI
PLATE 12; FIG. 14-4

Location:

   a. Irregularly shaped masses of gray matter within the medullary white matter of the cerebellum. They extend somewhat horizontally, with poorly defined separations.

I. Lateral (Dentate) Nucleus
1. Location and shape:
   a. This most lateral nucleus has a dorsal and a ventral eminence protruding laterally to give it a somewhat dentate form.
2. Afferents:
   a. Posterior and anterior lobes of the cerebellum via the axons of the Purkinje cells.
3. Efferents:
   a. Red nucleus and thalamus via the rostral (superior) cerebellar peduncle.

II. Nucleus Interpositus
1. Location:
   a. This is the middle gray mass which is as large as the lateral (dentate) nucleus.
2. Afferents:
   a. Paleocerebellum (rostral vermis plus pyramis, uvula, and paraflocculus).
   b. Neocerebellum (hemispheres).
3. Efferents:
   a. Red nucleus via the rostral (superior) cerebellar peduncle.

III. Fastigial Nucleus
1. Location and structure:
   a. A solid round gray mass about as large as the nucleus interpositus immediately lateral to the midline.
   b. Dorsal to the nodulus of the vermis.
2. Afferents:
   a. Paleocerebellum.
   b. Vestibular nuclei.
3. Efferents:
   a. Brain stem nuclei (especially vestibular and reticular).

IV. Special Comments
a. The dentate nucleus in man is a thin convoluted linear gray mass somewhat more dentate in outline.
b. The nucleus interpositus as described above for the dog is not present in man. Instead, this large gray mass is divided into a lateral *emboliform nucleus* close to the dentate nucleus, and a medial *globose nucleus* immediately lateral to the fastigial nucleus. Therefore, instead of the three deep cerebellar nuclei observed in the dog, in man there are four.

CEREBELLAR PEDUNCLES
PLATES 12 TO 19

I. Caudal (Inferior) Cerebellar Peduncle (Restiform body)
1. Location:
   a. Dorsolateral margin of the medulla.
   b. Lateral to the descending (spinal) vestibular nucleus.
   c. Dorsal to the descending root of the trigeminal nerve.
   d. Grossly between the other two peduncles in cross section.
2. Afferents:
   a. Olivocerebellar fibers.
   b. Dorsal spinocerebellar tract.
   c. External (superficial) arcuate fibers.
   d. Vestibulocerebellar fibers (primary and secondary).
   e. Reticulocerebellar fibers.
3. Efferents:
   a. Cerebellovestibular fibers.
   b. Cerebelloreticular fibers.

II. Middle Cerebellar Peduncle (Brachium pontis)
1. Location:

a. Grossly lateral to the other two peduncles at the level of the pons.
2. Afferents:
   a. Directly from nuclei pontis of opposite side to neocerebellum.
3. Function:
   a. Afferent to cerebellum for cortico-ponto-cerebellar system (best developed in man where cerebral influence over cerebellum is greatest).

III. Rostral (Superior) Cerebellar Peduncle (Brachium conjunctivum)
1. Location:
   a. Medial to the other two peduncles.
   b. Connects with the midbrain.
   c. Serves for attachment of the rostral (anterior) medullary velum.
2. Afferents:
   a. Ventral spinocerebellar tract.
3. Efferents:
   From lateral (dentate) and interpositus nuclei to:
   a. Red nucleus
   b. Thalamus and globus pallidus related to cerebral cortex.
   c. Reticular formation.

IV. Special Comments:
   a. In dog all three peduncles are about equal in diameter and grossly are oriented in a medial-lateral relationship.
   b. In man the middle cerebellar peduncle is more than twice the size of each of the others, with the rostral and caudal cerebellar peduncles appearing to be embedded within it.

CAUDAL (INFERIOR) COLLICULUS
PLATES 17 AND 18

1. Location:
   a. Caudal tectum of the mesencephalon.

b. Immediately caudolateral to the rostral (superior) colliculus. (The commissure of the caudal colliculi is longer than that of the rostral colliculi.)
2. Afferents:
   a. Lateral lemniscus.
   b. Caudal colliculus of opposite side via commissure of caudal colliculi.
3. Efferents:
   a. To medial geniculate body via brachium of the caudal colliculus.
   b. Opposite caudal colliculus via the commissure.
   c. Reticular formation via tectoreticular fibers.
   d. To the rostral colliculus.
4. Function:
   a. Auditory relay and reflexes.
5. Special Comments:
   a. Relatively few, if any, tectospinal fibers originate in the caudal colliculus.
   b. The two caudal colliculi plus the two rostral colliculi are collectively referred to as the corpora quadrigemina.

ROSTRAL (SUPERIOR) COLLICULUS
PLATES 19, 20, AND 21

1. Location:
   a. Tectum (roof) of midbrain.
   b. One on each side of the midline, rostromedial to the caudal (inferior) colliculi.
   c. Medial to the medial geniculate body, and dorsomedial to the brachium of the caudal colliculus.
   d. The pineal body lies at the rostral tips of the two colliculi, in the intercollicular groove.
2. Afferents:
   a. From the optic tract (those neurons which do not pass to the lateral geniculate body as origins of the optic radiations).

b. Spinotectal tract from the spinal cord.

c. Opposite rostral colliculus via commissure of the rostral colliculi.

3. Efferents (arise from the deeper layers):

    a. Tectoreticular fibers to the reticular formation of midbrain.

    b. Tectospinal fibers cross to the opposite side immediately ventral to the medial longitudinal fasciculus and descend to cervical and possibly upper thoracic levels in the ventral funiculus of the spinal cord.

    c. Opposite rostral colliculus via the commissure.

    d. Ascending projections to the pretectum and pulvinar.

4. Function:

    a. Center for visual reflexes concerned with involuntary eye movements, including perhaps pupillary adjustments, accommodation, convergence, and blinking.

LATERAL (EXTERNAL, ACCESSORY) CUNEATE NUCLEUS
PLATES 10 AND 11

1. Location: Caudal portion of medulla oblongata.

    a. A small mass of cells lateral to and continuous with the rostral part of the main cuneate nucleus. The accessory cuneate nucleus is bounded laterally by the caudal cerebellar peduncle.

2. Afferents:

    a. Neurons of the first order from the cervical region. These are actually a part of the fasciculus cuneatus.

3. Efferents:

    a. To the cerebellar cortex via the dorsal external arcuate fibers and caudal cerebellar peduncle.

4. Function:

    a. Prioprioceptives (GSA) from the neck region and upper limb to the lateral cuneate nucleus and then to the cerebellum.

5. Special Comments:

    a. Probably this nucleus is more important in lower animals than in man. It is very important in the righting reflex and functions synergistically with and contributes to the vestibular–cerebellar systems.

    b. This nucleus is sometimes regarded as a medullary equivalent of the nucleus dorsalis (of Clarke) of the spinal cord.

MEDIAL CUNEATE NUCLEUS
PLATES 8, 9, AND 10

1. Location:

    a. Caudal region of the medulla oblongata. Lateral to nucleus gracilis. Cuneatus extends farther rostrally than gracilis, that is, the caudal end is lateral to gracilis but the rostral end is lateral to the fourth ventricle and the vagal trigone.

    b. Caudally the medial cuneate nucleus is medial to the nucleus and tract of the descending root of the trigeminal nerve (V).

    c. More rostrally the medial cuneate nucleus is medial to the lateral cuneate nucleus.

    d. Caudally at the level of the pyramidal decussation this nucleus is dorsal to the somewhat diffuse fasciculi of the lateral corticospinal tract as they enter their decussation.

    e. More rostrally it is dorsal to the internal arcuate fibers within the reticular formation.

    f. Grossly, the fasciculus cuneatus can be followed from the dorsal funiculus of the spinal cord lateral to the fasciculus gracilis rostrally to the obex, where the cuneate nucleus and the lateral cuneate nucleus form a nodular protuberance, the cuneate tubercle.

2. Afferents:

    a. Neurons of the first order have their

cell bodies in the dorsal root ganglion. As part of the dorsal root these neurons enter the medial division of the dorsal root of the spinal nerves in the upper thoracic and cervical region (brachial plexus).

b. Upon entering the spinal cord, collaterals are given off from neurons of the first order to form the fasciculus cuneatus in the dorsal funiculus of the spinal cord and ascend directly to the cuneate nucleus within the medulla.

3. Efferents:

a. Axons from the cuneate nucleus follow the same course as those from the nucleus gracilis; that is, myelinated fibers of the second order sweep ventromedially as the internal arcuate fibers to cross over and enter the *medial lemniscus* of the opposite side.

4. Function:

a. Receives neurons of the first order from spinal nerves conveying primarily proprioceptive impulses from the forelimbs and upper thoracic region which ultimately reach the cerebral cortex for conscious proprioception (see Fig. 10-9, p. 161).

FACIAL (MOTOR) NUCLEUS AND NERVE (SVE)
PLATES 13, 14, AND 15

1. Location:

a. Directly caudal to the dorsal nucleus of the trapezoid body (superior olive) in the ventrolateral area of the reticular formation of the upper medulla. (In man, it is dorsal and somewhat lateral to the superior olive at the level of the facial nerve.)

b. Notice that the functional component of this nucleus (SVE) is the same as that of the nucleus ambiguus, which

is nearly in line caudally, also within the reticular formation.

2. Afferents:

a. The visceral afferent neurons of the facial nerve (GVA and SVA) both have cell bodies within the geniculate ganglion. These neurons terminate within the nucleus solitarius, from which internuncial neurons pass to the motor nucleus for reflexes.

b. Secondary and direct collaterals from the spinal nucleus of cranial nerve V for corneal and other trigeminofacial reflexes.

c. Superior olive and acoustic nuclei for acousticofacial reflexes (*e.g.,* closing the eyes when hearing a loud noise).

d. Fibers from the rostral colliculi for reflexes such as closing the eyes on seeing an approaching object.

e. Corticobulbar (corticonuclear) neurons from higher centers.

3. Efferents:

a. Axons from the cell bodies in the facial nucleus pass dorsomedially around the dorsal side of the abducens nucleus from medial to lateral as the internal genu of the facial nerve. The fibers pass ventrolaterally between the spinal tract of V and its own nucleus to make their exit from the brain for motor innervation of the muscles of facial expression. These muscles are embryonically derived from the second branchial arch and therefore are classified as branchiomeric, which are innervated by SVE neurons.

4. Special Comments:

a. What is commonly referred to as the 'facial nucleus' is the motor nucleus discussed above and classified as SVE. There are five other functional components of the facial nerve with nuclei of special names, some of which are shared with other nerves: GVE, parasympathetic nucleus of facial nerve (rostral salivatory nucleus); GVA and SVA (share the nucleus solitarius and its tract); GSA exteroceptive (spinal nucleus and tract of the trigeminal nerve); and GSA proprioceptive (probably the mesencephalic nucleus of the trigeminal nerve).

NUCLEUS GRACILIS
PLATES 8, 9, AND 10

1. Location:
    a. Caudal region of the medulla oblongata. Lateral to the dorsal median septum.
    b. Medial to the medial cuneate nucleus and the dorsal intermediate sulcus.
    c. Caudolateral to the vagal trigone and obex, extending from the obex caudally through the pyramidal decussation.
2. Afferents (Neuron cell bodies in the dorsal root ganglia):
    a. Axons enter the spinal cord via the medial division of the lumbosacral plexus dorsal roots and form the *fasciculus gracilis* in the dorsal funiculus of the spinal cord, as ipsilateral ascending axons of the first order which terminate in the nucleus gracilis within the medulla by synapsing with neurons of the second order.
3. Efferents (Neurons of the second order arising within the nucleus gracilis):
    a. These axons join those from the medial cuneate nucleus to form the *internal arcuate fibers* which decussate within the reticular formation to give rise to the *medial lemniscus,* which ascends on the opposite side to reach the thalamus and synapse with neurons of the third order.
4. System Concerned and Pathway:
    a. Primarily conscious kinesthetic sense (*i.e.,* sense of position and movement).
    b. The pathway is: Specialized nerve endings (*e.g.,* Pacinian corpuscles in subcutaneous area and free nerve endings in joint capsules) → dorsal root medial division → fasciculus

gracilis → nucleus gracilis (medulla, synapse with neuron II) → internal arcuate fibers (decussate) → medial lemniscus → thalamus (synapse with neuron III) → internal capsule → somesthetic area of cerebral cortex (see Fig. 10-9, p. 161).
5. Special Comments:
    In man, the fasciculi gracilis and cuneatus also transmit impulses of fine discriminatory touch, vibratory sense, and two-point tactile sense—which are not tested in animals during routine neurologic examinations, for obvious reasons.

HABENULA
PLATE 22

1. Structure and Location:
    a. A component of the epithalamus. Therefore dorsal to (upon) the thalamus.
    b. Two rounded eminences bordering the lateral edge of the third ventricle near its roof in the area of the pulvinar.
    c. At the caudal termination of the stria medullaris thalami.
    d. The right and left habenulae are joined by the habenular commissure.
    e. Each habenula consists of medial and lateral nuclei.
2. Afferents:
    a. Mainly via the stria medullaris thalami: possibly from hippocampus, olfactory area, septal area, amygdala, thalamus, hypothalamus, and tectum.
3. Efferents:
    a. Major efferent: the habenulopeduncular tract (fasciculus retroflexus of Meynert) to the interpeduncular nucleus.
    b. Other efferents: those to the tegmentum and tectum of the midbrain appear to be the most definite.
4. Functions:
    a. The functions are very poorly understood, but based on its connections it is considered to be an olfactory relay

station in olfacto-somatic correlations.

b. It seems to influence pituitary content and blood levels of TSH (Szentago-thai *et al.,* 1968).

c. It may be implicated with CNS control of body temperature and metabolism.

HYPOGLOSSAL NUCLEUS AND NERVE (SE)
PLATES 10 AND 11

1. Location:
   a. Near the midline in the central gray matter in the floor of the fourth ventricle.
   b. Most of this nucleus lies rostral to the nucleus gracilis.
   c. Medial to the vagal trigone, forming part of the hypoglossal trigone.
   d. Fibers from the hypoglossal nucleus pass ventrally to cross the internal arcuate fibers and emerge as several rootlets on the ventral surface dorsal to the caudal olive.
2. Afferents:
   a. Corticobulbar fibers—concerned with voluntary motion of tongue.
   b. Secondary fibers to complete reflexes for taste, pain, temperature, from the following cranial nerves: Vagus, Glossopharyngeal, Facial, and Trigeminal.
   c. From olfactory centers and tracts.
3. Efferents and Function:
   a. SE. This is the only voluntary motor supply to the tongue.
4. Special Comment:
   In man, with the large inferior olivary prominence, the hypoglossal rootlets emerge from the anterolateral (pre-olivary) sulcus between the pyramid and olive (ventral to olive).

HYPOTHALAMIC NUCLEI
PLATES 22 TO 25

1. Anatomic Organization:
   There are three nuclear areas of the hypothalamus in a rostro-caudal sequence based on the gross structures: the rostral, or supraoptic, area, which lies above the optic chiasm and fuses with the preoptic area; the intermediate, or tuberal, area, corresponding to the tuber cinereum; and the caudal hypothalamic, or mammillary, area, which is continuous caudally with the midbrain tegmentum. The specific nuclei within these areas are poorly defined.
   a. *The rostral, or supraoptic, area* contains two functionally well-known nuclei, the *supraoptic* and the *paraventricular.* The former is associated anatomically with the optic tract and the latter with the third ventricle. The *periventricular* (Gk. *peri-,* around) *nucleus is* in the wall of the third ventricle and is not restricted to any of the three rostro-caudal areas. The periventricular system connects the periventricular hypothalamic nucleus with the midline nuclei of the thalamus. The *rostral hypothalamic nucleus* infiltrates the preoptic area.
   b. *Intermediate, or tuberal, area.* Near the infundibular attachment to the tuber cinereum the ventral portion of the periventricular nucleus is indistinguishable from the arcuate nucleus, and both names are used collectively for the nuclear mass at the base of the third ventricle in this area (Plate 24).
   c. The *caudal hypothalamic, or mammillary, area* contains the mammillary bodies as the most prominent structures of this area (Plates 22 and 23). The area directly dorsal to the mammillary bodies is the *caudal hypothalamic nucleus.*
2. Major Afferent and Efferent Hypothalamic fibers:
   a. Afferents: from the hippocampus via the fornix, the periventricular system transmitting impulses from the thalamus, and from the retina via the optic tract.
   b. Efferents: The mammillary body gives

rise to the *mammillothalamic tract,* which extends to the anterior nuclear group of the thalamus, and the *mammillotegmental tract,* which courses caudally to reach the tegmentum of the midbrain. From the supraoptic and paraventricular nuclei axons form the *supraopticohypophyseal tract,* which passes to the neurohypophysis. At the base of the infundibulum in the area of the median eminence the supraopticohypophyseal tract is joined by the *tuberohypophyseal tract* from the tuberal area, to form the *hypothalamo-hypophyseal tract,* which extends to the neurohypophysis.

3. Hypothalamic Functions:

The hypothalamus strongly influences the physiologic activity of two important regulatory mechanisms, the autonomic nervous system and the pituitary gland.

   a. Hypothalamic–autonomic nervous system relations:

     (1) In general terms, the rostromedial hypothalamic area is parasympathetic dominant, whereas the caudolateral region is sympathetic dominant.

     (2) Stimulation of the caudolateral hypothalamic area produces sympathomimetic responses characteristic of emotional stress and anger similar to those in sham rage.

     (3) Destruction of the caudal hypothalamic area results in emotional lethargy, abnormal sleepiness, and fall in body temperature due to reduced visceral and somatic activity.

     (4) The ventromedial hypothalamic nucleus in the tuberal area is concerned with *satiety,* whereas the lateral hypothalamic area is regarded as a *feeding center.*

     (5) The hypothalamus regulates body temperature. The caudal hypothalamus is stimulated by low environmental temperature. Therefore this area is concerned with the conservation and the increased production of body heat. Lesions in this area produce a poikilothermic animal. The rostral hypothalamus (especially the preoptic area) is stimulated by increased environmental temperature. Therefore this area is concerned with dissipation of body heat. Localized lesions in this area result in hyperthermia (hyperpyrexia).

   b. Hypothalamic-pituitary relations

     (1) The supraoptic and paraventricular nuclei and the hypothalamohypophyseal tract are associated with the neurohypophysis in the production and transmission of two hormones: antidiuretic hormone (ADH) and oxytocin.

     (2) Rather than direct neural connections as for the neurohypophysis, there is a neurovascular mechanism for hormonal activity of the anterior pituitary. The hypothalamus produces an activating agent, a *releasing factor (neurohormone)* for each of the anterior pituitary hormones except prolactin, which has an inhibitory factor (PIF).

INTERPEDUNCULAR (INTERCRURAL) NUCLEUS
PLATES 20 AND 21

1. Location: Midbrain.

   a. In the midline between the substantiae nigrae of the cerebral peduncles.

   b. Ventral to the decussation of the rostral cerebellar peduncles.

   c. Dorsal to (at the apex of) the interpeduncular fossa.

2. Afferents:

   a. From the habenular nucleus as the habenulopeduncular tract or fasciculus retroflexus of Meynert.

   b. From the mammillary bodies.

3. Efferents:
   a. To the dorsal tegmental nucleus within the central gray matter dorsal to the MLF and the trochlear nucleus. The dorsal tegmental nucleus also receives fibers from the mammillary bodies as the *mammillotegmental tract.*
4. Function:
   a. This nucleus is a way-station in discharges from the hypothalamic olfacto-visceral and epithalamic olfacto-somatic correlation centers to the midbrain tegmentum.
5. Special Comments:
   This nucleus is relatively larger in carnivores than in man.

LATERAL LEMNISCUS
PLATES 16, 17, AND 18

1. Location:
   a. In the ventrolateral border of the pons lateral to the trapezoid body.
   b. Within the midbrain, a large fasciculus of lateral lemniscal fibers enters the ventrolateral border of the caudal colliculus and the brachium of the caudal colliculus.
2. Function and Course:
   a. Carries auditory impulses to higher centers as described under trapezoid body.
   b. The lateral lemniscus continues rostrally from the dorsal nucleus of the trapezoid body (superior olive) and the trapezoid body. Most of the fibers of the trapezoid body arise in the ventral cochlear nucleus and do not synapse in the trapezoid nuclei.
   c. Fibers of the lateral lemniscus enter the caudal colliculus where they may: (1) terminate by synapses for reflexes, (2) synapse with neurons projecting into the brachium of the caudal colliculus, or (3) pass through the colliculus without synapse to enter the brachium to the medial geniculate body.

LIMBIC SYSTEM
PLATES 19 TO 25

1. General Comments and Anatomic Relations:
   a. The limbic system is difficult to define. One reason is that various authors include different structures and areas in their definitions of limbic lobe and system. The consideration here is uncomplicated and basic, but seems to be adequate for study of the system.
   b. The limbic system is referred to as the 'visceral brain.' The association is with special visceral afferent (SVA) olfaction, and the general visceral efferent (GVE–autonomic nervous system).
   c. The literal meaning of the word 'limbic' is 'border,' in reference to the fact that the limbic lobe borders the rostral end of the brain stem.
   d. The limbic lobe consists of: the hippocampal, cingulate, and subcallosal gyri, plus the deeper hippocampus and dentate gyrus.
   e. The limbic system consists of the limbic lobe as defined in *d*, plus associated subcortical nuclei (*e.g.*, amygdaloid complex, hypothalamus, epithalamus, septal nuclei, and anterior thalamic nuclear area).
   f. The *Papez circuit* is a continuous neuronal pathway involving structures of the limbic system as follows: hippocampus to fornix to mammillary bodies to anterior nuclei of the thalamus to cingulate gyrus and back to the hippocampus.
2. Functions:
   a. The limbic system is in a position to receive and associate olfactory, visceral, oral, sexual, and basic sensory impulses—for example, olfactory (SVA), optic (SSA), auditory (SSA), exteroceptive (GSA), and interoceptive (GVA)—and project them to the hypothalamus. This phylogenetically

old system is connected to the neo-cortex.

b. The limbic system is involved with emotional and behavioral patterns. This is primarily because of its relations with the hypothalamus, which regulates the autonomic nervous system.

c. The Papez circuit referred to above was proposed as the basis for emotion underlying the psychic expression of the activity of neocortical areas. This phenomenon applies to animals as well as man. (See Chapter 16 of text for elaboration.)

d. Stimulation of the hippocampal formation results in cardiovascular and respiratory changes plus a general arousal response.

e. The Klüver-Bucy syndrome illustrates the significance of some limbic structures in emotional and behavioral qualities. This syndrome results from bilateral ablation of the rostral temporal lobe, including parts of the hippocampal formation and the amygdala. Some of the changes are: wild and aggressive monkeys become docile and show no emotional anger or fear. Dietary changes include excessive eating of unusual foods for monkeys (*e.g.,* fish and meat). Sexual aberrations include: attempted interspecies mating, and autosexual, homosexual, and excessive heterosexual activities.

f. Both 'pleasure centers' and 'punishing centers' have been localized within the limbic system.

MEDIAL LONGITUDINAL FASCICULUS (MLF)
PLATES 11, 15 TO 20; FIGS. 10-7 (p. 157) AND 12-4 (p. 189)

1. Location:
   a. From the level of the oculomotor nucleus and the rostral colliculi caudally into the lumbar segments of the spinal cord (dog).

   b. Adjacent to the midline at the dorso-lateral edge of the ventral median sulcus of the spinal cord.

   c. In the caudal medulla the MLF is ventromedial to the hypoglossal nucleus, and dorsal to the tectospinal fasciculus.

2. Afferents:
   a. All four vestibular nuclei on each side project axons medially to enter the MLF. These vestibular fibers are crossed and uncrossed, as well as ascending and descending. This afferent path is: labyrinth → vestibular nuclei → MLF.

   b. The MLF conveys indirect cerebellar influence by the afferent path: cerebellar cortex → fastigial nuclei → vestibular nuclei → MLF.

3. Efferents and Function:
   a. Ascending vestibular fibers (crossed and uncrossed) synapse with neurons in nuclei of cranial nerves III, IV, and VI, which innervate the extrinsic eye muscles to account for eye deviations and nystagmus resulting from abnormal vestibular stimulation.

   b. Intact secondary vestibular fibers in the MLF are essential for normal conjugate eye movements.

   c. Descending branches to the nucleus of cranial nerve XI and to motor neurons in the ventral horn of the upper segments of the spinal cord.

   d. Descending fibers, almost exclusively from the medial vestibular nucleus, give branches to the reticular formation and reach visceral motor and autonomic nuclei to contribute to reflex symptoms that occur with excessive labyrinthine stimulation (*e.g.,* nausea, vomiting, and vasomotor changes).

OCULOMOTOR NUCLEUS AND NERVE
PLATES 20 AND 21

1. Location:

a. Central gray matter of mesencephalon (ventral portion).

b. Ventral to aqueduct at the level of the rostral colliculus.

c. Dorsal to red nucleus.

2. Afferents:

a. Collaterals from the medial longitudinal fasciculus (MLF), mostly from the vestibular nuclei.

b. Reticular formation (which may relay corticobulbar fibers).

c. Direct corticobulbar fibers.

3. Function and Efferents:

a. Somatic motor for cranial nerve III.

b. Axons of III arch ventrolaterally through the medial portion of the red nucleus to emerge at the base of the midbrain.

c. This nerve passes rostrally lateral to the pituitary within the cavernous sinus to traverse the orbital fissure and enter the orbit.

d. Most of the extrinsic ocular muscles are innervated by III, namely, dorsal, medial, and ventral recti muscles and the ventral oblique muscle. In addition, this nerve innervates the levator palpebrae muscle.

e. The GVE preganglionic parasympathetic fibers of the oculomotor nerve arise in the Edinger-Westphal nucleus, located at the dorsomedial edge of the oculomotor (SE) nucleus. These synapse in the ciliary ganglion from which postganglionic neurons pass to the intrinsic ocular muscles (see Fig. 9-1, p. 131). Primary parasympathetic effect is pupillary constriction (miosis).

THE (INFERIOR) OLIVARY
NUCLEAR COMPLEX
PLATES 10, 11, AND 13

1. Location and Subdivisions:

a. Within the medulla rostral to the pyramidal decussation at the dorsolateral border of the pyramids.

b. The complex is composed of three bandlike nuclei (in the dog and cat): (i) the *principal olivary nucleus,* which is bordered on the medial side by (ii) the *medial accessory olivary nucleus* adjacent to the pyramid, and on the dorsolateral side by (iii) the *dorsal accessory olivary nucleus.*

c. In serial cross sections proceeding from caudal to rostral, the medial accessory olivary nucleus appears first.

2. Afferents:

a. Mostly supra-olivary descending connections from: cerebral cortex, caudate nucleus, globus pallidus, reticular formation, red nucleus, periaqueductal gray matter, and deep tegmental nuclei. Afferent fibers from the last three compose the *central tegmental tract.*

3. Efferents:

a. The main efferent fibers from the olivary complex are the olivocerebellar fibers which pass through the ventrolateral reticular formation to the contralateral cerebellum via the caudal cerebellar peduncle.

b. The existence of the olivospinal tract reported in many texts is not substantiated by experimental studies (Brodal, 1969).

4. Comments:

a. In man the inferior olivary complex consists of the same subdivisions as in the dog and cat, but they are much larger with more prominent fiber connections. In the human brain the inferior olive is grossly obvious as a nodular swelling on the ventrolateral side of the medulla, bordered ventrally and dorsally by well-defined pre-olivary and post-olivary sulci, respectively. Internally in the human medulla the myelinated axons which terminate in the olive form a heavy capsule, the *amiculum,* which surrounds the extremely large principal olivary nucleus.

b. Nomina Anatomica Veterinaria (1968) gives *nucleus olivaris* as the proper term, ". . . but 'inferior' is no longer necessary because the term Nucleus olivaris superior has been changed to Nucleus dorsalis corporis trapezoidei." 'Inferior' has been retained in this text because of the popular usage.

RED NUCLEUS
PLATES 20 AND 21

1. Location:
   a. Rostral collicular level of the mesencephalon. A rostral extension penetrates the subthalamus.
   b. Ventral to the motor nucleus of cranial nerve III. Emerging oculomotor fibers cross the medial side of the nucleus.
   c. Dorsal to the substantia nigra.
   d. Medial to the medial lemniscus.
2. Histologic Structure:
   a. Rostral portion—small cells (parvocellular), phylogenetically new, best demonstrated in man.
   b. Caudal portion—large cells (magnocellular), phylogenetically old, best demonstrated in lower animals (including the dog and cat).
3. Afferents:
   a. From cerebellar nuclei (lateral, or dentate, nucleus, and nucleus interpositus) via the rostral cerebellar peduncle, as a major route of discharge from the cerebellum.
   b. Corticorubral fibers from the cat's 'motor cortex' (corresponding approximately to the precruciate gyrus) to the red nucleus. These fibers are somatotopically arranged to terminate on the corresponding somatotopic organization of the rubrospinal cell bodies within the red nucleus. This may be considered an indirect corticospinal pathway with a relay in the red nucleus (Fig. 10-10, p. 166).
   c. From the corpus striatum.
4. Efferents:
   a. Rubrospinal tract. Fibers leave the red nucleus to cross immediately and descend within the brain stem and lateral funiculus of the spinal cord. In cat the fibers have been traced to lumbosacral levels (Hinman and Carpenter, 1959).
   b. Rubrobulbar tract. Fibers project from the red nucleus to the branchiomeric motor nuclei of the cranial nerves.
   c. Rubroreticular tract. Fibers pass mostly from the rostral portion of the red nucleus to be dispersed to nuclei in the reticular formation.
   d. Rubrothalamic fibers pass from the red nucleus to the thalamus as a segment of the cerebello-rubro-thalamic-cortical pathway given below.
5. Functions:
   a. Important way-station for proprioceptive impulses from cerebellum to cerebral cortex (pathway is cerebellum → red nucleus → thalamus → cerebrum).
   b. Clark (1965) considers the rubrospinal tract to be the most important motor pathway in animals. A review of the literature, however, does not seem to reveal full agreement with his view that voluntary motor paralysis in animals is a direct result of a lesion that severs the rubrospinal tract.
   c. The discharge to the red nucleus from the corpus striatum gives evidence of its connection with coordination of movements.
   d. There are conflicting reports concerning relation of the red nucleus to tonic and righting movements, that is, complex postural reflexes involved in changing body positions.
   e. Bilateral discrete lesions in the red nucleus (monkey) result in hypokinesis. The voluntary motor system is not impaired, but animals show disinclination to move.
   f. Stimulation of the red nucleus in the cat produces flexion in either the forelimb or the hindlimb of the opposite side.

THE RETICULAR FORMATION
(SUBSTANCE)
PLATES 9, 10, 11, 16, 19, 22, 23, AND 24

1. Structure and Location:
   a. As the name indicates, the reticular formation (RF) is a network-like structure which is generally considered to be anatomically poorly defined in that it has no sharp boundaries, nor is it composed of a single type of cell body or neuron process.
   b. It is a continuous meshwork of cell bodies and processes extending from the corpus striatum caudally throughout the brain stem and the spinal cord.
   c. It is a mixture of gray and white matter, ascending, descending, and transverse fibers, and sensory and motor neurons which pass on the same side or cross to the other.
   d. A retention of the primitive neuronal linkage is evident within the RF by the predominance of short-axon multisynaptic intrinsic paths.
   e. The RF serves as a matrix in which there are embedded specific brainstem pathways and nuclei of termination and origin for cranial nerves and other fibers associated with pathways connecting various parts of the brain and spinal cord. Some cell bodies in the reticular formation aggregate to form specific nuclei which are called reticular nuclei (Plate 9). In the spinal cord the RF lies close to the gray matter.
   f. Axons of neurons in the RF may project from the caudal medulla to the rostral thalamus or striatum, but a single neuron may produce a variety of anatomic patterns, that is, as axons penetrate through the RF they give rise to many collaterals which pass to various nuclei and generally do not join specific ascending pathways.
2. Functions:

   a. The RF acts to govern the behavior of an animal (including man). It is likened to an alarm for the cerebral cortex in the sense that the RF is the center of arousal for the cerebral cortex; that is, the cerebral cortex must be awakened and made ready for the sensory messages that are relayed from the thalamus, and for initiation of descending messages from the cortex. In this context the title 'reticular activating system' (RAS), which is more appropriate, is used.
   b. Related to the above concept, the RF serves as a 'general alarm' for the cortex in that it arouses the whole cortex rather than a specific receptive area for specific incoming messages for restricted areas. The specific ascending pathways, for example, the spinothalamic, send collaterals into the RF as they proceed to the thalamus and before entering the thalamic radiation.
   c. Somnolence, stupor, and coma represent different degrees of cerebral cortical suppression. Hypoactivity of the RF is usually thought to be directly responsible for these conditions and a destructive lesion of the RF may result in a state of coma for the animal (or man) from which recovery is impossible.
   d. Within the RF of the medulla there are scattered cells which constitute the respiratory and vasomotor centers. The existence of specific separate respiratory areas for initiation of expiration and inspiration is not substantiated by different authors.
   e. Besides the functions mentioned above, among many others the RF acts as a regulator for voluntary and reflex motor reactions. Certain areas of the RF are excitatory, whereas other areas are inhibitory, for example, in general the bulbar reticular area is inhibitory for extensor muscles and the lateral

reticular area functions as a facilitatory area for the same muscles.

SOLITARIUS FASCICULUS AND NUCLEUS
PLATES 11, 12, AND 13

1. Location:
   a. Deep in the ventrolateral floor of the fourth ventricle of the medulla. Indistinct in the area of the pons.
   b. Lateral to the nucleus of XII.
   c. Ventral to the medial and descending vestibular nuclei.
2. Afferents:
   a. Taste (SVA) from the anterior two-thirds of the tongue via VII.
   b. Taste from the posterior one-third of the tongue via IX.
   c. Taste from the root of the tongue via X.
   d. GVA from the thoracic and abdominal viscera innervated by X.
   e. GVA from carotid sinus (IX).
3. Function:
   a. SVA (taste) from cranial nerves VII, IX, and X.
   b. GVA from VII, IX, and X.
4. Special Comments:
   a. The fasciculus solitarius represents a descending bundle of visceral afferents, whereas the spinal tract of V constitutes a descending bundle of somatic afferents.
   b. The rostral portion of the nucleus solitarius, which receives SVA (taste) fibers from VII and IX, is also called the *gustatory nucleus.*

DORSAL SPINOCEREBELLAR TRACT (DIRECT CEREBELLAR TRACT)
PLATES 5 TO 11

1. Location:
   a. Extends through the spinal cord and medulla to the caudal cerebellar peduncle. This tract increases in size as it ascends.

   b. Near the surface of the lateral funiculus of the spinal cord.
   c. Lateral to the lateral corticospinal and rubrospinal tracts.
   d. Ventral to the entrance of the dorsal roots of the spinal nerves.
   e. Dorsal to the ventral spinocerebellar tract. This relationship extends through the lateral region of the medulla.
2. Function:
   Proprioceptives from the lower trunk and hindlimbs to the cerebellum.
3. Origin:
   This tract is composed of neurons of the second order which arise in the nucleus dorsalis (Clarke's nucleus).
4. Route of Impulses:
   Receptors (proprioceptive) → Neuron I (cell bodies in dorsal root ganglion) → Medial division of dorsal root of spinal cord levels T1 to L3 or L4 (cat) → Dorsal nucleus of Clarke (synapse with neuron II) → Dorsal spinocerebellar tract (ipsilateral) → Caudal cerebellar peduncle → Hindlimb and lower trunk regions of cerebellar cortex (Fig. 14-7, p. 211).
5. Comparison of the dorsal nucleus of Clarke and the lateral cuneate nucleus:

The cervical equivalent of the dorsal nucleus of Clarke is the external (accessory or lateral) cuneate nucleus (Plate 10). The pathway for the upper body is as follows:
   Receptors for spinal nerves C1 to T4 or T5 (cat) → Lateral cuneate nucleus → Cuneocerebellar tract → Caudal cerebellar peduncle (ipsilateral) → Forelimb, neck, and upper trunk regions of cerebellar cortex (Fig. 14-7, p. 211).

VENTRAL SPINOCEREBELLAR TRACT
PLATES 5 TO 11, 14, AND 15

1. Location and Course:
   The ventral spinocerebellar tract arises

from ipsilateral and contralateral dorsal horn cells in the lumbar segments of the spinal cord in the cat. Therefore, this tract transmits impulses which arise in the hindlimbs and lower trunk only. The tract is located at the periphery of the lateral funiculus of the spinal cord ventral to the dorsal spinocerebellar tract and lateral to the lateral spinothalamic tract (Fig. 10-7, p. 157).

In the medulla this tract ascends ventral to the spinal tract of the trigeminal nerve, continues to the level of the midbrain, where it enters the rostral cerebellar peduncle (Plate 15), which it follows to the cerebellum, and there terminates in the hindlimb areas of the cortex.

2. Function:
   Similar to the dorsal spinocerebellar tract, the ventral spinocerebellar tract transmits proprioceptive impulses (not conscious) from skeletal muscles, tendons, and ligaments of the caudal body area to the cerebellum.

3. Special Comments:
   Clinically, cerebellar influences are ipsilateral, and proprioceptive projections from spinal cord levels are ipsilateral. Those fibers which decussate in the cord cross again within the cerebellum, so that they too conform to the ipsilateral dominance (see Chapter 14).

TECTOSPINAL AND TECTOBULBAR TRACTS
PLATES 11, 15, AND 17

1. Location and Origin:
   a. The tectospinal and tectobulbar tracts are intimately related longitudinal bundles extending caudally from the rostral colliculus of the midbrain. The tectospinal tract descends to the lower cervical spinal cord, but on its way the tectobulbar fibers are given off primarily to motor nuclei of cranial nerves.

   b. The tectospinal fibers are best demonstrated in the medulla near the median raphe between the medial longitudinal fasciculus (MLF) dorsally and the medial lemniscus ventrally. At this level of the brain stem the bulbar portion of the tract may be referred to as the tectomedullary tract.

   c. The primary source of fibers is the rostral colliculus (optic reflex center), but perhaps other fibers are added from the caudal colliculus (auditory reflex center).

2. Function:
   a. To mediate reflex postural movements in response to visual and auditory stimuli (e.g., blinking at loud sounds and pricking up of the ears in animals).

3. Pathway:
   a. The cell bodies in the rostral colliculus give rise to axons which cross the midline of the mesencephalon as the *dorsal tegmental decussation* and descend within the brain stem as described above.

   b. The tectobulbar fibers leave the parent bundle to terminate by synapses with motor centers (e.g., with cranial nerve nuclei); the tectospinal fibers descend to the ventral funiculus of the (cervical) spinal cord.

THALAMIC NUCLEI
PLATES 22 TO 25

There are five groups of nuclei, based on their relative topography: midline, rostral (anterior), medial, lateral, and ventral. The internal medullary lamina divides the thalamus into the first three groups.

1. *Midline Nuclei:*
   a. Phylogenetically old with sensory relays through the reticular formation and rhinencephalon; efferents mainly to hypothalamus, basal ganglia, and amygdala.

   b. Connections with the hypothalamus

via periventricular system suggest involvement with visceral activity. An obscure function is the activation of the cerebral cortex in a vague general way.

 c. The most conspicuous nucleus is the *nucleus centralis medialis,* located directly in the midline between the right and left internal medullary laminae. Less distinct nuclei are: commissural, associated with the massa intermedia, and paraventricular and periventricular, related to the third ventricle.

2. *Rostral (Anterior) Group of Nuclei:*

 a. This group composes the rostral or anterior tuberculum at the extreme rostrodorsolateral thalamic pole separated from the caudate nucleus by the stria terminalis.

 b. This gray mass consists of three nuclei: the *rostrodorsal nucleus* as a dense cellular cap dorsal to the other two nuclei lateral to the stria medullaris; the *rostroventral nucleus* lies deep to the rostrodorsal nucleus and extends laterally as the largest of the three rostral nuclei; the *rostromedial nucleus* is in the deep ventromedial area of the rostral tuberculum beneath the other two nuclei.

3. *Medial Group of Nuclei:*

 a. According to Rioch (1929) there are nine nuclei in this group. Most, however, are poorly defined and their functional significance is questionable. The *medial dorsal nucleus* is the largest and includes most of the area between the internal medullary lamina and the midline nuclei.

4. *Lateral Group of Nuclei:*

 a. This group forms the dorsolateral area of the thalamus throughout its length between the internal and external medullary laminae. The most caudal portion of this group is the *pulvinar,* which is very poorly developed in the dog and cat when compared with that in man.

 b. The *reticular nucleus* is a thin gray area wedged between the external medullary lamina and the concave medial side of the internal capsule.

5. *Ventral Group of Nuclei:*

 a. This gray mass occupies the ventro-(lateral) area of the thalamus. The general boundaries are: dorsally, the lateral nuclear group; medially, the midline nuclei; rostrally, laterally, and ventrally, the external medullary lamina; caudally and laterocaudally, the medial geniculate body.

 b. The ventral nuclear group is subdivided into quadrants, each considered in general terms as a nucleus: the ventral rostromedial and rostrolateral nucleus, and the ventral caudomedial and caudolateral nucleus.

 c. The ventral caudomedial nucleus receives fibers conveying general somatic afferent (GSA) exteroceptive and proprioceptive impulses from the head area and relays these impulses to the sensory cerebral cortex. The ventral caudolateral nucleus is also a relay center for GSA exteroceptive and proprioceptive impulses from the body. This nucleus receives fibers from the lateral and ventral spinothalamic tracts plus the medial lemniscus and relays their impulses to the sensory cortex.

TRAPEZOID BODY AND AUDITORY PATHWAY
PLATE 15; FIG. 19-6 (p. 310)

1. Location and Function:

 a. In the rhombencephalon, transverse fibers forming an elevated band on the ventral surface between the medulla oblongata and the pons proper.

 b. The trapezoid fibers are a portion of the auditory pathway.

 c. Grossly, the abducens nerve emerges from the caudal portion of the trapezoid body near the midline. Laterally within the trapezium the facial nerve emerges directly behind the trigeminal, and the vestibulocochlear nerve is attached laterally immediately ventral to the flocculus of the cerebellum.

2. Afferents:
   a. The axons making up the trapezoid body arise mostly within the ventral cochlear nucleus.

3. Efferents:
   a. There are nuclei within the trapezoid body for synapses of some of the fibers.
   b. The fibers which decussate within the trapezoid body ascend within the lateral lemniscus.

4. Auditory Pathway:
   This is only one of the auditory pathways, and the one that includes the trapezoid body:

   Organ of Corti → cochlear nerve (Cell bodies in cochlear [spiral] ganglion) → ventral cochlear nucleus → trapezoid body (decussate) → superior olive (dorsal nucleus of the trapezoid body) → lateral lemniscus → nucleus of lateral lemniscus → caudal colliculus → medial geniculate body → auditory radiation → Sylvian gyrus (Dog).

5. Special Comments:
   In the human brain the trapezoid body is not evident grossly but is covered ventrally by the basilar portion of the pons (see Fig. 12-2, p. 187).

TRIGEMINAL NERVE COMPLEX

I. Components: Functional
GSA exteroceptive,
GSA proprioceptive,
SVE branchiomotor

1. Ganglion:
   Semilunar (Gasserian) located on the nerve near the brain on the cerebral surface of the petrous bone. This consists of unipolar neurons which give rise to GSA sensory fibers, ex-

cept the proprioceptives, which contribute to all three divisions of the trigeminal nerve. The central fibers from these neurons enter the brain via its main trunk (portio major), to divide into ascending and descending branches.

2. Nuclei:
   a. Sensory:
      Several sensory nuclei within the dorsolateral portion of the brain stem. There is a continuous longitudinal chain of sensory nuclei associated with the trigeminal nerve extending from the substantia gelatinosa of the spinal cord which is continuous cephalically with the spinal nucleus of V (in medulla), which is continuous with the main sensory nucleus (in pons) and extends into the mesencephalic nucleus of V in the midbrain.

      In the human it has been demonstrated that facial pain and temperature sense are handled entirely by the spinal tract and nucleus of V, whereas touch and two-point discrimination are centered in the main sensory nucleus of V, and proprioceptive neurons from the muscles of mastication are thought to have their cell bodies located in the mesencephalic nucleus of V.

   b. Motor:
      The motor (masticatory) nucleus of V is located within the pons medial to the main sensory nucleus of V (Plate 15).

II. Sensory Complex:

1. Spinal nucleus of V (Nucleus of the descending root of V)
   PLATES 7 TO 15
   Location:
   a. Caudal to the entry of nerve V,

extending caudad to cervical levels of the spinal cord.

b. Ventral to the caudal cerebellar peduncle near pons; ventromedial to caudal cerebellar peduncle in medulla.

c. Near the dorsolateral portion of the reticular formation in the medulla.

d. Continuous with the substantia gelatinosa in the dorsal gray column of the spinal cord.

Contents:

a. Multipolar cell bodies (second order) which serve as internuncial neurons between the trigeminal and other nuclei of the brain stem and cord.

GSA Afferents:

a. Descending branches of neurons entering from the semilunar ganglion.

b. It should be noted that, in addition, the spinal nucleus and tract of V also relay pain and temperature sense (GSA) from cranial nerves VII, IX, and X.

Efferents and Connections:

a. Axons pass ventromedially to cross within the reticular formation to enter the contralateral medial lemniscus.

b. Some fibers pass to the cerebellum via the caudal cerebellar peduncle.

c. Similar to the efferents of the main sensory nucleus of V.

2. Main (Pontine, Superior, Principal) Sensory Nucleus of V

PLATE 15

Location:

a. Within the pons medial to the middle cerebellar peduncle, in the lateral border of the reticular formation.

Contents:

a. Multipolar neurons which serve

as internuncials for GSA impulses from the trigeminal nerve. Neurons of the first order have cell bodies within the semilunar ganglion.

Efferents:

a. Axons from this nucleus and the spinal nucleus of V form the *trigeminal lemniscus,* fibers which cross over to join fibers from the medial lemniscus of the opposite side and thereby pass through the pons, midbrain, to enter the thalamus (Meyer, 1964).

3. Mesencephalic Nucleus and Tract of V

PLATES 15 TO 20

Location:

a. A thin column of neurons in the lateral border of the central gray, lateral to the cerebral aqueduct.

b. The caudal limitation is the rostral part of the fourth ventricle.

c. Rostral extent of this nucleus is the rostral limit of midbrain.

Contents:

a. Large unipolar neurons of the first order. This nucleus actually serves the function of a sensory ganglion and is an exception to the rule that sensory neurons of peripheral nerves have their cell bodies in a ganglion.

Afferents and Function:

a. Proprioceptives from muscles of mastication and structures associated with mouth (*i.e.,* masseter, temporal, and pterygoid muscles, hard palate, etc.). Proprioceptive afferents from these structures pass through the mandibular division of V, through the semilunar ganglion without cell bodies until reaching the mesencephalic nucleus of V.

b. According to some authorities, deep sensibility of the ocular, facial, and perhaps the lingual muscles is transmitted by fibers of the mesencephalic root of V.

Efferents and Connections:

a. Short fibers descend to synapse with motor nucleus of V, which completes the reflex for muscles of mastication.

b. Fibers connect with other motor nuclei via the secondary trigeminal lemniscus described under the main sensory nucleus of V.

Reflex Routes:

Receptors in muscles of mastication
↓
Mesencephalic nucleus of V (site of cell bodies of first order neurons)
↓ ↘
         Motor nucleus of V
         ↓
         Muscles of mastication (effectors)
↓
Trigeminal lemniscus
↓
Motor nuclei of brain stem and spinal cord

III. The Motor (Masticatory) Nucleus of V

PLATE 15

Location:

a. Medial to the main sensory nucleus of V.

Function:

a. Nucleus of origin for SVE component of the trigeminal nerve which sends axons through the portio minor with the mandibular division to supply muscles of mastication, plus tensor tympani and tensor veli palatini muscles. It should be noted that this is another example of a visceral component (SVE) supplying skeletal muscle.

Summary

Pathway of Exteroceptive Impulses from Head to Cerebral Cortex

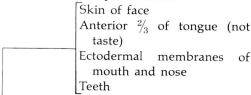

GSA exteroceptives from:
- Skin of face
- Anterior ⅔ of tongue (not taste)
- Ectodermal membranes of mouth and nose
- Teeth

→ Semilunar ganglion (cell bodies for neurons of first order) → Enter brain stem as central fibers of V → Ascending fibers → Chief sensory nucleus (termination of neurons of first order) → Neurons of second order → Trigeminal lemniscus → Thalamus (synapse with neurons of the third order) → Cerebral cortex.

The axons of neurons of the third order that arise in the thalamus enter the internal capsule to participate in the thalamic radiation to reach the cerebral cortex.

Additional Reflexes Utilizing Trigeminal Nerve as Afferent Component

In addition to the direct trigeminal lemniscus, there are collaterals given off to synapse with motor nuclei for the completion of important reflexes. Examples of such reflexes (with only the efferent portion of the reflexes given) are:

(1) Corneal reflex, trigeminal, and facial reflex—Motor nucleus VII.

(2) Tongue reflexes (not taste)—Motor nucleus XII.

(3) Salivary reflexes—Salivatory nuclei VII, IX.

(4) Vomiting reflex—Dorsal efferent nucleus of X, nucleus ambiguus, spinal cord, and motor nucleus of V.

TROCHLEAR NUCLEUS AND NERVE

PLATES 17, 18, AND 19

1. Location:

a. The trochlear nucleus lies within the ventral part of the central (periaqueductal) gray matter of the mesencephalon at the level of the caudal colliculus.

b. This nucleus is very close to the caudal border of the oculomotor nucleus. Both nuclei have the same motor functional component (SE) and therefore are in direct line with each other near the midline.

c. Dorsal to the decussation of the rostral cerebellar peduncle.

2. Function and Course of the Nerve:

a. The axons from the cell bodies in the trochlear nucleus pass dorsomedially to decussate within the rostral medullary velum, and emerge dorsally from the brain stem as the trochlear nerve.

b. The nerve passes from the dorsal surface of the brain stem lateroventrally between the cerebellum and the cerebral hemisphere.

c. The trochlear nerve makes its exit from the skull via the orbital fissure and enters the orbit along with the oculomotor and abducens nerves and the ophthalmic division of V. The trochlear nerve passes directly to the dorsal oblique muscle. This is the only target organ for this nerve, and the trochlear nerve provides the only somatic efferent innervation to this individual muscle.

d. The trochlear nucleus receives afferents from the cortex via the corticobulbar system and is intimately connected with cranial nerves III and VI via the medial longitudinal fasciculus (MLF) for the synergistic action of the extrinsic eye muscles. In addition, the trochlear nucleus is connected caudally via the MLF with the vestibular nuclei (VIII) and with the motor neurons of the upper cervical region of the spinal cord and the accessory nerve (XI) for reflex control of the movement of the head, neck, and eyes.

PARASYMPATHETIC NUCLEUS OF THE VAGUS (X), OR DORSAL EFFERENT NUCLEUS OF VAGUS
PLATE 11

1. Location:
a. Medulla, dorsolateral to hypoglossal nucleus.
b. Medial to tractus solitarius.
c. Somewhat longer than hypoglossal nucleus, and extends farther rostrad and caudad.

2. Afferents:
a. Sensory division of X.
b. Sensory division of IX.
c. Olfactory centers.
d. Medial vestibular nuclei.

3. Function and Efferents:
a. GVE preganglionic parasympathetic neuron cell bodies with axons contributing to the vagus nerve for innervation of thoracic and abdominal viscera. Postganglionics arise from intramural ganglia within the target organs. See Chapter 9 for functions.

VESTIBULAR NUCLEI
PLATES 12, 13, AND 14

1. Names and Location:
a. Four vestibular nuclei on each side of the midline in the floor of the fourth ventricle at the level of the upper medullary–pons junction.
b. The descriptive names of the nuclei indicate their relative positions. Three have commonly used eponyms: the lateral (Deiters'), medial (Schwalbe's), rostral, or superior (Bechterew's); the fourth is the caudal, spinal, or inferior, vestibular nucleus. Not all four can be cut in the same transection, although Plate 13 shows three.

2. Afferents:
a. Direct primary vestibular axons from the vestibular (Scarpa's) ganglion to all four nuclei via the vestibular nerve.
b. Cerebellovestibular fibers from the

cerebellar cortex and fastigial nucleus as a feedback mechanism between the vestibular nuclei and the cerebellum.

c. Proprioceptive impulses from the neck muscles, tendons, and joints permit head movements and position to direct the orientation of the body in space as the primary basic principle in the vestibular righting reflex.

3. Efferents:

a. The lateral vestibular nucleus projects descending axons ipsilaterally to the ventral funiculus of the lumbosacral spinal cord as the *lateral vestibulospinal tract.*

b. All four nuclei send axons medially to enter the medial longitudinal fasciculus (MLF). Descending MLF fibers from the medial vestibular nucleus form the *medial vestibulospinal tract* to the mid-thoracic level of the spinal cord.

c. Secondary vestibulocerebellar axons are relayed from nuclei to the cerebellar cortex.

d. The rostral (superior) vestibular nucleus projects fibers rostrally as a direct *vestibulomesencephalic tract* to the abducens, trochlear, and oculomotor nuclei.

4. Functions:

a. Descending fibers exert a facilitatory effect on spinal reflexes which control muscle tone, especially to maintain appropriate posture or strength of supporting and balancing movements.

b. Equilibrium is largely reflex activity which primarily is governed by general proprioceptive impulses from muscles, tendons, and joints in the trunk and limbs, and special proprioceptive nerve endings in the labyrinth which initiate impulses conveyed via the vestibular nerve to the vestibular nuclei.

c. The physiologic synergism of the vestibulospinal tract and the rhombencephalic reticular formation may be

demonstrated in decerebrate rigidity (see Chapter 11). Such an experimental preparation demonstrates the facilitatory effect of the vestibulospinal and reticulospinal tracts on the spinal cord lower motor neurons of particularly the antigravity (postural and extensor) muscles.

d. Labyrinthine overstimulation, disease, and irritative or destructive lesions may cause: disequilibrium, staggering, postural changes, falling or rolling to the same side, nystagmus, deviation of the eyes, nausea, vomiting, and vasomotor reactions.

Bibliography

Brodal, A.: *Neurological Anatomy in Relation to Clinical Medicine,* 2nd ed. New York, Oxford University Press, 1969.

Clark, C. H.: Basic Concepts in Neurologic Diagnostics. Chapter 2 in *Canine Neurology— Diagnosis and Treatment* (Hoerlein, B. F.). Philadelphia, W. B. Saunders Co., 1965. Pp. 7–24.

Crosby, E. C., Humphrey, T., and Lauer, E. W.: *Correlative Anatomy of the Nervous System.* New York, Macmillan Co., 1962.

Hinman, A., and Carpenter, M. B.: Efferent fiber projections of the red nucleus in the cat. J. Comp. Neurol. *113:* 61–82, 1959.

Meyer, H.: The Brain. Chapter 8 in *Anatomy of the Dog* (Miller, M. E., Christensen, G. C., and Evans, H. E.). Philadelphia, W. B. Saunders Co., 1964. Pp. 480–532.

Nomina Anatomica Veterinaria, Adopted by the General Assembly of the World Association of Veterinary Anatomists, Paris, 1967. Vienna, Adolf Holzhausen's Successors, 1968.

Rioch, D. McK.: Studies on the diencephalon of carnivora. I. The nuclear configuration of the thalamus, epithalamus, and hypothalamus of the dog and cat. J. Comp. Neurol. *49:* 1–119, 1929.

Szentagothai, J., Flerko, B., Mess, B., and Halasz, B.: *Hypothalamic Control of the Anterior Pituitary. An Experimental—Morphological Study,* 3rd ed. Budapest, Akademiai Kiado, 1968.

Part III

Atlas of Central Nervous System of the Dog

Atlas of Central Nervous System of the Dog

The 28 plates in this Atlas are photographs of transverse sections of the central nervous system of the dog taken in caudal-rostral sequence from the sacral level of the spinal cord to the olfactory tract. The sections were prepared by the usual histologic technique used for paraffin embedding (see Chapter 3). The tissue sections were stained with Luxol Fast Blue and counterstained with cresyl violet, according to the procedure of Klüver and Barrera (1953), mentioned in Chapter 3. This staining procedure permits study in the same section of both the white matter (stained with Luxol Fast Blue) and the gray matter (stained with cresyl violet). In the black-and-white plates the heavily myelinated fibers (white matter) are darkest, and the cell bodies (nuclei, or gray matter) are lightest.

Structures shown in the sections are identified by number. The Atlas is designed so that the student can turn the book clockwise to study the plates on the top page and attempt to recognize selected structures. He can then check his answers with the legend on the page directly below the plate. The solid black line in the small inset of the sagittal section of the brain indicates the level and plane of section for each plate from 8 through 28. The enlargement for each plate is indicated in the legend.

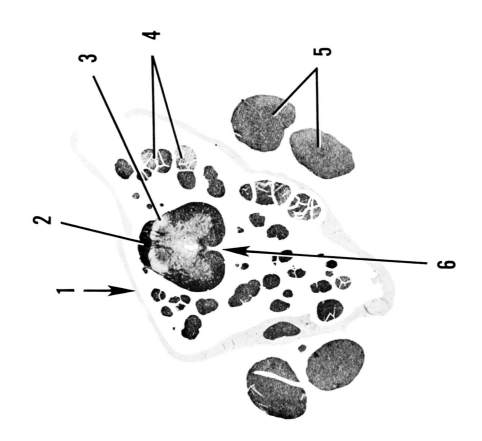

PLATE 1

Transverse section at sacral level of spinal cord.
× 15.

1. Dura mater
2. Dorsal funiculus of white matter
3. Substantia gelatinosa
4. Nerve roots
5. Ventral divisions of spinal nerves (outside dura mater)
6. Ventral median fissure

Characteristics of sacral level of spinal cord:

a. Small overall size.
b. Reduced amount of white matter.
c. Many nerve roots are descending inside the dura, entering into the formation of the cauda equina.

349

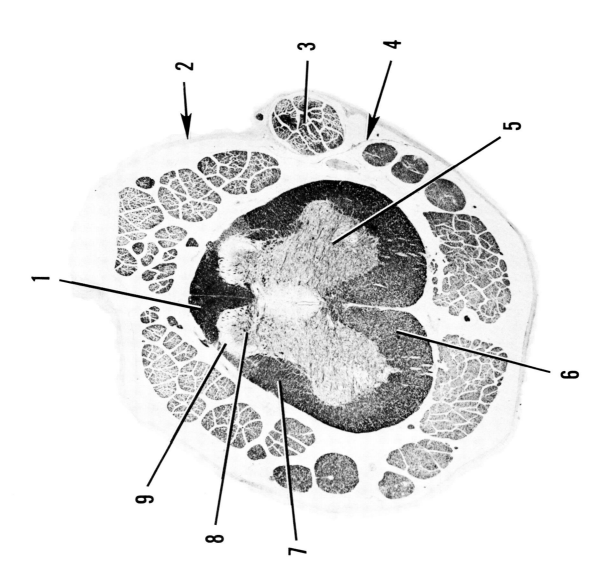

PLATE 2

Transverse section at low lumbar level of spinal cord. ×15.

1. Dorsal funiculus (fasciculus gracilis)
2. Dura mater
3. Nerve roots (descending to cauda equina)
4. Arachnoid
5. Ventral horn (GSE cell bodies)
6. Ventral funiculus
7. Lateral funiculus
8. Dorsal (sensory) horn of gray matter
9. Substantia gelatinosa

Characteristics of low lumbar levels of spinal cord:

a. Large dorsal and ventral gray horns compared with a relatively small amount of white matter.

b. Small size of cord with ventral area larger than the dorsal (dorsal funiculus consists of only fasciculus gracilis).

351

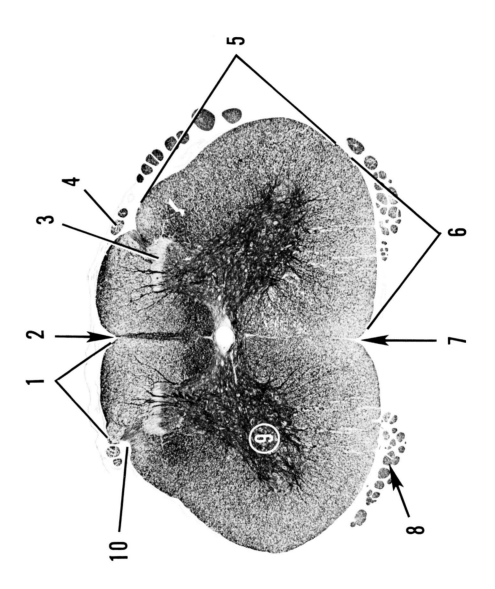

PLATE 3

Transverse section at mid-lumbar level of spinal cord.
× 15.

1. Dorsal funiculus
2. Dorsal median septum
3. Substantia gelatinosa
4. Dorsal (sensory) rootlets of spinal nerve
5. Lateral funiculus
6. Ventral funiculus
7. Ventral median fissure
8. Ventral (motor) rootlets of spinal nerve
9. Ventral horn (cell bodies of lower motor neurons)
10. Dorsolateral sulcus

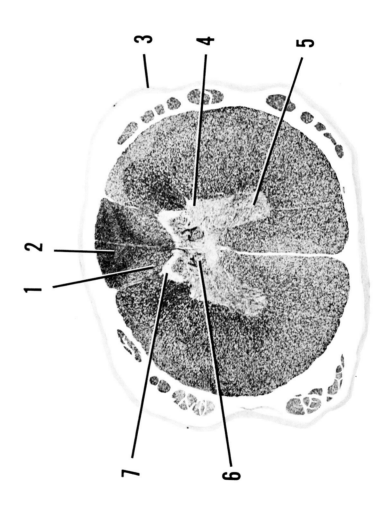

PLATE 4

Transverse section at mid-thoracic level of spinal cord. ×15.

1. Fasciculus cuneatus
2. Fasciculus gracilis
3. Dura mater
4. Intermediolateral horn (GVE cell bodies)
5. Ventral horn (GSE cell bodies)
6. Dorsal nucleus of Clarke
7. Substantia gelatinosa

Characteristics of thoracic levels of spinal cord:

a. Small amount of gray matter and a comparatively large amount of white matter.

b. Distinct intermediolateral cell column.

c. Circular outline of spinal cord.

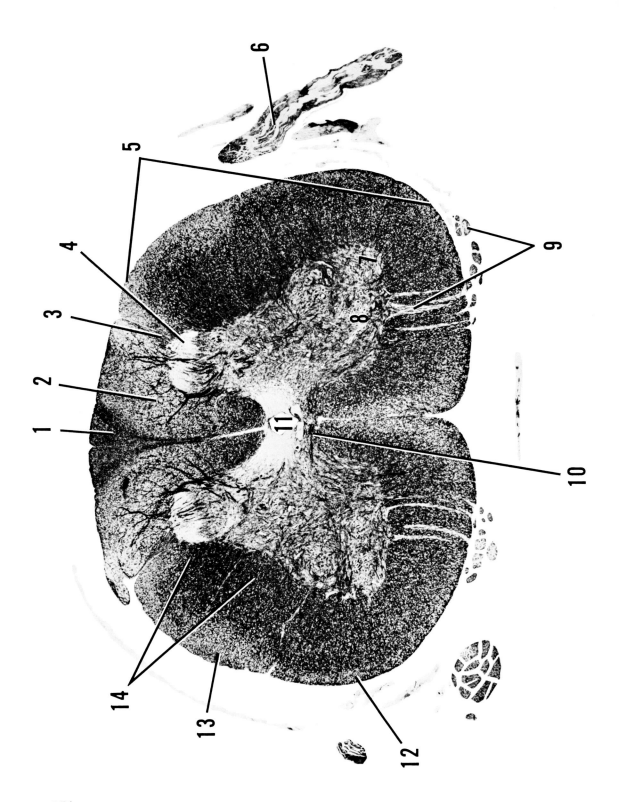

356

PLATE 5

Transverse section at low cervical level of spinal cord. ×15.

1. Fasciculus gracilis
2. Fasciculus cuneatus
3. Dorsolateral fasciculus of Lissauer
4. Substantia gelatinosa
5. Lateral funiculus
6. Dorsal root of spinal nerve
7. Lateral motor nucleus
8. Medial motor nucleus
9. Ventral (motor) rootlets
10. Ventral white commissure
11. Central canal
12. Ventral spinocerebellar tract
13. Dorsal spinocerebellar tract
14. Lateral corticospinal (pyramidal) and rubrospinal tracts

357

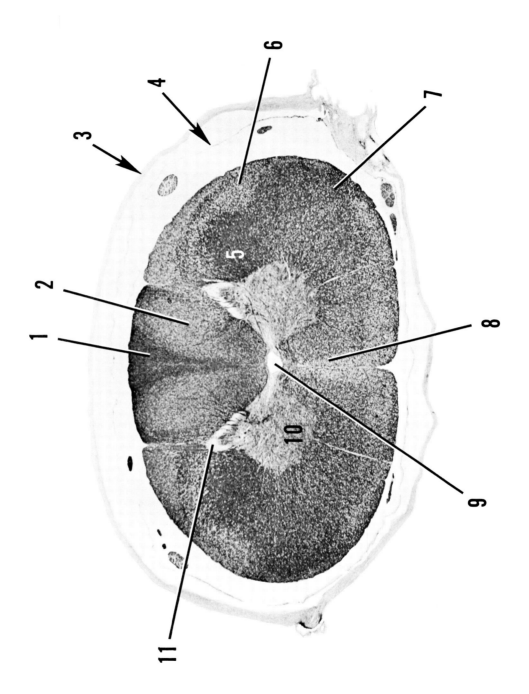

358

PLATE 6

Transverse section at mid-cervical level of spinal cord. ×15.

1. Fasciculus gracilis
2. Fasciculus cuneatus
3. Dura mater
4. Arachnoid
5. Lateral corticospinal (pyramidal) and rubrospinal tracts
6. Dorsal spinocerebellar tract
7. Ventral spinocerebellar tract
8. Ventral corticospinal (pyramidal) tract
9. Central canal
10. Ventral horn of gray matter
11. Substantia gelatinosa

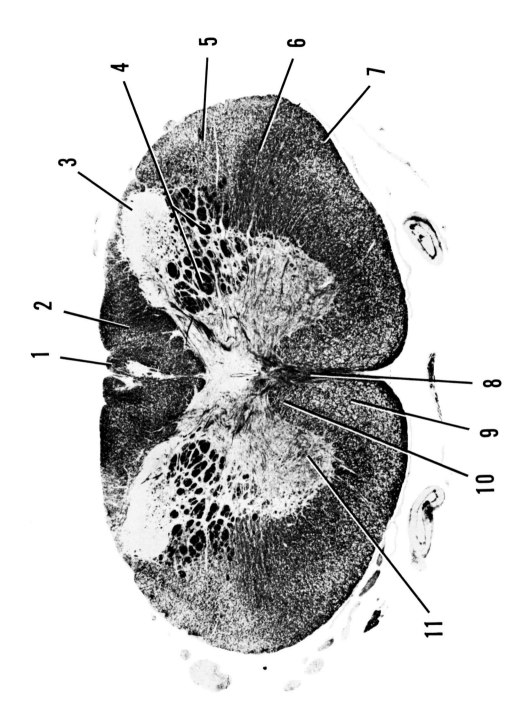

PLATE 7

Transverse section at the level of spinal cord–medulla junction. × 15.

1. Fasciculus gracilis
2. Fasciculus cuneatus
3. Nucleus of the spinal tract (descending root) of trigeminal nerve
4. Lateral corticospinal (pyramidal) tract
5. Dorsal spinocerebellar tract
6. Rubrospinal tract
7. Ventral spinocerebellar tract
8. Caudal portion of pyramidal decussation
9. Ventral corticospinal tract
10. Medial longitudinal fasciculus (MLF)
11. Cell bodies of lower motor neurons of accessory (XI) and first cervical nerves

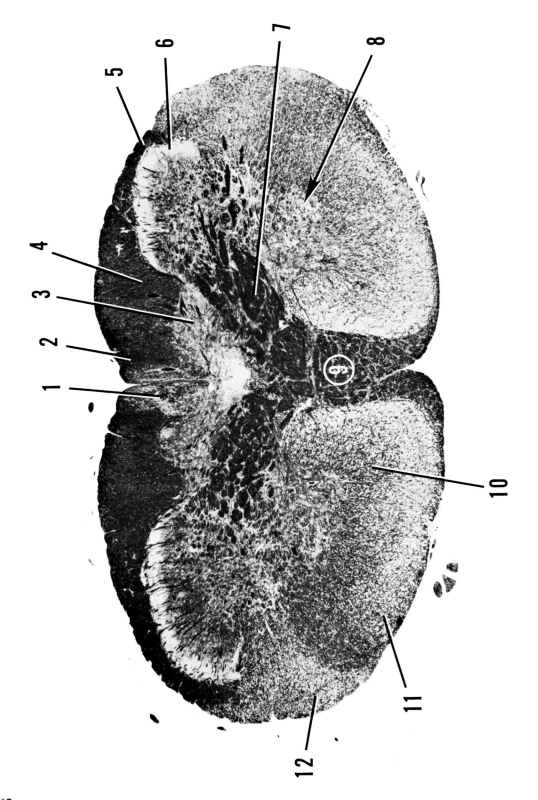

PLATE 8

Transverse section through the nucleus gracilis (1) and the pyramidal decussation (9). ×15.

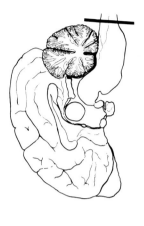

1. Nucleus gracilis
2. Fasciculus gracilis
3. Nucleus cuneatus
4. Fasciculus cuneatus
5. Spinal (descending) tract of trigeminal nerve (V)
6. Nucleus of the spinal tract (descending root) of V
7. Lateral corticospinal fibers
8. Nucleus of the accessory nerve (XI)
9. Pyramidal decussation
10. Vestibulospinal tract
11. Ventral spinocerebellar tract
12. Dorsal spinocerebellar tract

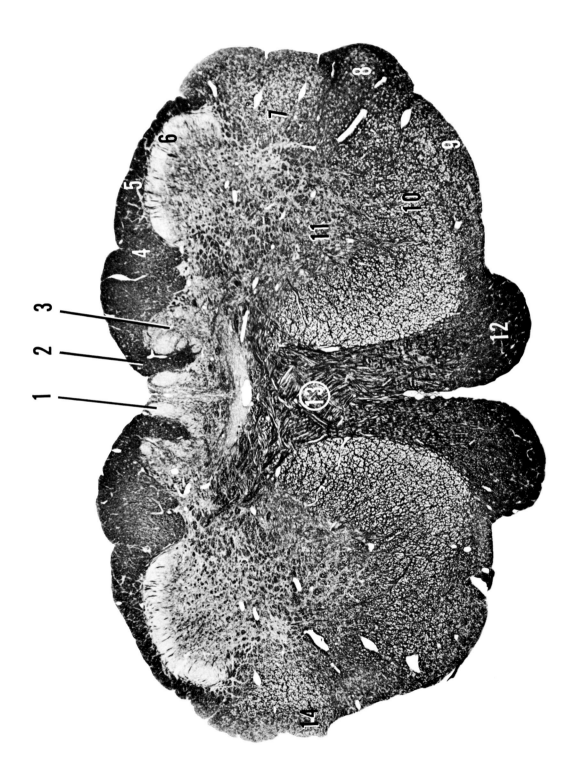

PLATE 9

Transverse section at the level of pyramidal decussation (13). ×15.

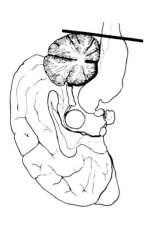

1. Nucleus gracilis
2. Fasciculus gracilis
3. Nucleus cuneatus
4. Fasciculus cuneatus
5. Spinal tract of trigeminal nerve (V)
6. Nucleus of the spinal tract of trigeminal nerve (V)
7. Rubrospinal tract
8. Lateral reticular nucleus
9. Ventral spinocerebellar tract
10. Spinothalamic tract
11. Ventral reticular nucleus
12. Pyramid
13. Pyramidal decussation
14. Dorsal spinocerebellar tract

Anatomic evidence that Plate 9 is rostral to Plate 8:

a. The dorsal median sulcus is beginning to widen (at obex).

b. The fasciculus gracilis is small.

c. The ventral median fissure is vertical and a distinct pyramid is present.

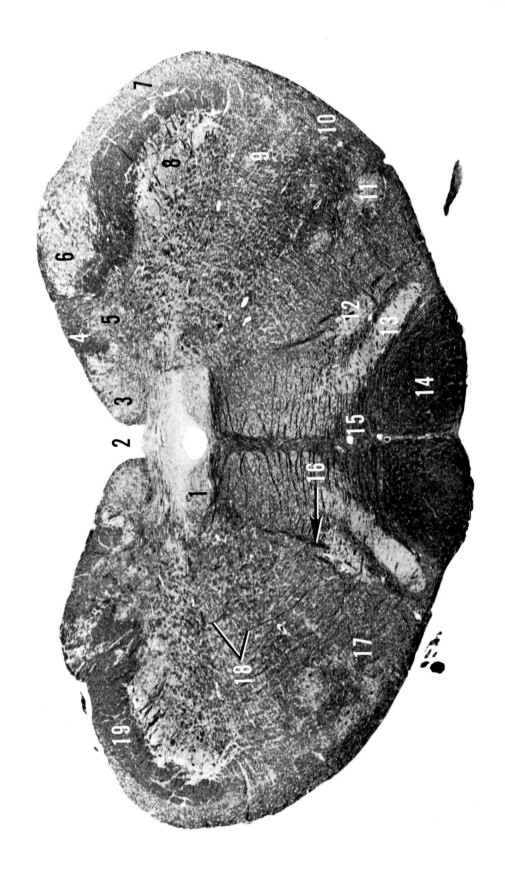

PLATE 10

Transverse section through the hypoglossal nucleus (1) near the level of the obex. ×15.

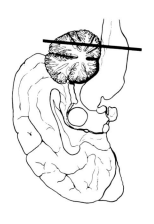

1. Hypoglossal nucleus
2. Dorsal median sulcus near the level of obex
3. Nucleus gracilis
4. Fasciculus cuneatus
5. Nucleus cuneatus
6. Accessory (lateral) cuneate nucleus
7. Dorsal spinocerebellar tract
8. Spinal nucleus of the trigeminal nerve
9. Nucleus ambiguus
10. Ventral spinocerebellar tract
11. Lateral reticular nucleus
12. Dorsal accessory olivary nucleus
13. Medial accessory olivary nucleus
14. Pyramid
15. Medial lemniscus (decussation)
16. Hypoglossal nerve
17. Spinothalamic tract
18. Internal arcuate fibers
19. Spinal tract of the trigeminal nerve

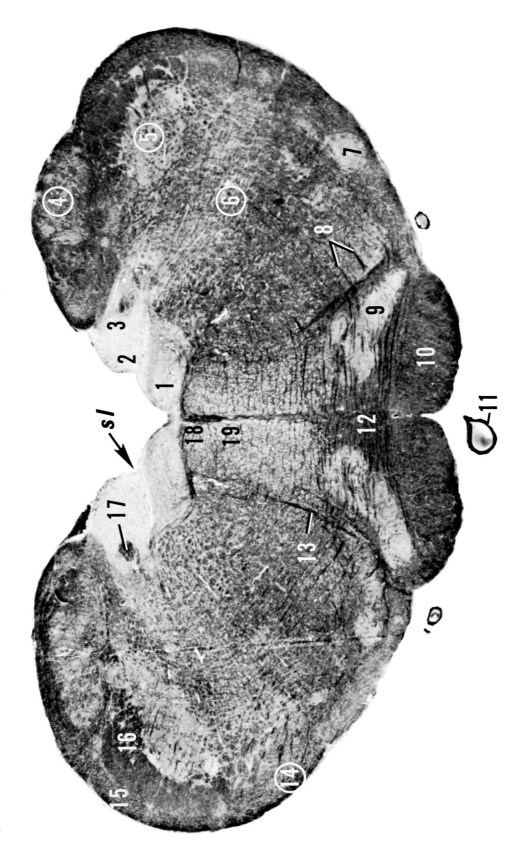

PLATE 11

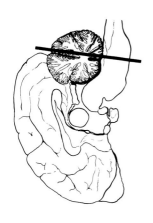

Transverse section at the level of the hypoglossal nucleus (1) and the decussation of the medial lemniscus (12). × 15.

1. Hypoglossal nucleus
2. Dorsal efferent nucleus of vagus (parasympathetic nucleus of X)
3. Nucleus solitarius
4. Accessory cuneate nucleus
5. Spinal nucleus of trigeminal nerve
6. Reticular formation (substance)
7. Lateral reticular nucleus
8. Internal arcuate fibers
9. Olive (inferior)
10. Pyramidal tract
11. Basilar artery
12. Decussation of medial lemniscus
13. Hypoglossal nerve
14. Ventral spinocerebellar tract
15. Dorsal spinocerebellar tract
16. Spinal tract of the trigeminal nerve
17. Tractus solitarius
18. Medial longitudinal fasciculus (MLF)
19. Tectospinal tract
sl Sulcus limitans

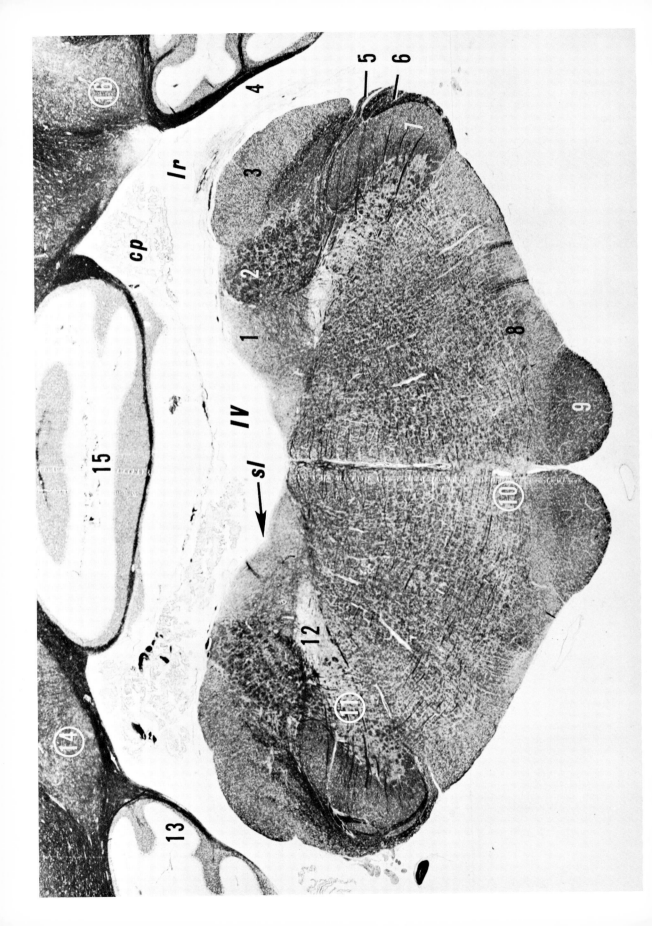

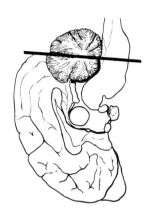

PLATE 12

Transverse section at the level of the medial (1) and descending (2) vestibular nuclei through the foramen of Luschka (4). × 14.

1. Medial vestibular nucleus
2. Descending (spinal) vestibular nucleus
3. Caudal (inferior) cerebellar peduncle
4. Foramen of Luschka
5. Sensory rootlets of vagus nerve
6. Motor rootlets of vagus nerve
7. Descending (spinal) tract of trigeminal nerve
8. Spinothalamic tract
9. Pyramid
10. Medial lemniscus
11. Spinal nucleus of trigeminal nerve
12. Nucleus solitarius
13. Flocculus of cerebellum
14. Nucleus interpositus of cerebellum
15. Nodulus of cerebellar vermis
16. Lateral (dentate) nucleus of cerebellum
cp Choroid plexus of fourth ventricle
lr Lateral recess of fourth ventricle
sl Sulcus limitans
IV Fourth ventricle

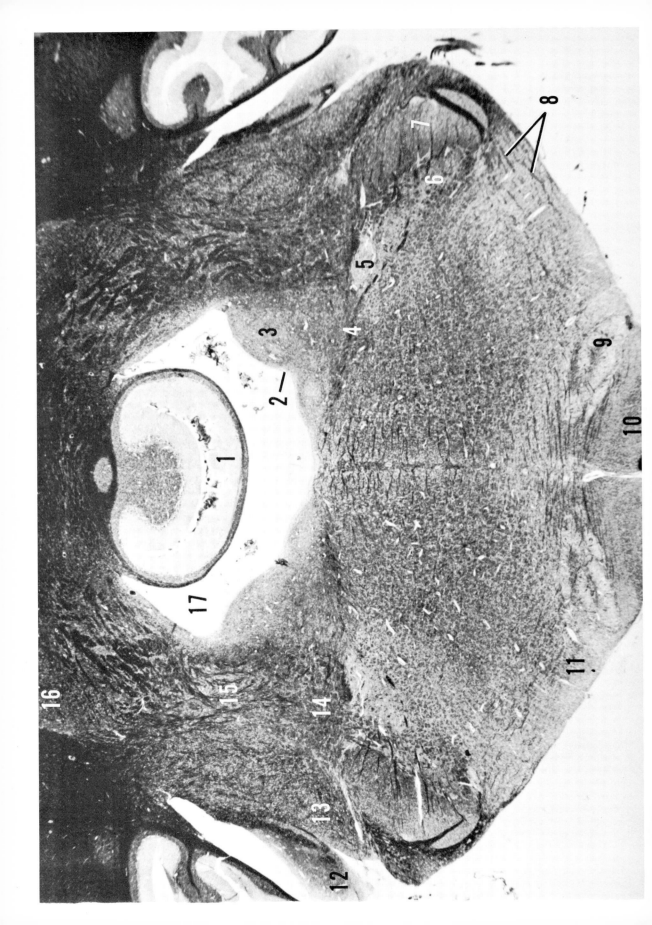

PLATE 13

Transverse section through the nodular lobule of the cerebellar vermis (1) and the caudal (inferior) olivary nucleus (9). × 14.

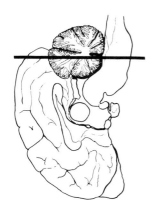

1. Nodular lobule of cerebellar vermis
2. Sulcus limitans
3. Medial vestibular nucleus
4. Facial nerve fibers
5. Nucleus solitarius
6. Nucleus of spinal tract of trigeminal nerve
7. Spinal tract of trigeminal nerve
8. Olivocerebellar fibers
9. Caudal (inferior) olivary nucleus
10. Pyramid
11. Spinothalamic tract
12. Cochlear nuclei and nerve
13. Caudal (inferior) cerebellar peduncle
14. Spinal vestibular nucleus
15. Lateral vestibular nucleus
16. Nucleus interpositus of cerebellum
17. Fourth ventricle

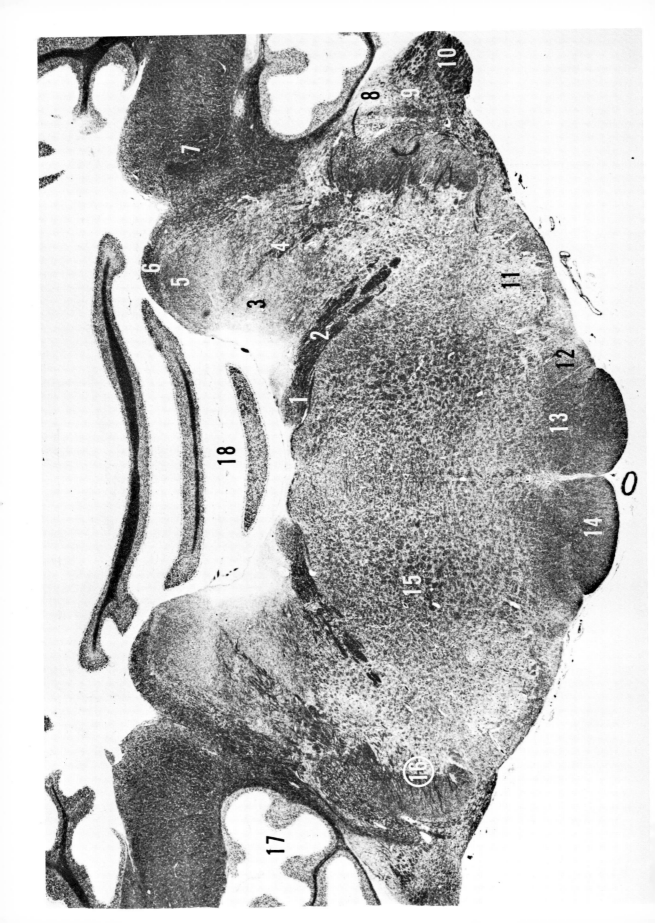

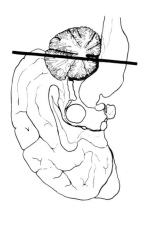

PLATE 14

Transverse section through the internal genu of facial nerve (1) and vestibulocochlear nerve (10). ×13.

1. Internal genu of facial nerve
2. Facial nerve
3. Rostral (superior) vestibular nucleus
4. Vestibular fibers
5. Rostral (superior) cerebellar peduncle
6. Ventral spinocerebellar fibers
7. Middle cerebellar peduncle
8. Dorsal cochlear nucleus
9. Ventral cochlear nucleus
10. Vestibulocochlear nerve (VIII)
11. Facial nucleus
12. Spinothalamic tract
13. Medial lemniscus
14. Pyramidal tract
15. Central tegmental tract
16. Spinal tract and nucleus of trigeminal nerve
17. Flocculus of cerebellum
18. Vermis (lingula) of cerebellum

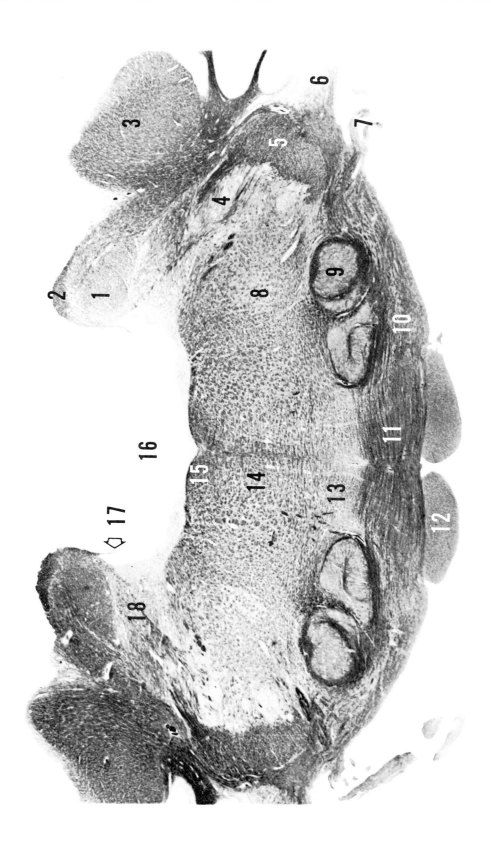

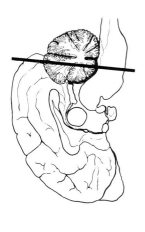

PLATE 15

Transverse section through the rostral cerebellar peduncle (1) and the dorsal nucleus of the trapezoid body (superior olive) (9). ×10.

1. Rostral (superior) cerebellar peduncle
2. Ventral spinocerebellar tract
3. Middle cerebellar peduncle
4. Main sensory nucleus of trigeminal nerve
5. Spinal tract of trigeminal nerve
6. Dorsal and ventral cochlear nuclei
7. Facial nerve
8. Motor nucleus of trigeminal nerve
9. Superior olivary nucleus (dorsal nucleus of trapezoid body)
10. Spinothalamic tract
11. Trapezoid body
12. Pyramidal tract (pyramid)
13. Medial lemniscus
14. Tectospinal tract
15. Medial longitudinal fasciculus (MLF)
16. Fourth ventricle
17. Attachment of rostral (anterior) medullary velum
18. Mesencephalic tract of trigeminal nerve

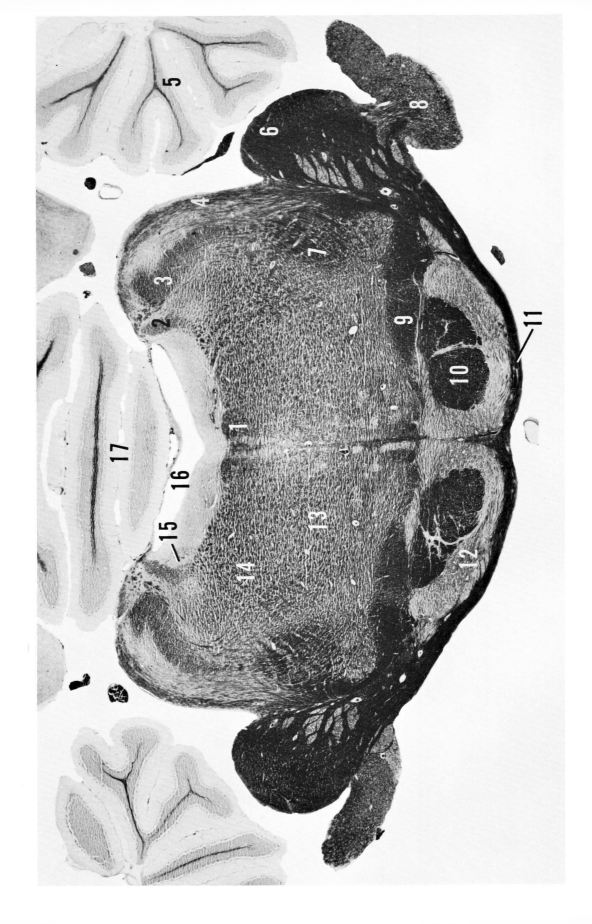

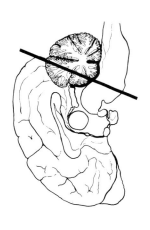

PLATE 16

Transverse section at the level of the trigeminal nerve (8). ×10.

1. Medial longitudinal fasciculus (MLF)
2. Mesencephalic tract of trigeminal nerve
3. Rostral (superior) cerebellar peduncle
4. Lateral lemniscus
5. Paraflocculus of cerebellum
6. Middle cerebellar peduncle
7. Rubrospinal fibers
8. Trigeminal nerve
9. Medial lemniscus
10. Corticospinal and corticobulbar fibers
11. Pontocerebellar fibers
12. Pontine nuclei (gray)
13. Reticular formation
14. Central tegmental fasciculus
15. Mesencephalic nucleus of trigeminal nerve
16. Fourth ventricle
17. Vermis (lingula) of cerebellum

379

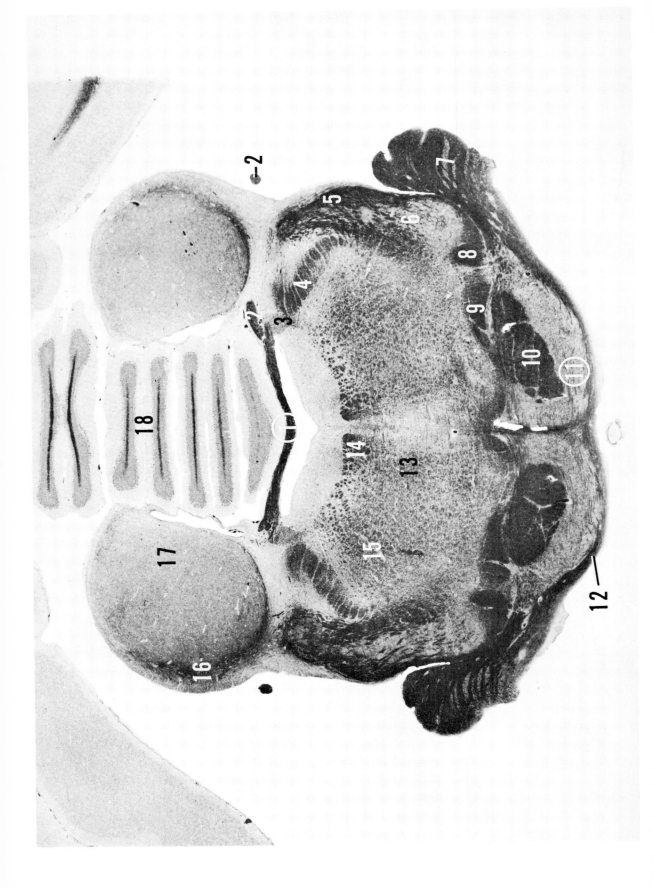

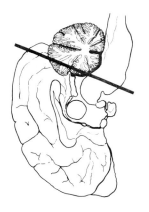

PLATE 17

Transverse section through the decussation of the trochlear nerve (1) and middle cerebellar peduncle (7). ×7.

1. Decussation of trochlear nerve (in rostral medullary velum)
2. Trochlear nerve
3. Mesencephalic tract of trigeminal nerve
4. Rostral (superior) cerebellar peduncle
5. Lateral lemniscus
6. Nuclei of lateral lemniscus
7. Middle cerebellar peduncle (containing trigeminal fibers).
8. Spinothalamic fibers
9. Medial lemniscus
10. Corticospinal and corticobulbar fibers
11. Pontine nuclei (gray)
12. Pontocerebellar fibers
13. Tectospinal fibers
14. Medial longitudinal fasciculus (MLF)
15. Central tegmental fasciculus
16. Brachium of caudal colliculus
17. Caudal colliculus
18. Vermis of cerebellum (rostral lobe)

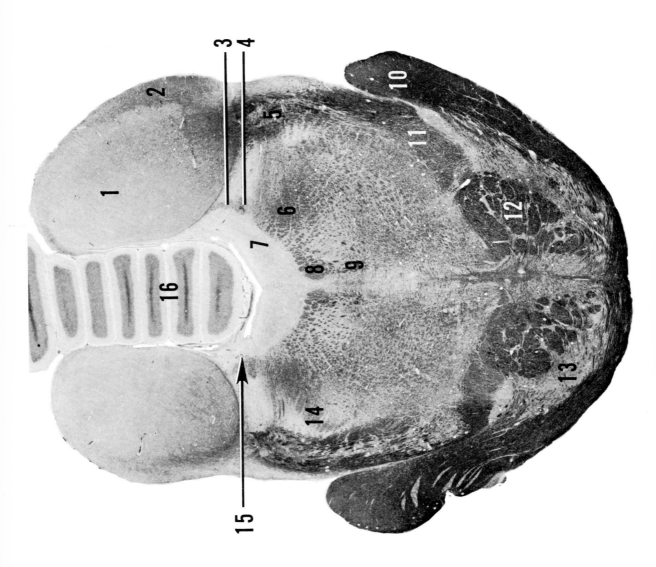

PLATE 18

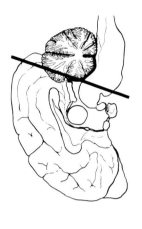

Transverse section through the caudal colliculus (1) and rostral edge of brachium pontis (10). × 7.

1. Caudal colliculus
2. Brachium of caudal colliculus
3. Mesencephalic tract of trigeminal nerve
4. Trochlear nerve
5. Lateral lemniscus and nuclei
6. Central tegmental tract
7. Dorsal longitudinal fasciculus
8. Medial longitudinal fasciculus (MLF)
9. Tectospinal tract
10. Middle cerebellar peduncle
11. Spinothalamic tract
12. Corticospinal and corticobulbar fibers
13. Pontine gray (nuclei)
14. Rostral superior cerebellar peduncle
15. Nucleus pigmentosus (ceruleus)
16. Vermis of cerebellum

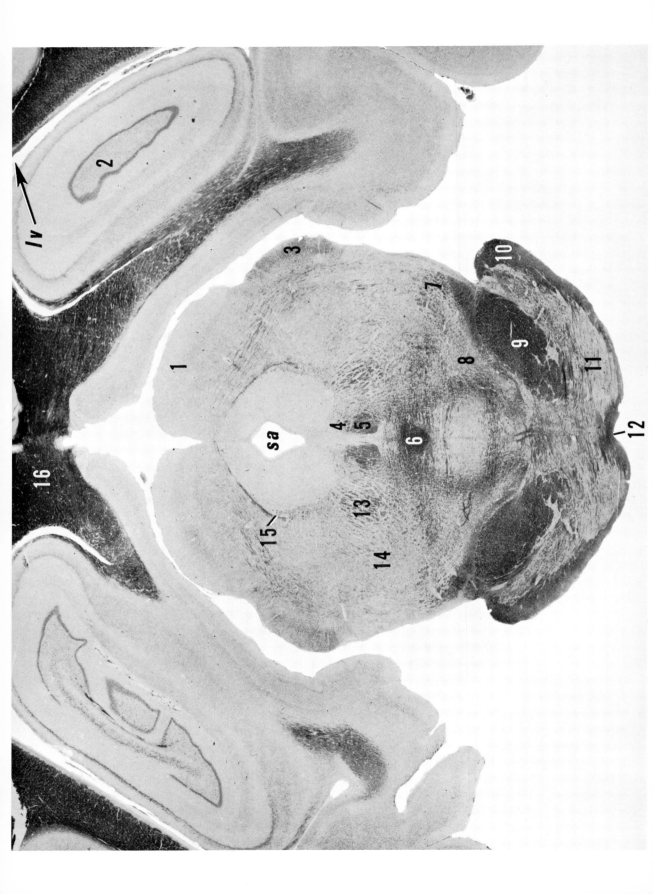

PLATE 19

Transverse section through the rostral colliculus (1) and rostral pons (11). × 7.

1. Rostral colliculus
2. Hippocampus and dentate gyrus
3. Brachium of caudal colliculus
4. Trochear nucleus
5. Medial longitudinal fasciculus (MLF)
6. Decussation of rostral (superior) cerebellar peduncle
7. Spinothalamic fibers
8. Medial lemniscus
9. Corticospinal and corticobulbar fibers
10. Middle cerebellar peduncle
11. Pontine nuclei (gray)
12. Ventral median sulcus
13. Central tegmental tract
14. Reticular formation (of mesencephalon)
15. Mesencephalic tract and nucleus of trigeminal nerve
16. Corpus callosum
lv   Lateral ventricle
sa   Sylvian aqueduct

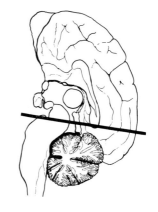

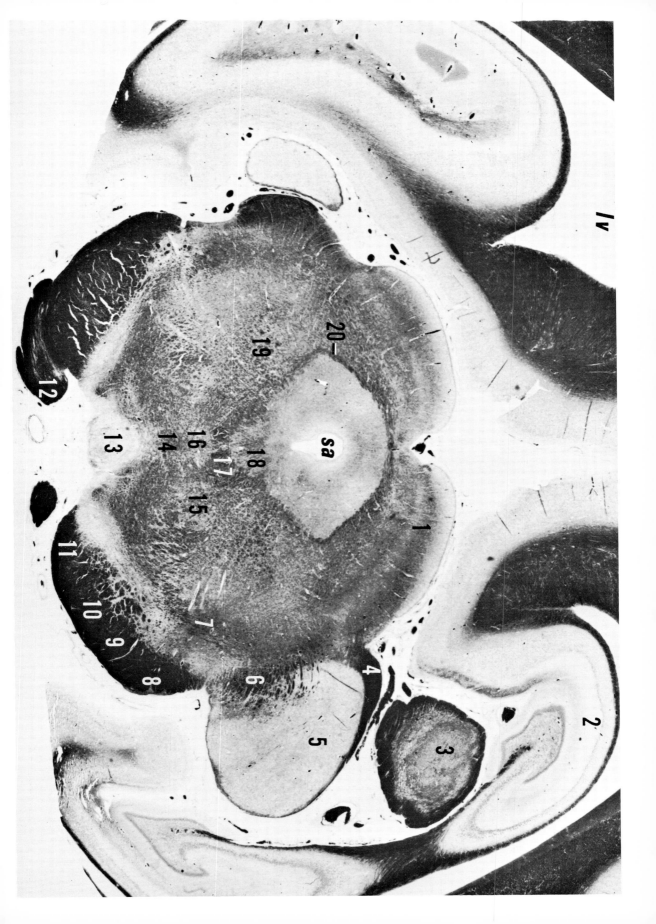

PLATE 20

Transverse section at the level of the rostral colliculus (1) and the interpeduncular nucleus (13). × 7.

1. Rostral colliculus
2. Hippocampus
3. Lateral geniculate nucleus (body)
4. Brachium of the rostral colliculus
5. Medial geniculate nucleus (body)
6. Brachium of caudal colliculus
7. Medial lemniscus
8. Parieto-occipito-temporal fibers
9. Cerebrospinal fibers
10. Cortico(cerebro)bulbar fibers
11. Frontopontine fibers
12. Pontine fibers
13. Interpeduncular nucleus
14. Ventral tegmental decussation
15. Red nucleus
16. Dorsal tegmental decussation
17. Medial longitudinal fasciculus (MLF)
18. Oculomotor nucleus
19. Central tegmental tract
20. Mesencephalic tract and nucleus of trigeminal nerve
lv Lateral ventricle
sa Sylvian aqueduct

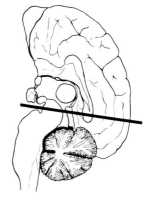

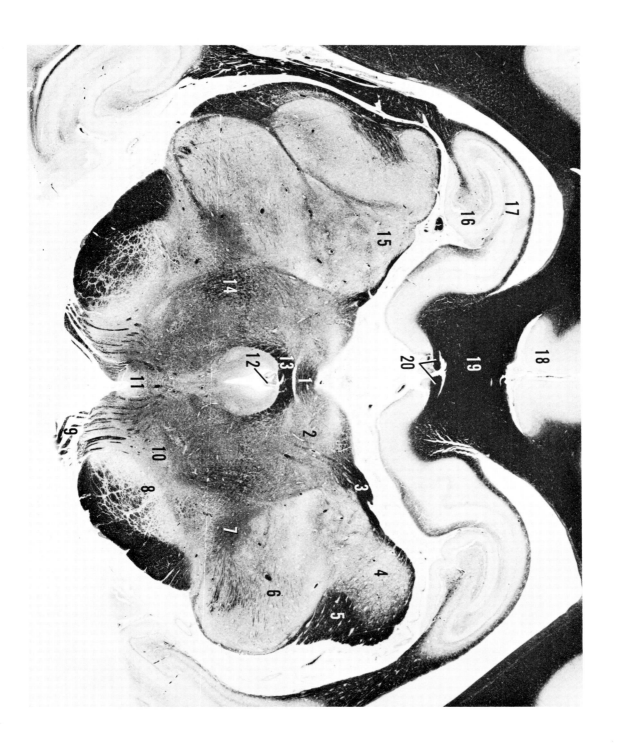

PLATE 21

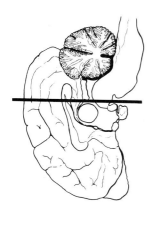

Transverse section at the level of the posterior commissure (13) and the exit of the oculomotor nerve (9). ×7.

1. Commissure of rostral colliculus
2. Pretectal region
3. Brachium of rostral colliculus
4. Lateral geniculate nucleus (body)
5. Optic tract
6. Medial geniculate nucleus (body)
7. Medial lemniscus and spinothalamic tract
8. Substantia nigra
9. Oculomotor nerve
10. Red nucleus
11. Interpeduncular nucleus
12. Subcommissural organ
13. Posterior commissure
14. Central tegmental tract
15. Pulvinar of thalamus
16. Dentate gyrus
17. Hippocampus
18. Cingulate gyrus
19. Corpus callosum
20. Commissure of fornices (hippocampi)

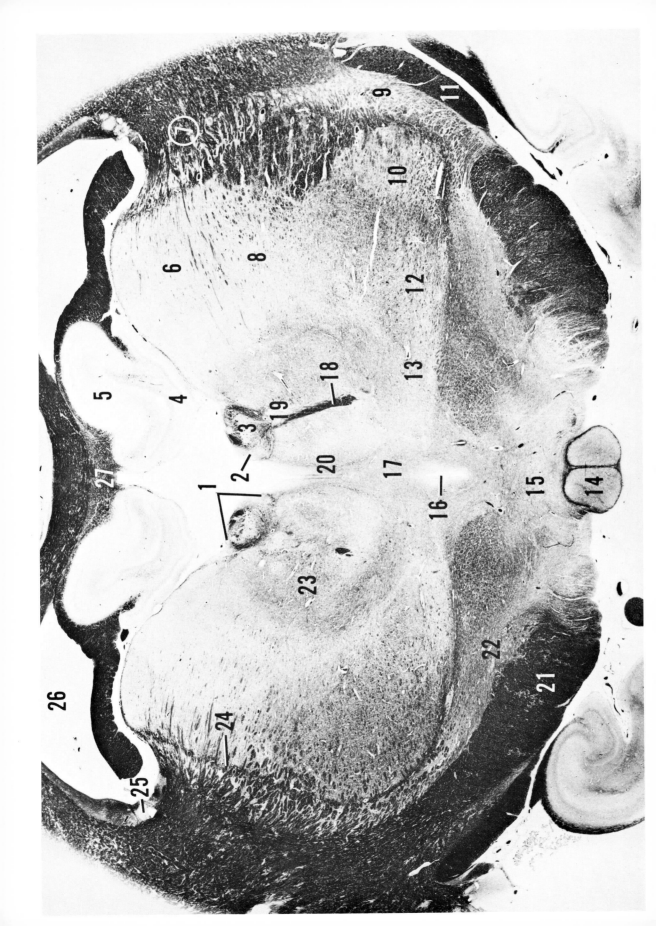

PLATE 22

Transverse section through the habenula (1) and the mammillary bodies (14). ×7.

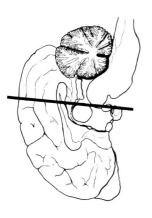

1. Habenular body
2. Medial habenular nucleus
3. Lateral habenular nucleus
4. Dentate gyrus
5. Hippocampus
6. Pulvinar
7. Optic radiation
8. Lateral nuclear area of thalamus (caudal port. of intermed. subdiv.)
9. Reticular nucleus of thalamus
10. Medial geniculate nucleus (body)
11. Optic tract
12. Ventral caudo(postero)lateral nucleus of thalamus
13. Ventral caudo(postero)medial nucleus of thalamus
14. Mammillary body
15. Caudal hypothalamic nucleus
16. Third ventricle
17. Central medial nucleus of thalamus
18. Habenulopeduncular tract
19. Nucleus of habenulopeduncular tract
20. Paraventricular nucleus
21. Crus cerebri
22. Subthalamic nucleus
23. Medial dorsal nucleus of thalamus
24. External medullary lamina
25. Caudate nucleus
26. Lateral ventricle
27. Commissure of fornices (hippocampi)

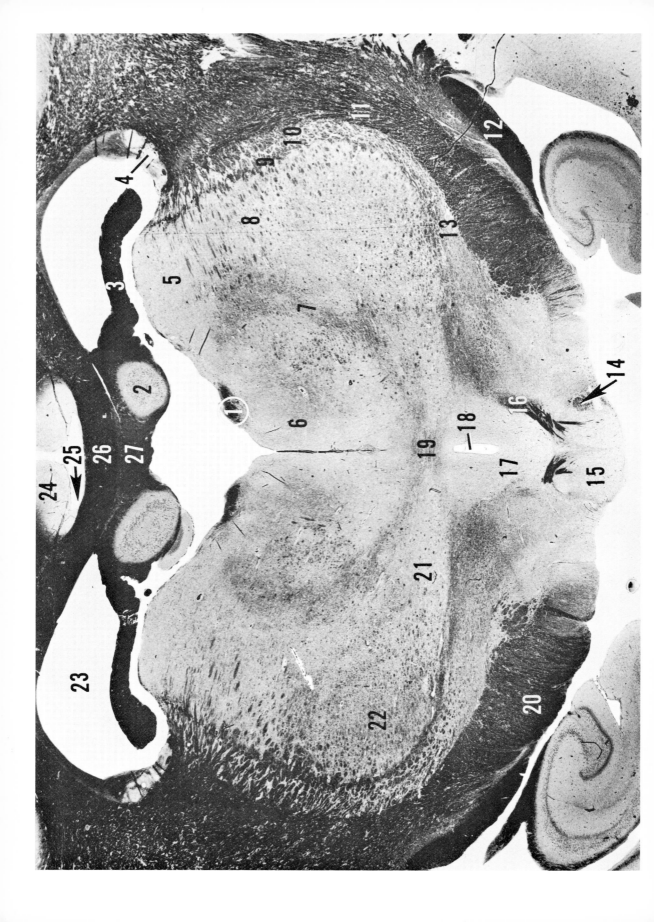

PLATE 23

Transverse section at the level of the mammillary bodies (15). ×7.

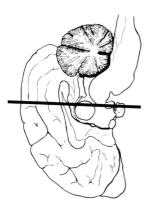

1. Stria medullaris thalami
2. Hippocampus
3. Fimbria
4. Caudate nucleus
5. Lateral nuclear area of thalamus (rostral portion)
6. Medial dorsal nucleus of thalamus
7. Internal medullary lamina
8. Lateral nuclear area of thalamus (intermed. portion)
9. External medullary lamina
10. Reticular nucleus of thalamus
11. Internal capsule (post. limb)
12. Optic tract
13. Zona incerta (of subthalamus)
14. Fornix
15. Mammillary body
16. Mammillothalamic tract
17. Caudal (post.) hypothalamic nucleus
18. Third ventricle
19. Central medial nucleus (of midline group)
20. Crus cerebri
21. Ventral caudo(postero)medial nucleus
22. Ventral caudo(postero)lateral nucleus
23. Lateral ventricle
24. Cingulate gyrus
25. Callosal sulcus
26. Corpus callosum
27. Commissure of fornices (hippocampi)

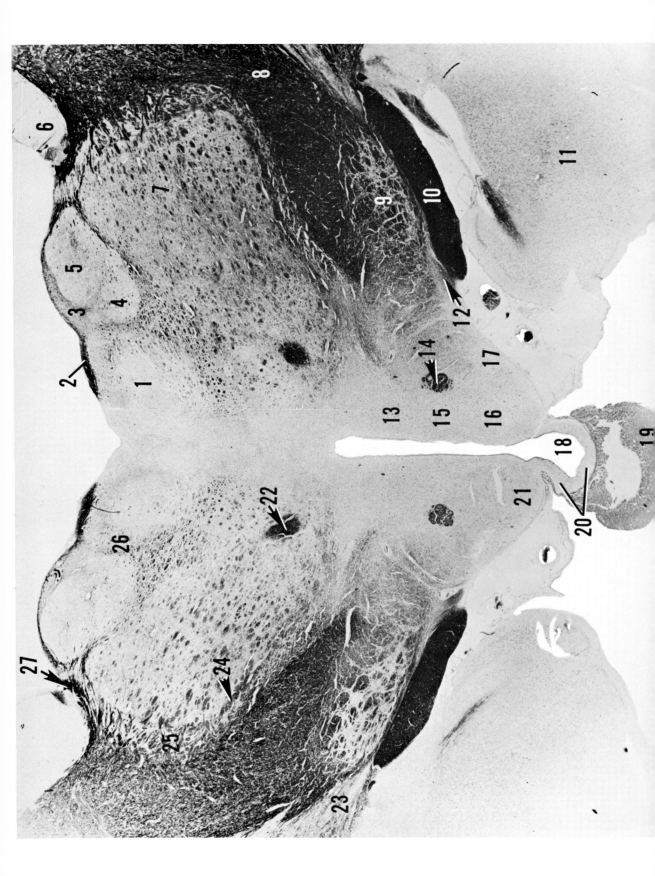

PLATE 24

Transverse section at the level of the medial dorsal nucleus of the thalamus (1) and the anterior lobe of the pituitary (19). ×10.

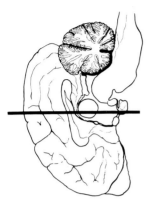

1. Medial dorsal nucleus of thalamus
2. Stria medullaris thalami
3. Rostro(antero)dorsal nucleus of thalamus
4. Rostro(antero)medial nucleus of thalamus
5. Rostro(antero)ventral nucleus of thalamus
6. Caudate nucleus
7. Ventral rostral (anterior) nucleus of thalamus
8. Genu of internal capsule
9. Entopeduncular nucleus
10. Optic tract
11. Amygdala
12. Ansa lenticularis
13. Paraventricular nucleus of thalamus
14. Column of fornix
15. Dorsomedial nucleus of hypothalamus
16. Ventromedial nucleus of hypothalamus
17. Lateral hypothalamic nucleus
18. Third ventricle (infundibular portion)
19. Anterior lobe of pituitary
20. Periventricular-arcuate nuclei

21. Supraoptic nucleus
22. Mammillothalamic tract
23. Globus pallidus
24. External medullary lamina
25. Reticular nucleus
26. Internal medullary lamina
27. Stria terminalis

395

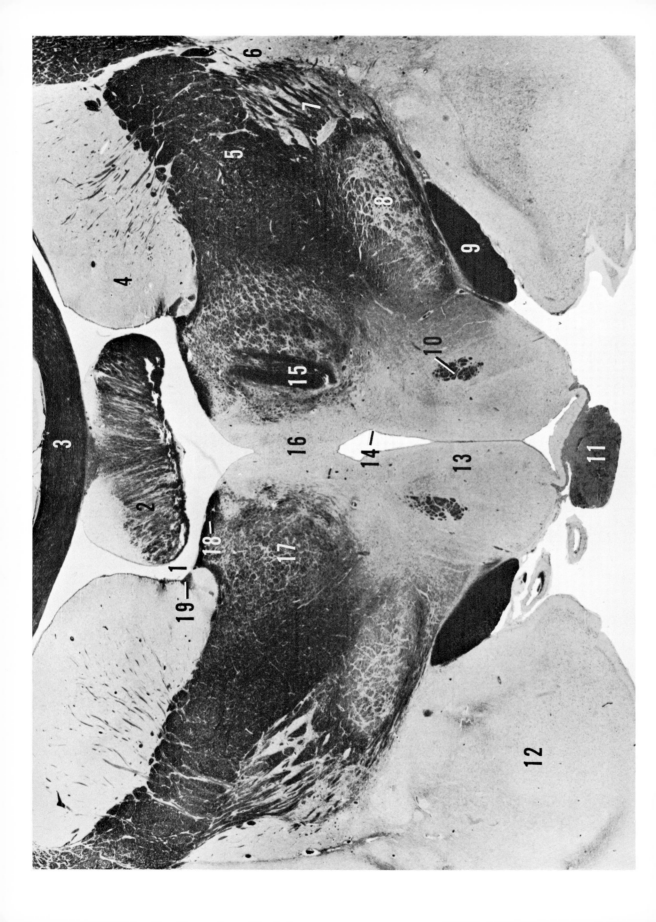

PLATE 25

Transverse section at the level of the interventricular foramen (1) and the anterior lobe of the pituitary (11). × 9.

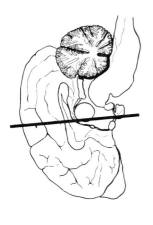

1. Interventricular foramen (of Monro)
2. Fornix
3. Corpus callosum
4. Caudate nucleus (head)
5. Genu of the internal capsule
6. Putamen
7. Globus pallidus
8. Entopeduncular nucleus
9. Optic tract
10. Column of fornix
11. Anterior lobe of pituitary
12. Amygdaloid nuclear complex
13. Paraventricular nucleus of hypothalamus
14. Hypothalamic sulcus
15. Mammillothalamic fasciculus
16. Thalamic adhesion (massa intermedia)
17. Ventral rostral (anterior) nucleus of thalamus
18. Stria medullaris thalami
19. Stria terminalis

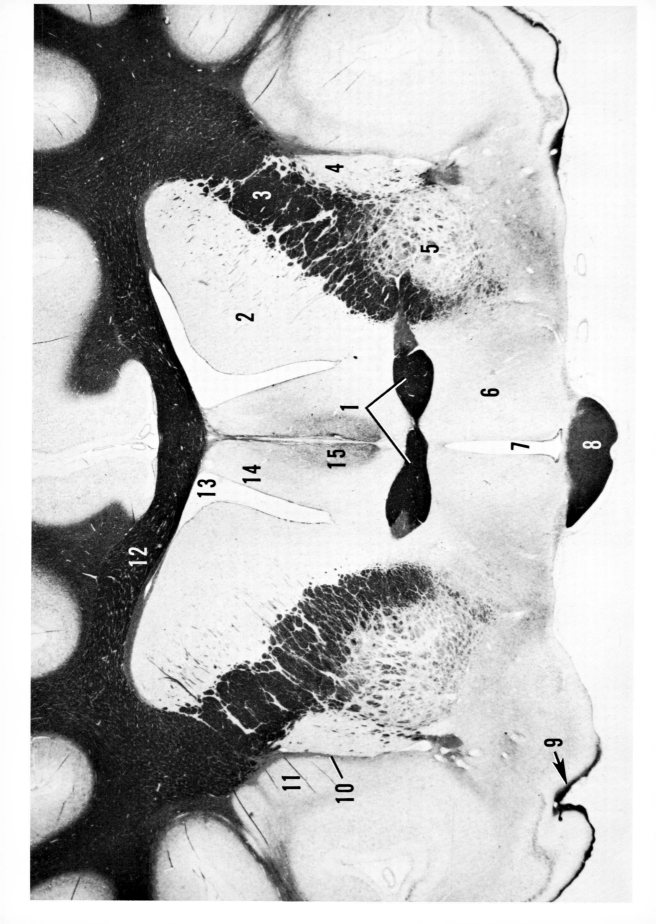

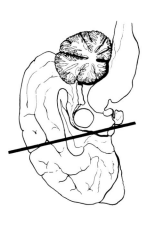

PLATE 26

Transverse section through the rostral (anterior) commissure (1) and the optic chiasm (8). × 7.

1. Rostral (anterior) commissure
2. Caudate nucleus (head)
3. Internal capsule (rostral limb)
4. Putamen
5. Globus pallidus
6. Supraoptic area of hypothalamus
7. Third ventricle
8. Optic chiasm
9. Lateral olfactory stria
10. External capsule
11. Claustrum
12. Corpus callosum
13. Lateral ventricle
14. Lateral septal nucleus
15. Medial septal nucleus

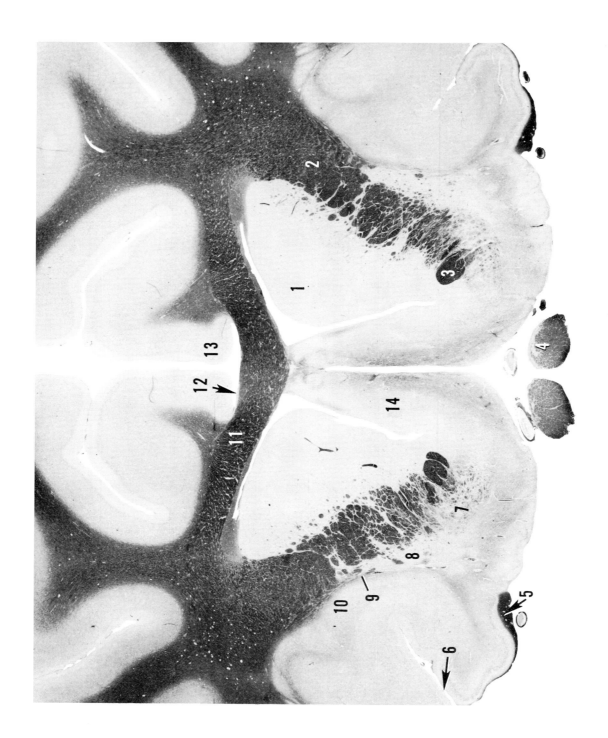

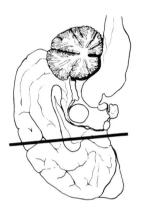

PLATE 27

Transverse section immediately rostral to the optic chiasm. ×7.

1. Caudate nucleus (head)
2. Internal capsule (rostral limb)
3. Rostral commissure (rostral part)
4. Optic nerve
5. Lateral olfactory stria
6. Rostral rhinal fissure
7. Globus pallidus
8. Putamen
9. External capsule
10. Claustrum
11. Corpus callosum
12. Indusium griseum
13. Cingulate gyrus
14. Septal nucleus

401

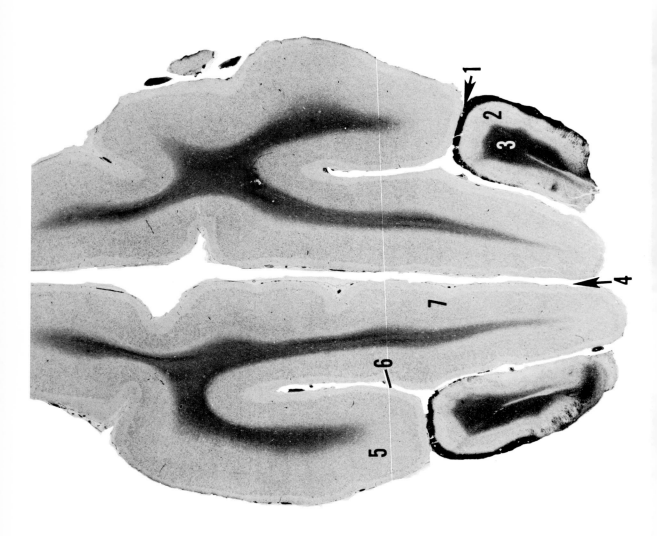

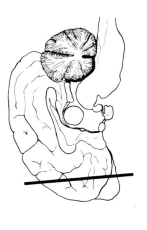

PLATE 28

Transverse section at the level of the olfactory tract (3). ×6.

1. Rostral (anterior) rhinal fissure
2. Rostral (anterior) olfactory nucleus
3. Olfactory tract
4. Ventral longitudinal fissure
5. Gyrus proreus
6. Olfactory sulcus
7. Gyrus rectus

Index

Page numbers in *italics* indicate references to the summary (Part II); page numbers preceded by P indicate Plates in the Atlas (Part III).

Dysmetria, 212
Dysphagia, 288
Dysphonia, 287
EAR, 302–317
  anatomy of, functional, 308, 309
  auditory pathways of, 308–311
  development of, 302
  divisions of, 302
  functions of, 302, 303, 305
  inner, 305–308
  middle, 303–305
  outer, 303
  transmission of sound from, 309
Ectoderm, superficial, 12
Effector, 102, 104
Efferents, basal ganglia, 235, 237
  mammillary body, 224
  reticular, 181
  rostral colliculus, 197
Embedding, nerve tissue, 61–62
Embryo, sensory functional components of, 111,
  112
Eminence, median, hypothalamal, 223
Emotion, mechanism of, 245
End-bulbs, 107
End-feet, 107
End-plate(s), motor, 102–104, 109–110
Endolymph, 306–307
Endoneurium, 44, 51, 94, 95
Enophthalmos, 300
Environment, internal, 7
Epimere, 96
Epineurium, 51, 94, 95
Epiphysis, 17, 31, 32, 172, 216–217, 225
Epithalamus, 17, 31, 215, 216–217
Epithelium, glandular, as target tissue, 132
  olfactory, 242
Equilibrium, 114, 302–303, 311–316
Eye, anatomy of, gross, 291–292
    microscopic, 293–295
  embryology of, 291
  histology of, 293–295
  section of, sagittal, 292
  and visual system, 290–301
FALX CEREBRI, 33, 79
Fasciculus(i), 153
  ascending, 157, 159–162
  of caudal medulla, 178–179
  cuneate, 32, 159–162, 172, 173, 178, P4–P10
  dorsolateral, 162, 179, P5
  Flechsig's, 162–163, 175
  gracilis, 32, 159–162, 172, 173, 178, P4–P9
  lenticular, 229
  of Lissauer, 157, 162, 179, P5
  longitudinal, dorsal, 29, 195, 199, 224, 225, 243,
    P18
    medial, 168–169, 188–190, 200, 312–314, 332,
    P7, P11, P15–P20
  mammillaris princeps, 224, 225
  proprius, dorsal, 157, 162, 169
    lateral, 164, 167
    ventral, 169
  retroflexus, of Meynert, 217, 243
  of Schutz, 29, 195, 199, 224, 225, 243, P18

  solitarius, 177, 336, P11
  spinal cord. See Tract, spinal
  thalamic, 229
Fat, in nervous system, 49
Feedback, 210
Feet, perivascular, 58, 59, 86
Fiber(s), 50
  A, 55
  afferent, to amygdala, 246
    cerebellar, 208
    cranial nerve, 276–277
    hypothalamic, 223–224
    proprioceptive, somatic, 200
    of red nucleus, 197
    of reticular formation, 181
    somatic, 100
    visceral, 100–101
  arcuate, dorsal external, 178, 208
    internal, 160, 176, 178, P10, P11
  association, 37–38, 258
  B, 55
  C, 55
  cerebellar cortex, 205–206, 207, 208–211
  cerebrospinal, 29
  climbing, 206, 207
  commissural, 38, 258
  corticifugal, 29, 256–257
  corticipetal, 257
  corticobulbar, 195, 257, P16–P18
  corticonigral, 236
  corticopontine, 195, 202
  corticoreticular, 181
  corticorubral, somatotopic patterns of, 166, 236,
    334
  corticospinal, 164, 165, 178, P8, P16–P20
  corticostriate, 235
  descending, 210
  efferent, cerebellar, 208
    cerebral, 256–257
    cranial nerve, 277
    hypothalamic, 223, 224
    reticular formation, 181
    gamma, 115
    somatic, 100
    visceral, 100–101, 130
  frontopontine, 195, 202, P20
  mammalian, conduction speed, 55
    diameter, 55
    function, 55
    properties of, 55
  mammilotegmental, 224, 225
  mossy, 206, 207
  nonmedullated, 50
  nonmyelinated, 50
  olivocerebellar, 179, 180, P13
  parallel, 207
  periventricular, 224
  pilomotor, 141
  pontine, P20
  pontocerebellar, P16, P17
  postganglionic, 50
  postsynaptic, 50
  projection, 37, 256–258
  Purkinje, 110, 111

Substance, anterior perforated, 241
  cerebral medullary, 33, 37–39
  chromidial. *See* Nissl bodies
  chromophil. *See* Nissl bodies
  reticular, 180–183
  rostral, 39
Substantia, gelatinosa Rolandi, 154–155, 156, 179, 286, P1–P6
  nigra, 29, 49, 195, 201, 230, 236, P20, P21
Subthalamus, 17, 33, 215, 229–230
  nuclei of, 229, P22
Sulcus(i), 251
  callosal, 25, 216, 256, 257, P23
  callosomarginal, 256
  cerebellar, 26
  cerebral, of dog, 252–253
  cingulate, 256
  coronal, 37, 253, 254
  cruciate, 37, 252, 253, 254
  definition, 18
  dorsal median, 23, 24, 32, 150, 152, 172, P3
  dorsolateral, 152, P3
  ectosylvian, 37, 253, 254
  genual, 256
  hippocampal, 256
  hypothalamic, 17, P25
  intermediate, 152
  limitans, 12, 14, 16, 18, 24, 149, 151, 173, 174, 176, P11, P12
  marginal, 37, 254
  median, 23, 24
  olfactory, P28
  postcruciate, 253, 254
  precruciate, 253, 254
  rhinal, 37, 254, P27, P28
  splenial, 256
  suprasylvian, 37, 254
  of telencephalon, 18–19, 34, 36–37
  ventral median, 150, 152, P1
Sympathetic chain, 131, 133, 134, 135, 138
Sympathomimetic effects, 136
Synapse(s), definition of, 107
  endings, 46
  evolution of, 4–5
  location of, 109
  morphology of, 107–108
  properties of, 108–109
  types of, 107
Syndrome, archicerebellar, 212
  Horner's, 146, 300
  Klüver-Bucy, 245–246, 261
  obstinate progression, 198
System, autonomic nervous. *See* Nervous system
  central nervous. *See* Nervous system
  corticopontocerebellar, 192
  corticospinal, function of, 165
  dentatorubrothalamic, 197
  endocrine, control of, 130
    function of, 7
  impulse-conducting, 110
  limbic, 245–250, *331*
    anatomy of, 245
    definition of, 245
    functions of, 245–256

  nervous. *See* Nervous system
  ocular, differences in, anatomic, 290–291
    physiologic, 290
  olfactory, degeneration of, 122
  reticulospinal fiber, 168
  vestibular, 187–188
  visual, 290–301
Tachycardia, 288
Tapetum lucidum, 292, 293
  nigrum, 292, 293
Target organ(s), correlation of, with functional components, 275–277
  efferent nerves of, 133–134
  innervation of, effects of, 136, 137
  of reflex arc, 102, 104
Target tissues, of autonomic nervous system, 132
Technique, celloidin, 62
  Golgi, 63
  Golgi-Cox, 63, 64
  Nauta-Gygax, 63
  neurohistologic, 60–64
Tectum, 14, 17, 23, 25, 29–30, 173, 195–197, 216, 225, 257
  optic, submammalian, 196
Tegmentum, 14, 17, 25, 29, 195, 197–201
Tela choroidea, 24, 60
Teleceptors, 114
Telencephalon, 8, 13, 14, 18–19, 33–39, 232–263
Telocoele(s), 15, 17, 18
Telodendria, 45, 46
Temperature change, 112
Tentorium, cerebelli, 79, 203
  osseous, 79
Thalamus, dorsal, 17, 18, 31–33, 189, 210, 216, 217–221, 227
  borders of, 217–218
  dog, 218
  functions of, 220–221
  nuclei of, 218–220, *337*, P22–P24
  ventral, 17, 33, 216, 229–230
Thermoreceptors, 113
Tissue, connective, of nerves, 94, 95
Tract(s). *See also* Fasciculus; Fiber(s); Pathway(s)
  auditory, 308–311
  comparison of, 8
  corticospinal, lateral, 20, 157, 164–165, 175, P5, P6, P11
    ventral, 157, 164, 165, 167, P6, P7
  crossed pyramidal, 164–165
  cuneatus, 159–162, 178, P4–P10
  cuneocerebellar, 163, 211, *336*
  dentatorubral, 197
  dentatothalamic, 210
  efferent, from red nucleus, 197
  habenular 198, 217, P22
  habenulopeduncular, 31, 217, 243, P22
  hypothalamohypophyseal, 224, 225
  mammillotegmental, 224, 225, 243
  mammillothalamic, 224, 225, 227, 243, P23, P24
  mesencephalic, trigeminal nerve, 186, 191, 192, 198, 286, *339*, P16, P18
  olfactohypothalamic, 223, 243
  olfactory, 39, 40, 241–242, P28
  olivospinal, 169